Petroleum Engineering

Petroleum Engineering

Edited by
James Cameron

⊟ Larsen & Keller
www.larsen-keller.com

Petroleum Engineering
Edited by James Cameron
ISBN: 978-1-63549-215-6 (Hardback)

🖥 Larsen & Keller

Published by Larsen and Keller Education,
5 Penn Plaza,
19th Floor,
New York, NY 10001, USA

Cataloging-in-Publication Data

Petroleum engineering / edited by James Cameron.
     p. cm.
Includes bibliographical references and index.
ISBN 978-1-63549-215-6
1. Petroleum engineering. 2. Petroleum--Prospecting.
3. Mining engineering. I. Cameron, James.
TN870 .P48 2017
665.5--dc23

The publisher's policy is to use permanent paper from mills that operate a sustainable forestry policy. Furthermore, the publisher ensures that the text paper and cover boards used have met acceptable environmental accreditation standards.

Printed and bound in the United States of America.

For more information regarding Larsen and Keller Education and its products, please visit the publisher's website www.larsen-keller.com

Table of Contents

Preface

Petroleum engineering deals with the production, development, extraction and management of hydrocarbons like crude oil and natural gas. It uses elements of petroleum geology, oil and gas facility engineering, and formation evaluation, etc. This book unfolds the innovative aspects of this subject, which will be crucial for the holistic understanding of this field. Some of the diverse topics covered in it address the branches and technologies that fall under this category. Those in search of information to further their knowledge will be greatly assisted by this textbook. This book will serve as a reference to a broad spectrum of readers.

A detailed account of the significant topics covered in this book is provided below:

Chapter 1- Petroleum engineering is a branch of engineering that is studies and researches the activities related to the production of natural gas or crude oil. This subject relates to a number of fields, such as drilling, economics, reservoir simulation and well engineering. This chapter will provide an integrated understanding of petroleum engineering.

Chapter 2- Petroleum engineering is an interdisciplinary subject. Some of the sub-fields related to this subject are reservoir engineering, drilling engineering, petroleum production engineering and subsurface engineering. Reservoir engineering is the branch of petroleum engineering that concerns itself with the issues caused because of the development and production of oil and gas reservoirs. This text will provide a glimpse of the related fields of petroleum engineering briefly.

Chapter 3- The Hubbert peak theory contents that the production of petroleum always tends to follow a bell-shaped curve. The following text also explains fractional distillation and octane rating. The topics discussed in the chapter are of great importance to broaden the existing knowledge on petroleum engineering.

Chapter 4- The process by which petroleum is drawn out of the Earth is known as extraction of petroleum. Shale oil extraction is the industrial process for oil production. Some of the topics discussed herein are unconventional oil, oil well, oil and gas well completion, drilling rig and peak oil. The topics discussed in this section help the reader in understanding petroleum extraction.

Chapter 5-Hydraulic fracturing is a technique of fracturing a rock by using pressuring liquid. The topics explained in this chapter are waterless fracturing, hydraulic fracturing proppants, regulation of hydraulic fracturing, the uses of radioactivity in oil and gas wells etc. The section on hydraulic fracturing is an overview of the subject matter incorporating all the major aspects of hydraulic fracturing.

Chapter 6- Petroleum refinery is an industrial process plant; it is used in processing crude oil. Some of the useful products produced by this process are diesel fuel, asphalt base, heating oil and kerosene. This chapter helps the readers in understanding the basic concepts and processes of petroleum refinery.

Chapter 7- Formation evaluation is the evaluation of a borehole to measure its ability to produce petroleum. Well logging, core sample, mud logging, gamma ray logging and formation evaluation neutron porosity are some of the chapters explained in this section. The text has been carefully written to provide an easy understanding of the formation evaluation of petroleum.

Chapter 8- The environmental concerns caused by petroleum are high, for it is poisonous for every form of life. It immensely damages the ecosystem and has resulted in climate change. The other aspects of the environmental concerns are oil spill, the environmental impact of the oil shale industry and the environmental impact of hydraulic fracturing.

I would like to make a special mention of my publisher who considered me worthy of this opportunity and also supported me throughout the process. I would also like to thank the editing team at the back-end who extended their help whenever required.

Editor

Introduction to Petroleum Engineering

Petroleum engineering is a branch of engineering that is studies and researches the activities related to the production of natural gas or crude oil. This subject relates to a number of fields, such as drilling, economics, reservoir simulation and well engineering. This chapter will provide an integrated understanding of petroleum engineering.

Petroleum Engineering

Example of a map used by reservoir engineers to determine where to drill a well. This screenshot is of a structure map generated by contour map software for an 8500 ft deep gas and oil reservoir in the Erath field, Vermilion Parish, Erath, Louisiana. The left-to-right gap, near the top of the contour map indicates a fault line. This fault line is between the blue/green contour lines and the purple/red/yellow contour lines. The thin red circular contour line in the middle of the map indicates the top of the oil reservoir. Because gas floats above oil, the thin red contour line marks the gas/oil contact zone

Petroleum engineering is a field of engineering concerned with the activities related to the production of hydrocarbons, which can be either crude oil or natural gas. Exploration and Production are deemed to fall within the *upstream* sector of the oil and gas industry. Exploration, by earth scientists, and petroleum engineering are the oil and gas industry's two main subsurface disciplines, which focus on maximizing economic recovery of hydrocarbons from subsurface reservoirs. Petroleum geology and geophysics focus on provision of a static description of the hydrocarbon reservoir rock, while petroleum engineering focuses on estimation of the recoverable volume of this resource using a detailed understanding of the physical behavior of oil, water and gas within porous rock at very high pressure.

The combined efforts of geologists and petroleum engineers throughout the life of a hydrocarbon accumulation determine the way in which a reservoir is developed and depleted, and usually they have the highest impact on field economics. Petroleum engineering requires a good knowledge of many other related disciplines, such as geophysics, petroleum geology, formation evaluation (well logging), drilling, economics, reservoir simulation, reservoir engineering, well engineering, artificial lift systems, completions and oil and gas facilities engineering.

Recruitment to the industry has historically been from the disciplines of physics, chemical engineering and mining engineering. Subsequent development training has usually been done within oil companies.

Overview

The profession got its start in 1914 within the American Institute of Mining, Metallurgical and

Petroleum Engineers (AIME). The first Petroleum Engineering degree was conferred in 1915 by the University of Pittsburgh. Since then, the profession has evolved to solve increasingly difficult situations, as much of the "low hanging fruit" of the world's oil fields have been found and depleted. Improvements in computer modeling, materials and the application of statistics, probability analysis, and new technologies like horizontal drilling and enhanced oil recovery, have drastically improved the toolbox of the petroleum engineer in recent decades.

Deep-water, arctic and desert conditions are usually contended with. High Temperature and High Pressure (HTHP) environments have become increasingly commonplace in operations and require the petroleum engineer to be savvy in topics as wide ranging as thermo-hydraulics, geomechanics, and intelligent systems.

The Society of Petroleum Engineers (SPE) is the largest professional society for petroleum engineers and publishes much information concerning the industry. Petroleum engineering education is available at 17 universities in the United States and many more throughout the world - primarily in oil producing regions - and some oil companies have considerable in-house petroleum engineering training classes.

Petroleum engineering has historically been one of the highest paid engineering disciplines, although there is a tendency for mass layoffs when oil prices decline. In a June 4, 2007 article, Forbes.com reported that petroleum engineering was the 24th best paying job in the United States. The 2010 National Association of Colleges and Employers survey showed petroleum engineers as the highest paid 2010 graduates at an average $125,220 annual salary. For individuals with experience, salaries can go from $170,000 to $260,000 annually. They make an average of $112,000 a year and about $53.75 per hour.

Types

Petroleum engineers divide themselves into several types:

- Reservoir engineers work to optimize production of oil and gas via proper well placement, production rates, and enhanced oil recovery techniques.
- Drilling engineers manage the technical aspects of drilling exploratory, production and injection wells.
- Production engineers, including subsurface engineers, manage the interface between the reservoir and the well, including perforations, sand control, downhole flow control, and downhole monitoring equipment; evaluate artificial lift methods; and also select surface equipment that separates the produced fluids (oil, natural gas, and water)..

Petroleum

Petroleum (from Latin: petra: "rock" + oleum: "oil".) is a naturally occurring, yellow-to-black liquid found in geological formations beneath the Earth's surface, which is commonly refined into various types of fuels. Components of petroleum are separated using a technique called fractional distillation.

It consists of hydrocarbons of various molecular weights and other organic compounds. The name *petroleum* covers both naturally occurring unprocessed crude oil and petroleum products that are made up of refined crude oil. A fossil fuel, petroleum is formed when large quantities of dead organisms, usually zooplankton and algae, are buried underneath sedimentary rock and subjected to both intense heat and pressure.

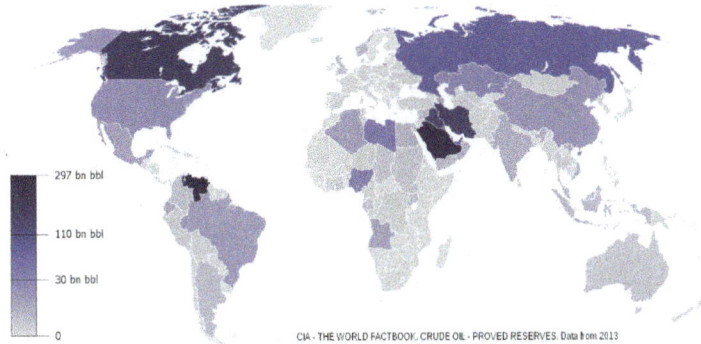

Proven world oil reserves, 2013. Unconventional reservoirs such as natural heavy oil and oil sands are included.

Pumpjack pumping an oil well near Lubbock, Texas

Petroleum has mostly been recovered by oil drilling (natural petroleum springs are rare). Drilling is carried out after studies of structural geology (at the reservoir scale), sedimentary basin analysis, and reservoir characterization (mainly in terms of the porosity and permeability of geologic reservoir structures) have been completed. It is refined and separated, most easily by distillation, into a large number of consumer products, from gasoline (petrol) and kerosene to asphalt and chemical reagents used to make plastics and pharmaceuticals. Petroleum is used in manufacturing a wide variety of materials, and it is estimated that the world consumes about 95 million barrels each day.

An oil refinery in Mina-Al-Ahmadi, Kuwait

Concern over the depletion of the earth's finite reserves of oil, and the effect this would have on a society dependent on it, is a concept known as peak oil. The use of fossil fuels, such as petroleum will have a negative impact on Earth's biosphere, damaging ecosystems through events such as oil spills and releasing a range of pollutants into the air including ground-level ozone and sulfur dioxide from sulfur impurities in fossil fuels. The burning of fossil fuels plays the major role in the current episode of global warming.

Natural petroleum spring in Korňa, Slovakia

Etymology

The term was found (in the spelling "petraoleum") in 10th-century Old English sources. It was used in the treatise De Natura Fossilium, published in 1546 by the German mineralogist Georg Bauer, also known as Georgius Agricola. In the 19th century, the term *petroleum* was often used to refer to mineral oils produced by distillation from mined organic solids such as cannel coal (and later oil shale), and refined oils produced from them; in the United Kingdom, storage (and later transport) of these oils were regulated by a series of Petroleum Acts, from the *Petroleum Act 1863* onwards. Petroleum.

History

Early History

Oil derrick in Okemah, Oklahoma, 1922

Petroleum, in one form or another, has been used since ancient times, and is now important across society, including in economy, politics and technology. The rise in importance was due to the invention of the internal combustion engine, the rise in commercial aviation, and the importance of petroleum to industrial organic chemistry, particularly the synthesis of plastics, fertilizers, solvents, adhesives and pesticides.

More than 4000 years ago, according to Herodotus and Diodorus Siculus, asphalt was used in the construction of the walls and towers of Babylon; there were oil pits near Ardericca (near Babylon), and a pitch spring on Zacynthus. Great quantities of it were found on the banks of the river Issus, one of the tributaries of the Euphrates. Ancient Persiantablets indicate the medicinal and lighting uses of petroleum in the upper levels of their society. By 347 AD, oil was produced from bamboo-drilled wells in China. Early British explorers to Myanmar documented a flourishing oil extraction industry based in Yenangyaung that, in 1795, had hundreds of hand-dug wells under production. The mythological origins of the oil fields at Yenangyaung, and its hereditary monopoly control by 24 families, indicate very ancient origins.

Pechelbronn (Pitch fountain) is said to be the first European site where petroleum has been explored and used. The still active Erdpechquelle, a spring where petroleum appears mixed with water has been used since 1498, e.g. for medical purposes. Oil sands have been mined since the 18th century.

In Wietze in lower Saxony, natural asphalt/bitumen has been explored since the 18th century. Both in Pechelbronn as in Wietze, the coal industry dominated the petroleum technologies.

Modern History

Scottish chemist James Young noticed a natural petroleum seepage in the Riddingscolliery at Alfreton, Derbyshire from which he distilled a light thin oil suitable for use as lamp oil, at the same time obtaining a thicker oil suitable for lubricating machinery. In 1848 Young set up a small business refining the crude oil.

Shale bings near Broxburn, 3 of a total of 19 in West Lothian

Young eventually succeeded, by distilling cannel coal at a low heat, in creating a fluid resembling petroleum, which when treated in the same way as the seep oil gave similar products. Young found that by slow distillation he could obtain a number of useful liquids from it, one of which he named "paraffine oil" because at low temperatures it congealed into a substance resembling paraffin wax.

The production of these oils and solid paraffin wax from coal formed the subject of his patent dated 17 October 1850. In 1850 Young & Meldrum and Edward William Binney entered into partnership under the title of E.W. Binney & Co. at Bathgate in West Lothian and E. Meldrum & Co. at Glasgow; their works at Bathgate were completed in 1851 and became the first truly commercial oil-works in the world with the first modern oil refinery, using oil extracted from locally mined torbanite, shale, and bituminous coal to manufacture naphtha and lubricating oils; paraffin for fuel use and solid paraffin were not sold until 1856.

The world's first oil refinery was built in 1856 by Ignacy Łukasiewicz. His achievements also included the discovery of how to distill kerosene from seep oil, the invention of the modern kerosene lamp (1853), the introduction of the first modern street lamp in Europe (1853), and the construction of the world's first modern oil well (1854).

The demand for petroleum as a fuel for lighting in North America and around the world quickly grew.Edwin Drake's 1859 well near Titusville, Pennsylvania, is popularly considered the first modern well. Already 1858 Georg Christian Konrad Hunäus had found a significant amount of petroleum while drilling for lignite 1858 in Wietze, Germany. Wietze later provided about 80% of the German consumption in the Wilhelminian Era. The production stopped in 1963, but Wietze has hosted a Petroleum Museum since 1970.

Drake's well is probably singled out because it was drilled, not dug; because it used a steam engine; because there was a company associated with it; and because it touched off a major boom. However, there was considerable activity before Drake in various parts of the world in the mid-19th century. A group directed by Major Alexeyev of the Bakinskii Corps of Mining Engineers hand-drilled a well in the Baku region in 1848. There were engine-drilled wells in West Virginia in the same year as Drake's well. An early commercial well was hand dug in Poland in 1853, and another in nearby Romania in 1857. At around the same time the world's first, small, oil refinery was opened at Jasło in Poland, with a larger one opened at Ploiesti in Romania shortly after. Romania is the first country in the world to have had its annual crude oil output officially recorded in international statistics: 275 tonnes for 1857.

The first commercial oil well in Canada became operational in 1858 at Oil Springs, Ontario (then Canada West). Businessman James Miller Williams dug several wells between 1855 and 1858 before discovering a rich reserve of oil four metres below ground. Williams extracted 1.5 million litres of crude oil by 1860, refining much of it into kerosene lamp oil. Williams's well became commercially viable a year before Drake's Pennsylvania operation and could be argued to be the first commercial oil well in North America. The discovery at Oil Springs touched off an oil boom which brought hundreds of speculators and workers to the area. Advances in drilling continued into 1862 when local driller Shaw reached a depth of 62 metres using the spring-pole drilling method. On January 16, 1862, after an explosion of natural gas Canada's first oil gusher came into production, shooting into the air at a recorded rate of 3,000 barrels per day. By the end of the 19th century the Russian Empire, particularly the Branobel company in Azerbaijan, had taken the lead in production.

Access to oil was and still is a major factor in several military conflicts of the twentieth century, including World War II, during which oil facilities were a major strategic asset and were extensively bombed. The German invasion of the Soviet Union included the goal to capture the Baku oilfields, as it would provide much needed oil-supplies for the German military which was suffering from blockades. Oil exploration in North America during the early 20th century later led to the US becoming the leading producer by mid-century. As petroleum production in the US peaked during the 1960s, however, the United States was surpassed by Saudi Arabia and the Soviet Union.

Today, about 90 percent of vehicular fuel needs are met by oil. Petroleum also makes up 40 percent of total energy consumption in the United States, but is responsible for only 1 percent of electricity generation.Petroleum's worth as a portable, dense energy source powering the vast majority of vehicles and as the base of many industrial chemicals makes it one of the world's most important commodities. Viability of the oil commodity is controlled by several key parameters, number of vehicles in the world competing for fuel, quantity of oil exported to the world market (Export Land Model), Net Energy Gain (economically useful energy provided minus energy consumed), political stability of oil exporting nations and ability to defend oil supply lines.

The top three oil producing countries are Russia, Saudi Arabia and the United States. About 80 percent of the world's readily accessible reserves are located in the Middle East, with 62.5 percent coming from the Arab 5: Saudi Arabia, UAE, Iraq, Qatar and Kuwait. A large portion of the world's total oil exists as unconventional sources, such as bitumen in Canada and extra heavy oil in Venezuela. While significant volumes of oil are extracted from oil sands, particularly in Canada, logistical and technical hurdles remain, as oil extraction requires large amounts of heat and water, making its net energy content quite low relative to conventional crude oil. Thus, Canada's oil sands are not expected to provide more than a few million barrels per day in the foreseeable future.

Composition

In its strictest sense, petroleum includes only crude oil, but in common usage it includes all liquid, gaseous and solid hydrocarbons. Under surface pressure and temperature conditions, lighter hydrocarbons methane, ethane, propane and butane occur as gases, while pentane and heavier hydrocarbons are in the form of liquids or solids. However, in an underground oil reservoir the proportions of gas, liquid, and solid depend on subsurface conditions and on the phase diagram of the petroleum mixture.

An oil well produces predominantly crude oil, with some natural gas dissolved in it. Because the pressure is lower at the surface than underground, some of the gas will come out of solution and be recovered (or burned) as *associated gas* or *solution gas*. A gas well produces predominantly natural gas. However, because the underground temperature and pressure are higher than at the surface, the gas may contain heavier hydrocarbons such as pentane, hexane, and heptane in the gaseous state. At surface conditions these will condense out of the gas to form natural gas condensate, often shortened to *condensate*. Condensate resembles gasoline in appearance and is similar in composition to some volatilelight crude oils.

The proportion of light hydrocarbons in the petroleum mixture varies greatly among different oil fields, ranging from as much as 97 percent by weight in the lighter oils to as little as 50 percent in the heavier oils and bitumens.

The hydrocarbons in crude oil are mostly alkanes, cycloalkanes and various aromatic hydrocarbons, while the other organic compounds contain nitrogen, oxygen and sulfur, and trace amounts of metals such as iron, nickel, copper and vanadium. Many oil reservoirs contain live bacteria. The exact molecular composition of crude oil varies widely from formation to formation but the proportion of chemical elements varies over fairly narrow limits as follows:

Total World Oil Reserves

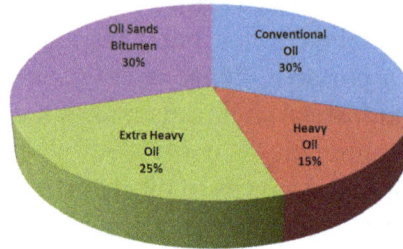

Most of the world's oils are non-conventional.

Composition by weight	
Element	**Percent range**
Carbon	83 to 85%
Hydrogen	10 to 14%
Nitrogen	0.1 to 2%
Oxygen	0.05 to 1.5%
Sulfur	0.05 to 6.0%
Metals	< 0.1%

Four different types of hydrocarbon molecules appear in crude oil. The relative percentage of each varies from oil to oil, determining the properties of each oil.

Composition by weight		
Hydrocarbon	**Average**	**Range**
Alkanes (paraffins)	30%	15 to 60%
Naphthenes	49%	30 to 60%
Aromatics	15%	3 to 30%
Asphaltics	6%	remainder

Crude oil varies greatly in appearance depending on its composition. It is usually black or dark brown (although it may be yellowish, reddish, or even greenish). In the reservoir it is usually found in association with natural gas, which being lighter forms a "gas cap" over the petroleum, and saline water which, being heavier than most forms of crude oil, generally sinks beneath it. Crude oil may also be found in a semi-solid form mixed with sand and water, as in the Athabasca oil sands in Canada, where it is usually referred to as crude bitumen. In Canada, bitumen is considered a sticky, black, tar-like form of crude oil which is so thick and heavy that it must be heated or diluted before it will flow. Venezuela also has large amounts of oil in the Orinoco oil sands, although the hydrocarbons trapped in them are more fluid than in Canada and are usually called extra heavy oil. These oil sands resources are called unconventional oil to distinguish them from oil which can be extracted using traditional oil well methods. Between them, Canada and Venezuela contain an estimated 3.6 trillion barrels (570×10^9 m³) of bitumen and extra-heavy oil, about twice the volume of the world's reserves of conventional oil.

Petroleum is used mostly, by volume, for producing fuel oil and gasoline, both important "primary energy" sources. 84 percent by volume of the hydrocarbons present in petroleum is converted into energy-rich fuels (petroleum-based fuels), including gasoline, diesel, jet, heating, and other fuel oils, and liquefied petroleum gas. The lighter grades of crude oil produce the best yields of these products, but as the world's reserves of light and medium oil are depleted, oil refineries are increasingly having to process heavy oil and bitumen, and use more complex and expensive methods to produce the products required. Because heavier crude oils have too much carbon and not enough hydrogen, these processes generally involve removing carbon from or adding hydrogen to the molecules, and using fluid catalytic cracking to convert the longer, more complex molecules in the oil to the shorter, simpler ones in the fuels.

Due to its high energy density, easy transportability and relative abundance, oil has become the world's most important source of energy since the mid-1950s. Petroleum is also the raw material for many chemical products, including pharmaceuticals, solvents, fertilizers, pesticides, and plastics; the 16 percent not used for energy production is converted into these other materials. Petroleum is found in porousrock formations in the upper strata of some areas of the Earth's crust. There is also petroleum in oil sands (tar sands). Known oil reserves are typically estimated at around 190 km³ (1.2 trillion(short scale)barrels) without oil sands, or 595 km³ (3.74 trillion barrels) with oil sands. Consumption is currently around 84 million barrels (13.4×10^6 m³) per day, or 4.9 km³ per year, yielding a remaining oil supply of only about 120 years, if current demand remains static.

Chemistry

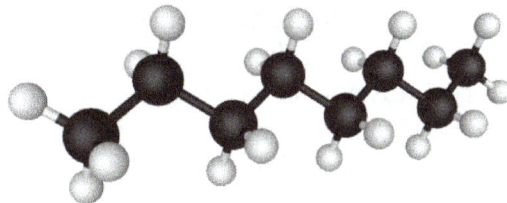

Octane, a hydrocarbon found in petroleum. Lines represent single bonds; black spheres represent carbon; white spheres represent hydrogen.

Petroleum is a mixture of a very large number of different hydrocarbons; the most commonly found molecules are alkanes (paraffins), cycloalkanes (naphthenes), aromatic hydrocarbons, or more complicated chemicals like asphaltenes. Each petroleum variety has a unique mix of molecules, which define its physical and chemical properties, like color and viscosity.

The *alkanes*, also known as *paraffins*, are saturated hydrocarbons with straight or branched chains which contain only carbon and hydrogen and have the general formula C_nH_{2n+2}. They generally have from 5 to 40 carbon atoms per molecule, although trace amounts of shorter or longer molecules may be present in the mixture.

The alkanes from pentane (C_5H_{12}) to octane (C_8H_{18}) are refined into gasoline, the ones from nonane (C_9H_{20}) to hexadecane ($C_{16}H_{34}$) into diesel fuel, kerosene and jet fuel. Alkanes with more than 16 carbon atoms can be refined into fuel oil and lubricating oil. At the heavier end of the range, paraffin wax is an alkane with approximately 25 carbon atoms, while asphalt has 35 and up, although these are usually cracked by modern refineries into more valuable products. The shortest molecules, those with four or fewer carbon atoms, are in a gaseous state at room temperature. They are the petroleum gases. Depending on demand and the cost of recovery, these gases are either flared off, sold as liquefied petroleum gas under pressure, or used to power the refinery's own burners. During the winter, butane (C_4H_{10}), is blended into the gasoline pool at high rates, because its high vapor pressure assists with cold starts. Liquified under pressure slightly above atmospheric, it is best known for powering cigarette lighters, but it is also a main fuel source for many developing countries. Propane can be liquified under modest pressure, and is consumed for just about every application relying on petroleum for energy, from cooking to heating to transportation.

The *cycloalkanes*, also known as *naphthenes*, are saturated hydrocarbons which have one or more carbon rings to which hydrogen atoms are attached according to the formula C_nH_{2n}. Cycloalkanes have similar properties to alkanes but have higher boiling points.

The *aromatic hydrocarbons* are unsaturated hydrocarbons which have one or more planar six-carbon rings called benzene rings, to which hydrogen atoms are attached with the formula C_nH_{2n-6}. They tend to burn with a sooty flame, and many have a sweet aroma. Some are carcinogenic.

These different molecules are separated by fractional distillation at an oil refinery to produce gasoline, jet fuel, kerosene, and other hydrocarbons. For example, 2,2,4-trimethylpentane (isooctane), widely used in gasoline, has a chemical formula of C_8H_{18} and it reacts with oxygen exothermically:

$$2\ C8H18_{(l)} + 25\ O2_{(g)} \rightarrow 16\ CO2_{(g)} + 18\ H2O_{(g)}\ (\Delta H = -5.51\ \text{MJ/mol of octane})$$

The number of various molecules in an oil sample can be determined by laboratory analysis. The molecules are typically extracted in a solvent, then separated in a gas chromatograph, and finally determined with a suitable detector, such as a flame ionization detector or a mass spectrometer. Due to the large number of co-eluted hydrocarbons within oil, many cannot be resolved by traditional gas chromatography and typically appear as a hump in the chromatogram. This unresolved complex mixture (UCM) of hydrocarbons is particularly apparent when analysing weathered oils and extracts from tissues of organisms exposed to oil.

Incomplete combustion of petroleum or gasoline results in production of toxic byproducts. Too little oxygen results in carbon monoxide. Due to the high temperatures and high pressures involved, exhaust gases from gasoline combustion in car engines usually include nitrogen oxides which are responsible for creation of photochemical smog.

Empirical Equations for Thermal Properties

Heat of Combustion

At a constant volume the heat of combustion of a petroleum product can be approximated as follows:

$$Q_v = 12400, -2,100d^2.$$

where Q_v is measured in cal/gram and d is the specific gravity at 60 °F (16 °C).

Thermal Conductivity

The thermal conductivity of petroleum based liquids can be modeled as follows:

$$K = \frac{1.62}{API}[1 - 0.0003(t - 32)]$$

where K is measured in $BTU \cdot °F^{-1}hr^{-1}ft^{-1}$, t is measured in °F and API is degrees API gravity.

Specific Heat

The specific heat of petroleum oils can be modeled as follows:

$$c = \frac{1}{d}[0.388 + 0.00046t],$$

where c is measured in BTU/lbm-°F, t is the temperature in Fahrenheit and d is the specific gravity at 60 °F (16 °C).

In units of kcal/(kg·°C), the formula is:

$$c = \frac{1}{d}[0.4024 + 0.00081t],$$

where the temperature t is in Celsius and d is the specific gravity at 15 °C.

Latent Heat of Vaporization

The latent heat of vaporization can be modeled under atmospheric conditions as follows:

$$L = \frac{1}{d}[110.9 - 0.09t],$$

where L is measured in BTU/lbm, t is measured in °F and d is the specific gravity at 60 °F (16 °C).

In units of kcal/kg, the formula is:

$$L = \frac{1}{d}[194.4 - 0.162t],$$

where the temperature t is in Celsius and d is the specific gravity at 15 °C.

Formation

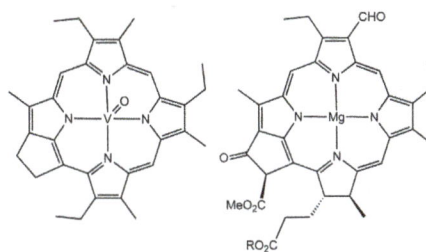

Structure of a vanadium porphyrin compound (left) extracted from petroleum by Alfred E. Treibs, father of organic geochemistry. Treibs noted the close structural similarity of this molecule and chlorophyll a (right).

Petroleum is a fossil fuel derived from ancient fossilizedorganic materials, such as zooplankton and algae. Vast quantities of these remains settled to sea or lake bottoms, mixing with sediments and being buried under anoxic conditions. As further layers settled to the sea or lake bed, intense heat and pressure build up in the lower regions. This process caused the organic matter to change, first into a waxy material known as kerogen, which is found in various oil shales around the world, and then with more heat into liquid and gaseous hydrocarbons via a process known as catagenesis. Formation of petroleum occurs from hydrocarbon pyrolysis in a variety of mainly endothermic reactions at high temperature and/or pressure.

There were certain warm nutrient-rich environments such as the Gulf of Mexico and the ancient Tethys Sea where the large amounts of organic material falling to the ocean floor exceeded the rate at which it could decompose. This resulted in large masses of organic material being buried under subsequent deposits such as shale formed from mud. This massive organic deposit later became heated and transformed under pressure into oil.

Geologists often refer to the temperature range in which oil forms as an "oil window"—below the minimum temperature oil remains trapped in the form of kerogen, and above the maximum temperature the oil is converted to natural gas through the process of thermal cracking. Sometimes, oil formed at extreme depths may migrate and become trapped at a much shallower level. The Athabasca Oil Sands are one example of this.

An alternative mechanism was proposed by Russian scientists in the mid-1850s, the hypothesis of abiogenic petroleum origin, but this is contradicted by geological and geochemical evidence.

Abiogenic (formed by inorganic means) sources of oil have been found, but never in commercially profitable amounts. The controversy isn't over whether abiogenic oil reserves exist, said Larry Nation of the American Association of Petroleum Geologists. The controversy is over how much they contribute to Earth's overall reserves and how much time and effort geologists should devote to seeking them out.

Reservoirs

Crude Oil Reservoirs

Hydrocarbon trap

Three conditions must be present for oil reservoirs to form: a source rock rich in hydrocarbon material buried deep enough for subterranean heat to cook it into oil, a porous and permeable reservoir rock for it to accumulate in, and a cap rock (seal) or other mechanism that prevents it from escaping to the surface. Within these reservoirs, fluids will typically organize themselves like a three-layer cake with a layer of water below the oil layer and a layer of gas above it, although the different layers vary in size between reservoirs. Because most hydrocarbons are less dense than rock or water, they often migrate upward through adjacent rock layers until either reaching the surface or becoming trapped within porous rocks (known as reservoirs) by impermeable rocks above. However, the process is influenced by underground water flows, causing oil to migrate hundreds of kilometres horizontally or even short distances downward before becoming trapped in a reservoir. When hydrocarbons are concentrated in a trap, an oil field forms, from which the liquid can be extracted by drilling and pumping.

The reactions that produce oil and natural gas are often modeled as first order breakdown reactions, where hydrocarbons are broken down to oil and natural gas by a set of parallel reactions, and oil eventually breaks down to natural gas by another set of reactions. The latter set is regularly used in petrochemical plants and oil refineries.

Wells are drilled into oil reservoirs to extract the crude oil. "Natural lift" production methods that rely on the natural reservoir pressure to force the oil to the surface are usually sufficient for a while after reservoirs are first tapped. In some reservoirs, such as in the Middle East, the natural pressure is sufficient over a long time. The natural pressure in most reservoirs, however, eventually dissipates. Then the oil must be extracted using "artificial lift" means. Over time, these "primary" methods become less effective and "secondary" production methods may be used. A common secondary method is "waterflood" or injection of water into the reservoir to increase pressure and force the oil to the drilled shaft or "wellbore." Eventually "tertiary" or "enhanced" oil recovery methods may be used to increase the oil's flow characteristics by injecting steam, carbon dioxide and other gases or chemicals into the reservoir. In the United States, primary production methods account for less than 40 percent of the oil produced on a daily basis, secondary methods account for about half, and tertiary recovery the remaining 10 percent. Extracting oil (or "bitumen") from oil/tar sand and oil shale deposits requires mining the sand or shale and heating it in a vessel or retort, or using "in-situ" methods of injecting heated liquids into the deposit and then pumping out the oil-saturated liquid.

Unconventional Oil Reservoirs

Oil-eating bacteria biodegrade oil that has escaped to the surface. Oil sands are reservoirs of partially biodegraded oil still in the process of escaping and being biodegraded, but they contain so much migrating oil that, although most of it has escaped, vast amounts are still present—more than can be found in conventional oil reservoirs. The lighter fractions of the crude oil are destroyed first, resulting in reservoirs containing an extremely heavy form of crude oil, called crude bitumen in Canada, or extra-heavy crude oil in Venezuela. These two countries have the world's largest deposits of oil sands.

On the other hand, oil shales are source rocks that have not been exposed to heat or pressure long enough to convert their trapped hydrocarbons into crude oil. Technically speaking, oil shales are not always shales and do not contain oil, but are fined-grain sedimentary rocks containing an insoluble organic solid called kerogen. The kerogen in the rock can be converted into crude oil using heat and pressure to simulate natural processes. The method has been known for centuries and was patented in 1694 under British Crown Patent No. 330 covering, "A way to extract and make great quantities of pitch, tar, and oil out of a sort of stone." Although oil shales are found in many countries, the United States has the world's largest deposits.

Classification

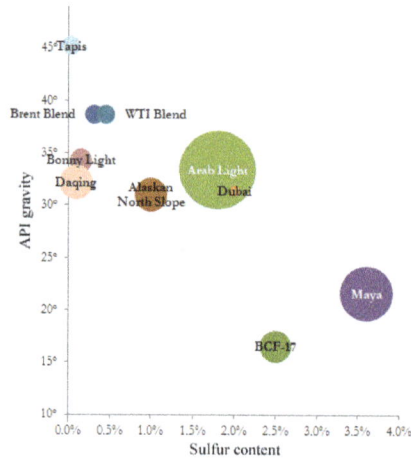

Some marker crudes with their sulfur content (horizontal) and API gravity (vertical) and relative production quantity

The petroleum industry generally classifies crude oil by the geographic location it is produced in (e.g. West Texas Intermediate, Brent, or Oman), its API gravity (an oil industry measure of density), and its sulfur content. Crude oil may be considered light if it has low density or heavy if it has high density; and it may be referred to as sweet if it contains relatively little sulfur or sour if it contains substantial amounts of sulfur.

The geographic location is important because it affects transportation costs to the refinery. *Light* crude oil is more desirable than *heavy* oil since it produces a higher yield of gasoline, while *sweet* oil commands a higher price than *sour* oil because it has fewer environmental problems and requires less refining to meet sulfur standards imposed on fuels in consuming countries. Each crude oil has unique molecular characteristics which are revealed by the use of Crude oil assay analysis in petroleum laboratories.

Barrels from an area in which the crude oil's molecular characteristics have been determined and the oil has been classified are used as pricing references throughout the world. Some of the common reference crudes are:

- West Texas Intermediate (WTI), a very high-quality, sweet, light oil delivered at Cushing, Oklahoma for North American oil

- Brent Blend, consisting of 15 oils from fields in the Brent and Ninian systems in the East Shetland Basin of the North Sea. The oil is landed at Sullom Voe terminal in Shetland. Oil production from Europe, Africa and Middle Eastern oil flowing West tends to be priced off this oil, which forms a benchmark

- Dubai-Oman, used as benchmark for Middle East sour crude oil flowing to the Asia-Pacific region

- Tapis (from Malaysia, used as a reference for light Far East oil)

- Minas (from Indonesia, used as a reference for heavy Far East oil)

- The OPEC Reference Basket, a weighted average of oil blends from various OPEC (The Organization of the Petroleum Exporting Countries) countries

- Midway Sunset Heavy, by which heavy oil in California is priced

- Western Canadian Select the benchmark crude oil for emerging heavy, high TAN (acidic) crudes.

There are declining amounts of these benchmark oils being produced each year, so other oils are more commonly what is actually delivered. While the reference price may be for West Texas Intermediate delivered at Cushing, the actual oil being traded may be a discounted Canadian heavy oil—Western Canadian Select— delivered at Hardisty, Alberta, and for a Brent Blend delivered at Shetland, it may be a discounted Russian Export Blend delivered at the port of Primorsk.

Petroleum Industry

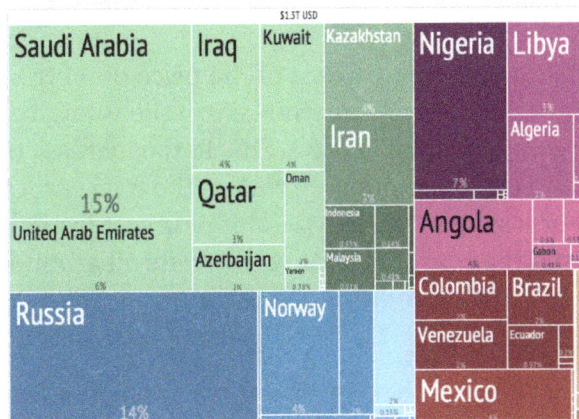

Crude oil export treemap (2012) from Harvard Atlas of Economic Complexity

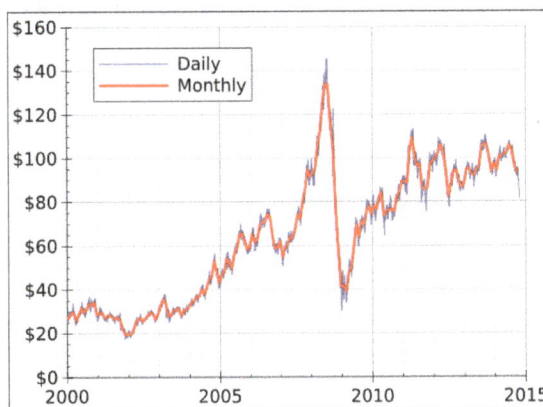

New York Mercantile Exchange prices ($/bbl) for West Texas Intermediate 2000 through Oct 2014

The petroleum industry is involved in the global processes of exploration, extraction, refining, transporting (often with oil tankers and pipelines), and marketing petroleum products. The largest volume products of the industry are fuel oil and gasoline. Petroleum is also the raw material for many chemical products, including pharmaceuticals, solvents, fertilizers, pesticides, and plastics. The industry is usually divided into three major components: upstream, midstream and downstream. Midstream operations are usually included in the downstream category.

Petroleum is vital to many industries, and is of importance to the maintenance of industrialized civilization itself, and thus is a critical concern to many nations. Oil accounts for a large percentage of the world's energy consumption, ranging from a low of 32 percent for Europe and Asia, up to a high of 53 percent for the Middle East, South and Central America (44%), Africa (41%), and North America (40%). The world at large consumes 30 billion barrels (4.8 km³) of oil per year, and the top oil consumers largely consist of developed nations. In fact, 24 percent of the oil consumed in 2004 went to the United States alone, though by 2007 this had dropped to 21 percent of world oil consumed.

In the US, in the states of Arizona, California, Hawaii, Nevada, Oregon and Washington, the Western States Petroleum Association (WSPA) represents companies responsible for producing, distributing, refining, transporting and marketing petroleum. This non-profit trade association was founded in 1907, and is the oldest petroleum trade association in the United States.

Shipping

In the 1950s, shipping costs made up 33 percent of the price of oil transported from the Persian Gulf to USA, but due to the development of supertankers in the 1970s, the cost of shipping dropped to only 5 percent of the price of Persian oil in USA. Due to the increase of the value of the crude oil during the last 30 years, the share of the shipping cost on the final cost of the delivered commodity was less than 3% in 2010. For example, in 2010 the shipping cost from the Persian Gulf to the USA was in the range of 20 $/t and the cost of the delivered crude oil around 800 $/t.

Price

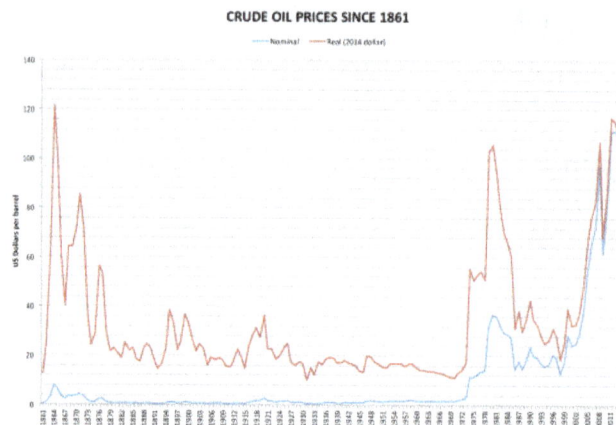

Nominal and inflation-adjusted US dollar price of crude oil, 1861–2015

After the collapse of the OPEC-administered pricing system in 1985, and a short-lived experiment with netback pricing, oil-exporting countries adopted a market-linked pricing mechanism. First adopted by PEMEX in 1986, market-linked pricing was widely accepted, and by 1988 became and still is the main method for pricing crude oil in international trade. The current reference, or pricing markers, are Brent, WTI, and Dubai/Oman.

Uses

The chemical structure of petroleum is heterogeneous, composed of hydrocarbon chains of different lengths. Because of this, petroleum may be taken to oil refineries and the hydrocarbon chemicals separated by distillation and treated by other chemical processes, to be used for a variety of purposes. The total cost of a plant is about 9 billion dollars per plant.

Fuels

A poster used to promote carpooling as a way to ration gasoline during World War II

The most common distillation fractions of petroleum are fuels. Fuels include (by increasing boiling temperature range):

Common fractions of petroleum as fuels	
Fraction	**Boiling range °C**
Liquefied petroleum gas (LPG)	−40
Butane	−12 to −1
Gasoline	−1 to 110
Jet fuel	150 to 205
Kerosene	205 to 260
Fuel oil	205 to 290
Diesel fuel	260 to 315

Petroleum classification according to chemical composition.

Class of petroleum	Composition of 250–300 °C fraction, wt. %				
	Par.	**Napth**	**Arom.**	**Wax**	**Asph.**
Paraffinic	46—61	22—32	12—25	1.5—10	0—6
Paraffinic-naphtenic	42—45	38—39	16—20	1—6	0—6
Naphthenic	15—26	61—76	8—13	Trace	0—6
Paraffinic-naphtenic-aromatic	27—35	36—47	26—33	0.5—1	0—10
Aromatic	0—8	57—78	20—25	0—0.5	0—20

Other Derivatives

Certain types of resultant hydrocarbons may be mixed with other non-hydrocarbons, to create other end products:

- Alkenes (olefins), which can be manufactured into plastics or other compounds

- Lubricants (produces light machine oils, motor oils, and greases, adding viscosity stabilizers as required)

- Wax, used in the packaging of frozen foods, among others

- Sulfur or sulfuric acid. These are useful industrial materials. Sulfuric acid is usually prepared as the acid precursor oleum, a byproduct of sulfur removal from fuels.

- Bulk tar

- Asphalt

- Petroleum coke, used in speciality carbon products or as solid fuel

- Paraffin wax

- Aromaticpetrochemicals to be used as precursors in other chemical production

Agriculture

Since the 1940s, agricultural productivity has increased dramatically, due largely to the increased use of energy-intensive mechanization, fertilizers and pesticides.

Petroleum by Country

Consumption Statistics

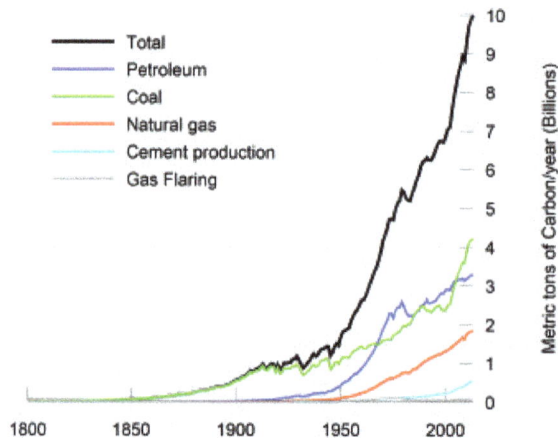

Global fossil carbon emissions, an indicator of consumption, for 1800–2007.

Total

Oil

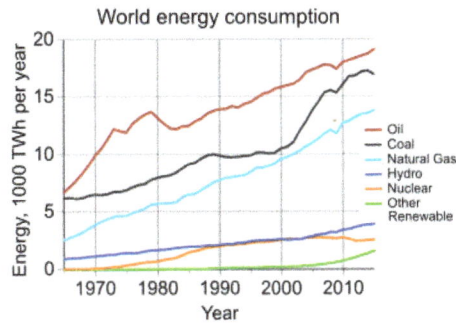

Rate of world energy usage per day, from 1970 to 2010. 1000 TWh = 1 PWh.

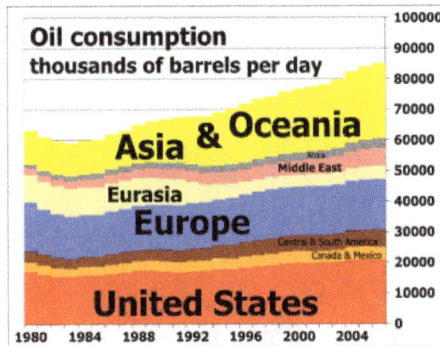

Daily oil consumption from 1980 to 2006

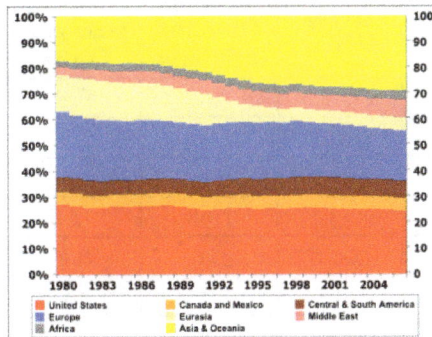

Oil consumption by percentage of total per region from 1980 to 2006:

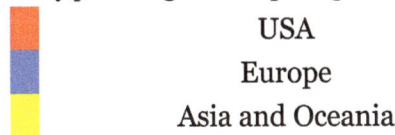

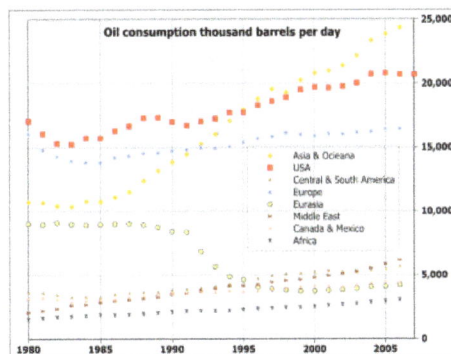

USA

Europe

Asia and Oceania

Oil consumption 1980 to 2007 by region

Consumption

According to the US Energy Information Administration (EIA) estimate for 2011, the world consumes 87.421 million barrels of oil each day.

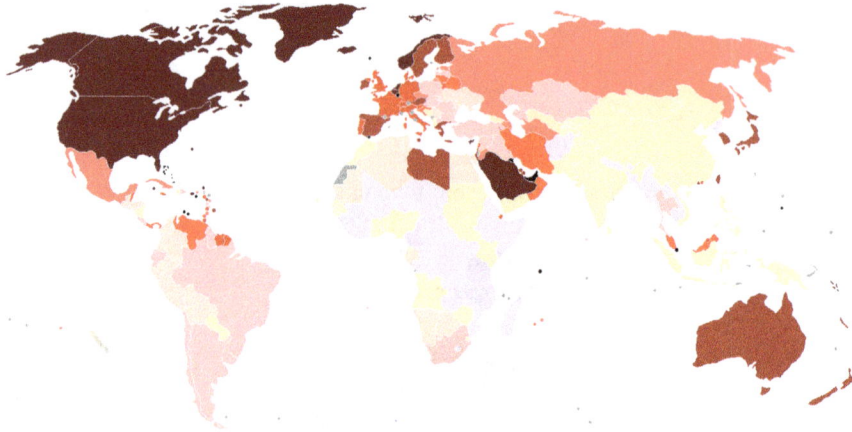

Oil consumption per capita (darker colors represent more consumption, gray represents no data)

This table orders the amount of petroleum consumed in 2011 in thousand barrels (1000 bbl) per day and in thousand cubic metres (1000 m³) per day:

Consuming nation 2011	(1000 bbl/ day)	(1000 m³/ day)	Population in millions	bbl/ year per capita	m³/year per capita	National production/ consumption
United States [1]	18,835.5	2,994.6	314	21.8	3.47	0.51
China	9,790.0	1,556.5	1345	2.7	0.43	0.41
Japan [2]	4,464.1	709.7	127	12.8	2.04	0.03
India [2]	3,292.2	523.4	1198	1	0.16	0.26
Russia [1]	3,145.1	500.0	140	8.1	1.29	3.35
Saudi Arabia (OPEC)	2,817.5	447.9	27	40	6.4	3.64
Brazil	2,594.2	412.4	193	4.9	0.78	0.99
Germany [2]	2,400.1	381.6	82	10.7	1.70	0.06
Canada	2,259.1	359.2	33	24.6	3.91	1.54
South Korea [2]	2,230.2	354.6	48	16.8	2.67	0.02
Mexico [1]	2,132.7	339.1	109	7.1	1.13	1.39
France [2]	1,791.5	284.8	62	10.5	1.67	0.03
Iran (OPEC)	1,694.4	269.4	74	8.3	1.32	2.54
United Kingdom [1]	1,607.9	255.6	61	9.5	1.51	0.93
Italy [2]	1,453.6	231.1	60	8.9	1.41	0.10

Source: US Energy Information Administration

Population Data:

1 peak production of oil already passed in this state

2 This country is not a major oil producer

Production

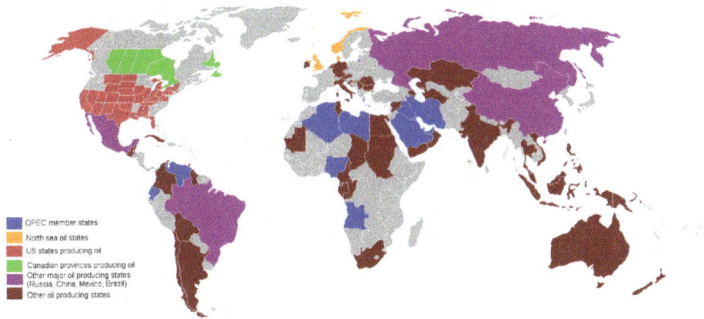

Oil producing countries

Selected Producers, 1973–2015

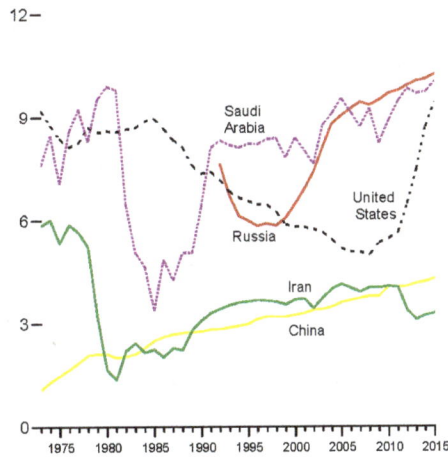

Top oil-producing countries(million barrels per day)

In petroleum industry parlance, *production* refers to the quantity of crude extracted from reserves, not the literal creation of the product.

#	Producing nation	10³bbl/d (2006)	10³bbl/d (2007)	10³bbl/d (2008)	10³bbl/d (2009)	Present share
1	Russia[1]	9,677	9,876	9,789	9,934	12.0%
2	Saudi Arabia **(OPEC)**	10,665	10,234	10,782	9,760	11.8%
3	United States[1]	8,331	8,481	8,514	9,141	11.1%
4	Iran **(OPEC)**	4,148	4,043	4,174	4,177	5.1%
5	China	3,846	3,901	3,973	3,996	4.8%
6	Canada[2]	3,288	3,358	3,350	3,294	4.0%
7	Mexico[1]	3,707	3,501	3,185	3,001	3.6%
8	United Arab Emirates **(OPEC)**	2,945	2,948	3,046	2,795	3.4%
9	Kuwait **(OPEC)**	2,675	2,613	2,742	2,496	3.0%
10	Venezuela **(OPEC)** [1]	2,803	2,667	2,643	2,471	3.0%
11	Norway[1]	2,786	2,565	2,466	2,350	2.8%
12	Brazil	2,166	2,279	2,401	2,577	3.1%

#	Producing nation	10³bbl/d (2006)	10³bbl/d (2007)	10³bbl/d (2008)	10³bbl/d (2009)	Present share
13	Iraq (OPEC) [3]	2,008	2,094	2,385	2,400	2.9%
14	Algeria (OPEC)	2,122	2,173	2,179	2,126	2.6%
15	Nigeria (OPEC)	2,443	2,352	2,169	2,211	2.7%
16	Angola (OPEC)	1,435	1,769	2,014	1,948	2.4%
17	Libya (OPEC)	1,809	1,845	1,875	1,789	2.2%
18	United Kingdom	1,689	1,690	1,584	1,422	1.7%
19	Kazakhstan	1,388	1,445	1,429	1,540	1.9%
20	Qatar (OPEC)	1,141	1,136	1,207	1,213	1.5%
21	Indonesia	1,102	1,044	1,051	1,023	1.2%
22	India	854	881	884	877	1.1%
23	Azerbaijan	648	850	875	1,012	1.2%
24	Argentina	802	791	792	794	1.0%
25	Oman	743	714	761	816	1.0%
26	Malaysia	729	703	727	693	0.8%
27	Egypt	667	664	631	678	0.8%
28	Colombia	544	543	601	686	0.8%
29	Australia	552	595	586	588	0.7%
30	Ecuador (OPEC)	536	512	505	485	0.6%
31	Sudan	380	466	480	486	0.6%
32	Syria	449	446	426	400	0.5%
33	Equatorial Guinea	386	400	359	346	0.4%
34	Thailand	334	349	361	339	0.4%
35	Vietnam	362	352	314	346	0.4%
36	Yemen	377	361	300	287	0.3%
37	Denmark	344	314	289	262	0.3%
38	Gabon	237	244	248	242	0.3%
39	South Africa	204	199	195	192	0.2%
40	Turkmenistan	No data	180	189	198	0.2%
41	Trinidad and Tobago[4]	181	179	176	174	0.1%

Source: U.S. Energy Information Administration

[1] Peak production of conventional oil already passed in this state

[2] Although Canada's conventional oil production is declining, its total oil production is increasing as oil sands production grows. When oil sands are included, Canada has the world's second largest oil reserves after Saudi Arabia.

[3] Though still a member, Iraq has not been included in production figures since 1998

[4] Trinidad and Tobago has the worlds third largest pitch lake situated La Brea south Trinidad

In 2013, the United States will produce an average of 11.4 million barrels a day, which would make it the second largest producer of hydrocarbons, and is expected to overtake Saudi Arabia before 2020.

Export

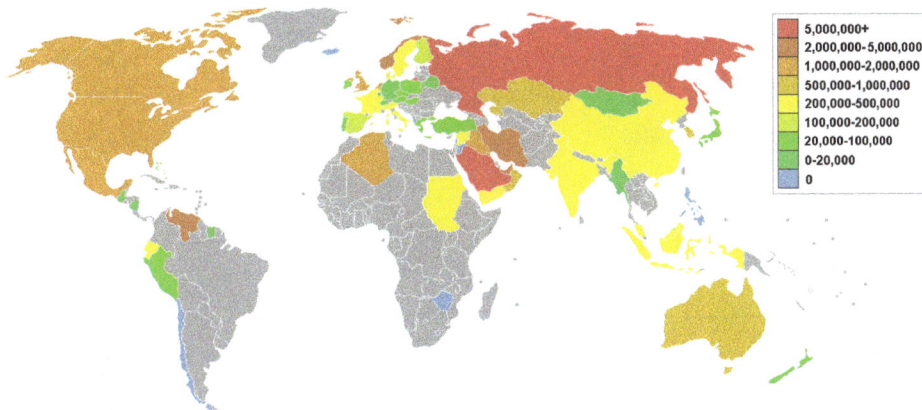

Oil exports by country

In order of net exports in 2011, 2009 and 2006 in thousand bbl/d and thousand m³/d:

#	Exporting nation	10^3bbl/d (2011)	10^3m³/d (2011)	10^3bbl/d (2009)	10^3m³/d (2009)	10^3bbl/d (2006)	10^3m³/d (2006)
1	Saudi Arabia (OPEC)	8,336	1,325	7,322	1,164	8,651	1,376
2	Russia[1]	7,083	1,126	7,194	1,144	6,565	1,044
3	Iran (OPEC)	2,540	403	2,486	395	2,519	401
4	United Arab Emirates (OPEC)	2,524	401	2,303	366	2,515	400
5	Kuwait (OPEC)	2,343	373	2,124	338	2,150	342
6	Nigeria (OPEC)	2,257	359	1,939	308	2,146	341
7	Iraq (OPEC)	1,915	304	1,764	280	1,438	229
8	Angola (OPEC)	1,760	280	1,878	299	1,363	217
9	Norway[1]	1,752	279	2,132	339	2,542	404
10	Venezuela (OPEC) [1]	1,715	273	1,748	278	2,203	350
11	Algeria (OPEC) [1]	1,568	249	1,767	281	1,847	297
12	Qatar (OPEC)	1,468	233	1,066	169	–	–
13	Canada[2]	1,405	223	1,168	187	1,071	170
14	Kazakhstan	1,396	222	1,299	207	1,114	177
15	Azerbaijan[1]	836	133	912	145	532	85
16	Trinidad and Tobago[1]	177	112	167	160	155	199

Source: US Energy Information Administration

1peak production already passed in this state

[2] Canadian statistics are complicated by the fact it is both an importer and exporter of crude oil, and refines large amounts of oil for the U.S. market. It is the leading source of U.S. imports of oil and products, averaging 2,500,000 bbl/d (400,000 m³/d) in August 2007.

Total world production/consumption (as of 2005) is approximately 84 million barrels per day (13,400,000 m³/d).

Import

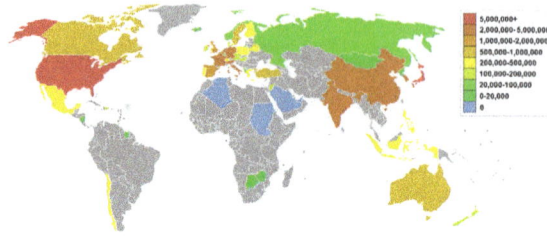

Oil imports by country

In order of net imports in 2011, 2009 and 2006 in thousand bbl/d and thousand m³/d:

#	Importing nation	10³bbl/day (2011)	10³m³/day (2011)	10³bbl/day (2009)	10³m³/day (2009)	10³bbl/day (2006)	10³m³/day (2006)
1	United States [1]	8,728	1,388	9,631	1,531	12,220	1,943
2	China [2]	5,487	872	4,328	688	3,438	547
3	Japan	4,329	688	4,235	673	5,097	810
4	India	2,349	373	2,233	355	1,687	268
5	Germany	2,235	355	2,323	369	2,483	395
6	South Korea	2,170	345	2,139	340	2,150	342
7	France	1,697	270	1,749	278	1,893	301
8	Spain	1,346	214	1,439	229	1,555	247
9	Italy	1,292	205	1,381	220	1,558	248
10	Singapore	1,172	186	916	146	787	125
11	Republic of China (Taiwan)	1,009	160	944	150	942	150
12	Netherlands	948	151	973	155	936	149
13	Turkey	650	103	650	103	576	92
14	Belgium	634	101	597	95	546	87
15	Thailand	592	94	538	86	606	96

Source: US Energy Information Administration

1 peak production of oil expected in 2020

2 Major oil producer whose production is still increasing

Oil Imports to the USA by Country 2010

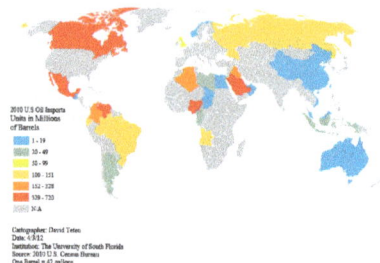

Oil imports to US 2010

Non-producing Consumers

Countries whose oil production is 10% or less of their consumption.

#	Consuming nation	(bbl/day)	(m³/day)
1	Japan	5,578,000	886,831
2	Germany	2,677,000	425,609
3	South Korea	2,061,000	327,673
4	France	2,060,000	327,514
5	Italy	1,874,000	297,942
6	Spain	1,537,000	244,363
7	Netherlands	946,700	150,513
8	Turkey	575,011	91,663

Source: CIA World Factbook

Environmental Effects

Diesel fuel spill on a road

Because petroleum is a naturally occurring substance, its presence in the environment need not be the result of human causes such as accidents and routine activities (seismic exploration, drilling, extraction, refining and combustion). Phenomena such as seeps and tar pits are examples of areas that petroleum affects without man's involvement. Regardless of source, petroleum's effects when released into the environment are similar.

Ocean Acidification

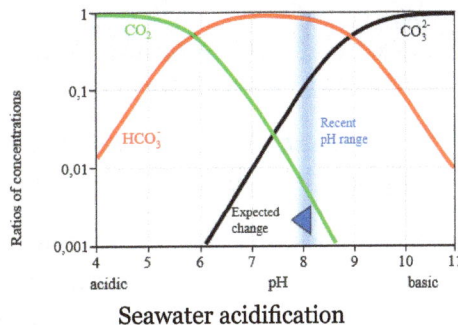

Seawater acidification

Ocean acidification is the increase in the acidity of the Earth's oceans caused by the uptake of carbon dioxide (CO_2) from the atmosphere. This increase in acidity inhibits all marine life - having a greater impact on smaller organisms as well as shelled organisms .

Global Warming

When burned, petroleum releases carbon dioxide, a greenhouse gas. Along with the burning of coal, petroleum combustion may be the largest contributor to the increase in atmospheric CO_2. Atmospheric CO_2 has risen over the last 150 years to current levels of over 390 ppmv, from the 180 – 300 ppmv of the prior 800 thousand years This rise in temperature may have reduced the Arctic ice cap to 1,100,000 sq mi (2,800,000 km²), smaller than ever recorded. Because of this melt, more oil reserves have been revealed. It is estimated by the International Energy Agency that about 13 percent of the world's undiscovered oil resides in the Arctic.

Extraction

Oil extraction is simply the removal of oil from the reservoir (oil pool). Oil is often recovered as a water-in-oil emulsion, and specialty chemicals called demulsifiers are used to separate the oil from water. Oil extraction is costly and sometimes environmentally damaging. Offshore exploration and extraction of oil disturbs the surrounding marine environment.

Oil Spills

Oil slick from the Montara oil spill in the Timor Sea, September, 2009

Volunteers cleaning up the aftermath of the Prestige oil spill

Crude oil and refined fuel spills from tanker ship accidents have damaged natural ecosystems in Alaska, the Gulf of Mexico, the Galapagos Islands, France and many other places.

The quantity of oil spilled during accidents has ranged from a few hundred tons to several hundred thousand tons (e.g., Deepwater Horizon Oil Spill, Atlantic Empress, Amoco Cadiz). Smaller spills have already proven to have a great impact on ecosystems, such as the Exxon Valdez oil spill

Oil spills at sea are generally much more damaging than those on land, since they can spread for hundreds of nautical miles in a thin oil slick which can cover beaches with a thin coating of oil. This can kill sea birds, mammals, shellfish and other organisms it coats. Oil spills on land are more readily containable if a makeshift earth dam can be rapidly bulldozed around the spill site before most of the oil escapes, and land animals can avoid the oil more easily.

Control of oil spills is difficult, requires ad hoc methods, and often a large amount of manpower. The dropping of bombs and incendiary devices from aircraft on the SS Torrey Canyon wreck produced poor results; modern techniques would include pumping the oil from the wreck, like in the Prestige oil spill or the Erika oil spill.

Though crude oil is predominantly composed of various hydrocarbons, certain nitrogen heterocylic compounds, such as pyridine, picoline, and quinoline are reported as contaminants associated with crude oil, as well as facilities processing oil shale or coal, and have also been found at legacy wood treatment sites. These compounds have a very high water solubility, and thus tend to dissolve and move with water. Certain naturally occurring bacteria, such as Micrococcus, Arthrobacter, and Rhodococcus have been shown to degrade these contaminants.

Tarballs

A tarball is a blob of crude oil which has been weathered after floating in the ocean. Tarballs are an aquatic pollutant in most environments, although they can occur naturally, for example in the Santa Barbara Channel of California or in the Gulf of Mexico off Texas. Their concentration and features have been used to assess the extent of oil spills. Their composition can be used to identify their sources of origin, and tarballs themselves may be dispersed over long distances by deep sea currents. They are slowly decomposed by bacteria, including Chromobacterium violaceum, *Cladosporium resinae, Bacillus submarinus, Micrococcus varians*, Pseudomonas aeruginosa, *Candida marina* and *Saccharomyces estuari.*

Whales

James S. Robbins has argued that the advent of petroleum-refined kerosene saved some species of great whales from extinction by providing an inexpensive substitute for whale oil, thus eliminating the economic imperative for open-boat whaling.

Alternatives to Petroleum

In the United States in 2007 about 70 percent of petroleum was used for transportation (e.g. gasoline, diesel, jet fuel), 24 percent by industry (e.g. production of plastics), 5 percent for residential and commercial uses, and 2 percent for electricity production. Outside of the US, a higher proportion of petroleum tends to be used for electricity.

Alternatives to Petroleum-based Vehicle Fuels

Alternative fuel vehicles refers to both:

- Vehicles that use alternative fuels used in standard or modified internal combustion engines such as natural gas vehicles, neat ethanol vehicles, flexible-fuel vehicles, biodiesel-powered vehicles, and hydrogen vehicles.

- Vehicles with advanced propulsion systems that reduce or substitute petroleum use such as battery electric vehicles, plug-in hybrid electric vehicles, hybrid electric vehicles, and hydrogenfuel cell vehicles.

Brazilian fuel station with four alternative fuels for sale: diesel (B3), gasohol (E25), neat ethanol (E100), and compressed natural gas (CNG)

Alternatives to Using Oil in Industry

Biological feedstocks do exist for industrial uses such as Bioplastic production. Hydroelectric energy could also be used instead of gasoline for cars, reducing the amount of oil needed for the world. Another, is the electric powered cars currently in production.

Alternatives to Burning Petroleum for Electricity

In oil producing countries with little refinery capacity, oil is sometimes burned to produce electricity. Renewable energy technologies such as solar power, wind power, micro hydro, biomass and biofuels are used, but the primary alternatives remain large-scale hydroelectricity, nuclear and coal-fired generation.

Future of Petroleum Production

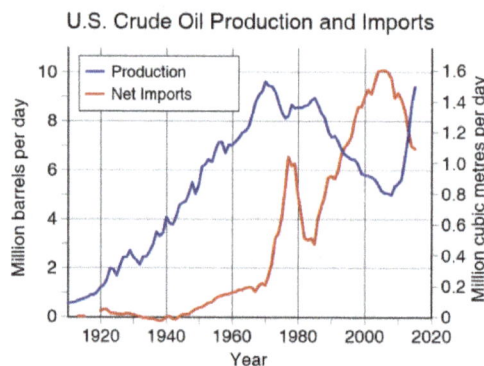

US oil production and imports, 1910-2012

Consumption in the twentieth and twenty-first centuries has been abundantly pushed by automobile growth; the 1985–2003 oil glut even fueled the sales of low economy vehicles in OECD countries. The 2008 economic crisis seems to have had some impact on the sales of such vehicles; still, the 2008 oil consumption shows a small increase. In 2016 however Goldman Sachs predicts lower demand due to emerging economies concerns, especially China. The BRICS (Brasil, Russia, India, China, South Africa) countries might also kick in, as China briefly was the first automobile market in December 2009. The immediate outlook still hints upwards. In the long term, uncertainties linger; the OPEC believes that the OECD countries will push low consumption policies at some point in the future; when that happens, it will definitely curb oil sales, and both OPEC and EIA kept lowering their 2020 consumption estimates during the past 5 years. Oil products are more and more in competition with alternative sources, mainly coal and natural gas, both cheaper sources. Production will also face an increasingly complex situation; while OPEC countries still have large reserves at low production prices, newly found reservoirs often lead to higher prices; offshore giants such as Tupi, Guara and Tiber demand high investments and ever-increasing technological abilities. Subsalt reservoirs such as Tupi were unknown in the twentieth century, mainly because the industry was unable to probe them. Enhanced Oil Recovery (EOR) techniques (example: DaQing, China) will continue to play a major role in increasing the world's recoverable oil. The reach of available petroleum ressources has always been around 35 years or even less since the start of the modern exploration. The oil constant, a insider pun in the German industry refers to that effect.

Peak Oil

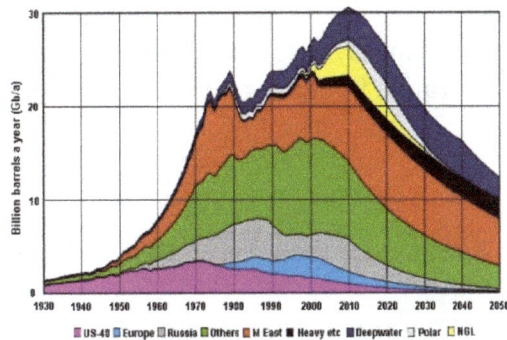

Global peak oil forecast

Peak oil is the projection that future petroleum production (whether for individual oil wells, entire oil fields, whole countries, or worldwide production) will eventually peak and then decline at a similar rate to the rate of increase before the peak as these reserves are exhausted. The peak of oil discoveries was in 1965, and oil production per year has surpassed oil discoveries every year since 1980. However, this does not mean that potential oil production has surpassed oil demand.

Hubbert applied his theory to accurately predict the peak of U.S. conventional oil production at a date between 1966 and 1970. This prediction was based on data available at the time of his publication in 1956. In the same paper, Hubbert predicts world peak oil in "half a century" after his publication, which would be 2006.

It is difficult to predict the oil peak in any given region, due to the lack of knowledge and/or transparency in accounting of global oil reserves. Based on available production data, proponents have previously predicted the peak for the world to be in years 1989, 1995, or 1995–2000. Some of these predictions date from before the recession of the early 1980s, and the consequent reduction in global consumption, the effect of which was to delay the date of any peak by several years. Just as the 1971 U.S. peak in oil production was only clearly recognized after the fact, a peak in world production will be difficult to discern until production clearly drops off. The peak is also a moving target as it is now measured as "liquids", which includes synthetic fuels, instead of just conventional oil.

The International Energy Agency (IEA) said in 2010 that production of conventional crude oil had peaked in 2006 at 70 MBBL/d, then flattened at 68 or 69 thereafter. Since virtually all economic sectors rely heavily on petroleum, peak oil, if it were to occur, could lead to a "partial or complete failure of markets". In the mid-2000s, widespread fears of an imminent peak led to the "peak oil movement," in which over one hundred thousand Americans prepared, individually and collectively, for the "post-carbon" future.

Unconventional Production

The calculus for peak oil has changed with the introduction of unconventional production methods. In particular, the combination of horizontal drilling and hydraulic fracturing has resulted in a significant increase in production from previously uneconomic plays. Analysts expect that $150 billion will be spent on further developing North American tight oil fields in 2015. The large increase in tight oil production is one of the reasons behind the price drop in late 2014. Certain rock strata contain hydrocarbons but have low permeability and are not thick from a vertical perspective. Conventional vertical wells would be unable to economically retrieve these hydrocarbons. Horizontal drilling, extending horizontally through the strata, permits the well to access a much greater volume of the strata. Hydraulic fracturing creates greater permeability and increases hydrocarbon flow to the wellbore.

References

- Longmuir, Marilyn V. "Oil in Burma: The Extraction of "Earth Oil" to 1914". Bangkok: White Lotus (2001) ISBN 974-7534-60-6 pp.329

- Rainer Karlsch, Raymond G. Stokes: Faktor Öl. Die Mineralölwirtschaft in Deutschland 1859–1974. Verlag C. H. Beck, München, 2003. ISBN 3-406-50276-8

- Russell, Loris S. (2003). A Heritage of Light: Lamps and Lighting in the Early Canadian Home. University of Toronto Press. ISBN 0-8020-3765-8.

- Frank, Alison Fleig (2005). Oil Empire: Visions of Prosperity in Austrian Galicia (Harvard Historical Studies). Harvard University Press. ISBN 0-674-01887-7.

- Polar Prospects:A minerals treaty for Antarctica. United States, Office of Technology Assessment. September 1989. p. 104. ISBN 978-1-4289-2232-7.

- Natural Resources Canada (May 2011). Canadian Crude Oil, Natural Gas and Petroleum Products: Review of 2009 & Outlook to 2030 (PDF) (Report). Ottawa, ON: Government of Canada. p. 9. ISBN 978-1-100-16436-6.

- Simanzhenkov, Vasily; Idem, Raphael (2003). Crude Oil Chemistry. CRC Press,. p. 33. ISBN 0203014049. Retrieved 10 November 2014.

- Schneider-Mayerson Matthew (2015). Peak Oil: Apocalyptic Environmentalism and Libertarian Political Culture. University of Chicago Press. ISBN 978-0-226-28543-6.

- Hume, Neil; Editor, Commodities (2016-03-08). "Goldman Sachs says commodity rally is unlikely to last". Financial Times. ISSN 0307-1766. Retrieved 2016-03-08.

- Ovale, Peder. "Her ser du hvorfor oljeprisen faller" In English Teknisk Ukeblad, 11 December 2014. Accessed: 11 December 2014.

- "Gasoline as Fuel – History of Word Gasoline – Gasolin and Petroleum Origins". Alternativefuels.about.com. 2013-07-12. Retrieved 2013-08-27.

- "A liquid market: Thanks to LNG, spare gas can now be sold the world over". The Economist. 14 July 2012. Retrieved 6 January 2013.

- Mark Thompson (12 November 2012). "U.S. to become biggest oil producer – IEA". CNN. Retrieved 9 February 2013.

Sub-fields of Petroleum Engineering

Petroleum engineering is an interdisciplinary subject. Some of the sub-fields related to this subject are reservoir engineering, drilling engineering, petroleum production engineering and subsurface engineering. Reservoir engineering is the branch of petroleum engineering that concerns itself with the issues caused because of the development and production of oil and gas reservoirs. This text will provide a glimpse of the related fields of petroleum engineering briefly.

Reservoir Engineering

Screenshot of a structure map generated by Contour map software for an 8500ft deep gas &Oil reservoir in the Earth field, Vermilion Parish, Erath, Louisiana. The left-to-right gap, near the top of the contour map indicates a fault. This fault line is between the blue/green contour lines and the purple/red/yellow contour lines. The thin red circular contour line in the middle of the map indicates the top of the oil reservoir. Because gas floats above oil, the thin red contour line marks the gas/oil contact zone. Reservoir engineers could use the map as a part of their well drill planning.

Reservoir engineering is a branch of petroleum engineering that applies scientific principles to the drainage problems arising during the development and production of oil and gas reservoirs so as to obtain a high economic recovery. The working tools of the reservoir engineer are subsurface geology, applied mathematics, and the basic laws of physics and chemistry governing the behavior of liquid and vapor phases of crude oil, natural gas, and water in reservoir rock. Of particular interest to reservoir engineers is generating accurate reserves estimates for use in financial reporting to the SEC and other regulatory bodies. Other job responsibilities include numerical reservoir modeling, production forecasting, well testing, well drilling and workover planning, economic modeling, and PVT analysis of reservoir fluids. Reservoir engineers also play a central role in field development planning, recommending appropriate and cost effective reservoir depletion schemes such as waterflooding or gas injection to maximize hydrocarbon recovery. Due to legislative changes in many hydrocarbon producing countries, they are also involved in the design and implementation of carbon sequestration projects in order to minimise the emission of greenhouse gases.

Types

Reservoir engineers often specialize in two areas:

- Surveillance (or production) engineering, i.e. monitoring of existing fields and optimization of production and injection rates. Surveillance engineers typically use analytical and empirical techniques to perform their work, including decline curve analysis, material balance modeling, and inflow/outflow analysis.

- Simulation modeling, i.e. the conduct of reservoir simulation studies to determine optimal development plans for oil and gas reservoirs. Also, reservoir engineers perform and integrate well tests into their data for reservoirs in geothermal drilling.

Drilling Engineering

Drilling engineering is a subset of petroleum engineering.

Drilling engineers design and implement procedures to drill wells as safely and economically as possible. They work closely with the drilling contractor, service contractors, and compliance personnel, as well as with geologists and other technical specialists. The drilling engineer has the responsibility for ensuring that costs are minimized while getting information to evaluate the formations penetrated, protecting the health and safety of workers and other personnel, and protecting the environment.

Overview

The planning phases involved in drilling an oil or gas well typically involve estimating the value of sought reserves, estimating the costs to access reserves, acquiring property by a mineral lease, a geological survey, a well bore plan, and a layout of the type of equipment required to reach the depth of the well. Drilling engineers are in charge of the process of planning and drilling the wells. Their responsibilities include:

1. Designing well programs (e.g., casing sizes and setting depths) to prevent blowouts (uncontrolled well-fluid release) while allowing adequate formation evaluation.
2. Designing or contributing to the design of casing strings and cementing plans, directional drilling plans, drilling fluids programs, and drill string and drill bit programs.
3. Specifying equipment, material and ratings and grades to be used in the drilling process.
4. Providing technical support and audit during the drilling process.
5. Performing cost estimates and analysis.
6. Developing contracts with vendors.

Drilling engineers are often degreed as petroleum engineers, although they may come from other technical disciplines (e.g., mechanical engineering or geology) and subsequently be trained by an oil and gas company. They also may have practical experience as a rig hand or mudlogger or mud engineer.

Petroleum Production Engineering

Petroleum production engineering is a subset of petroleum engineering.

Petroleum production engineers design and select subsurface equipment to produce oil and gas well fluids. They often are degreed as petroleum engineers, although they may come from other technical disciplines (e.g., mechanical engineering, chemical engineering, physics) and subsequently be trained by an oil and gas company.

Overview

Petroleum production engineers' responsibilities include:

1. Evaluating inflow and outflow performance between the reservoir and the wellbore.

2. Designing completion systems, including tubing selection, perforating, sand control, matrix stimulation, and hydraulic fracturing.

3. Selecting artificial lift equipment, including sucker-rod lift (typically beam pumping), gas lift, electrical submersible pumps, subsurface hydraulic pumps, progressing-cavity pumps, and plunger lift.

4. Selecting (not design) equipment for surface facilities that separate and measure the produced fluids (oil, natural gas, water, and impurities), prepare the oil and gas for transportation to market, and handle disposal of any water and impurities.

Note: Surface equipments are designed by Chemical engineers and Mechanical engineers according to data provided by the production engineers.

Subsurface Engineer

Subsurface engineers (also known as Completion engineers) are a subset within Petroleum Engineering and typically work closely with Drilling engineers. The job of a Subsurface Engineer is to effectively select equipment that will best suit the subsurface environment in order to best produce the hydrocarbon reserves. Once the hardware has been selected, a Subsurface Engineer will monitor and adjust the equipment to ensure the well and reservoir produces under ideal circumstances.

Overview

Subsurface engineers must design a successful well completion system by selecting equipment that is adequate for both downhole environments and applications. Considerations must be given to the various functions under which the completion equipment must operate and the effects any changes in temperatures or differential pressure will have on the equipment. The completion system must also be efficient and cost effective to achieve maximum production and financial goals. Another factor in the selection of specific completion equipment is the production rates of the well. The typical job duties of a Subsurface engineer include managing the interface between the reservoir and the well, including perforations, sand control, artificial lift, downhole flow control, and downhole monitoring equipment. Additional responsibilities of a Subsurface engineer include: performing a cost and risk analysis on the design, contacting vendors for the rental, purchase, and shipment of equipment, and working closely with fellow employees (geologists, reservoir engineers, drilling engineers, and production engineers).

Design Components

The design components considered to perform a well completion may include:

- Cost and risk analysis

- Determining the specifications for the wellbore clean-out

- Use of specific Packer assemblies

- Determining specific tool selection to operate equipment within the well

- Assess possible equipment load specifications and incorporation of safety factors

- Best use of flow control accessories (sliding sleeves and safety valves)

- Determining the appropriate perforating shots per foot and charges based on the target formations

- Acidifying the formation to inhibit flow of hydrocarbons

- Sand Control operations to increase production

- Prevention of formation sand production with the use of wire screens

- Review Well logs to determine equipment placement within the well

- Determination of specific production pipe regarding well flow rates

- Selection of equipment to maintain well stability

- Oversee completion operations

References

- Clegg, Joe Dunn, ed. (2007). Petroleum-Engineering-Handbook-Volume-IV-Production-Operations-Engineering. Dallas, Texas: Society of Petroleum Engineers. p. 900. ISBN 978-1-55563-118-5.

Theories and Concepts Related to Petroleum Engineering

The Hubbert peak theory contents that the production of petroleum always tends to follow a bell-shaped curve. The following text also explains fractional distillation and octane rating. The topics discussed in the chapter are of great importance to broaden the existing knowledge on petroleum engineering.

Hubbert Peak Theory

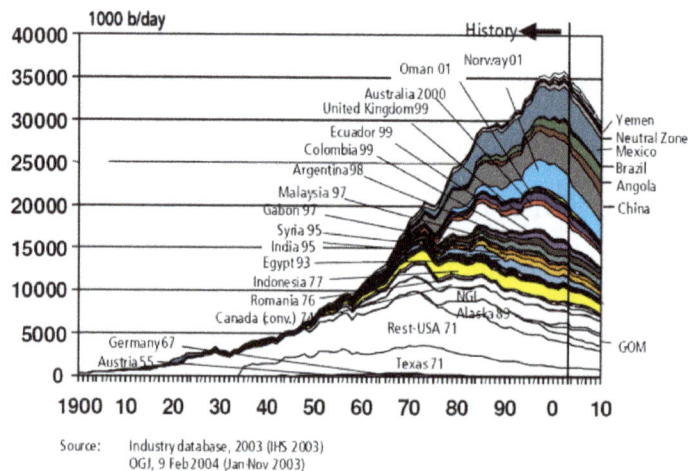

2004 U.S. government predictions for oil production other than in OPEC and the former Soviet Union

The Hubbert peak theory says that for any given geographical area, from an individual oil-producing region to the planet as a whole, the rate of petroleum production tends to follow a bell-shaped curve. It is one of the primary theories on peak oil.

Choosing a particular curve determines a point of maximum production based on discovery rates, production rates and cumulative production. Early in the curve (pre-peak), the production rate increases because of the discovery rate and the addition of infrastructure. Late in the curve (post-peak), production declines because of resource depletion.

The Hubbert peak theory is based on the observation that the amount of oil under the ground in any region is finite, therefore the rate of discovery which initially increases quickly must reach a maximum and decline. In the US, oil extraction followed the discovery curve after a time lag of 32 to 35 years. The theory is named after American geophysicist M. King Hubbert, who created a method of modeling the production curve given an assumed ultimate recovery volume.

Hubbert's Peak

"Hubbert's peak" can refer to the peaking of production of a particular area, which has now been observed for many fields and regions.

Hubbert's Peak was thought to have been achieved in the United States contiguous 48 states (that is, excluding Alaska) in the early 1970s. Oil production peaked at 10,200,000 barrels per day (1,620,000 m³/d) and then declined for several years since. Yet, recent advances in extraction technology, particularly those that led to the extraction of tight oil and oil from shale,have drastically changed the picture. A decline in production followed the 1970s peak, but it has been succeeded by a major increase in production.

Peak oil as a proper noun, or "Hubbert's peak" applied more generally, refers to a predicted event: the peak of the entire planet's oil production. After Peak Oil, according to the Hubbert Peak Theory, the rate of oil production on Earth would enter a terminal decline. On the basis of his theory, in a paper he presented to the American Petroleum Institute in 1956, Hubbert correctly predicted that production of oil from conventional sources would peak in the continental United States around 1965–1970. His prediction of inevitable decline has been incorrect, but the 1970 peak has yet not been surpassed. Hubbert further predicted a worldwide peak at "about half a century" from publication and approximately 12 gigabarrels (GB) a year in magnitude. In a 1976 TV interview Hubbert added that the actions of OPEC might flatten the global production curve but this would only delay the peak for perhaps 10 years. The development of new technologies has provided access to large quantities of unconventional resources, and the boost of production has largely discounted Huppert's prediction. In the future, pressure to limit the use of fossil fuels (and so reduce the release of greenhouse gasses) will curb production, not exhaustion of resources.

Hubbert's Theory

Hubbert Curve

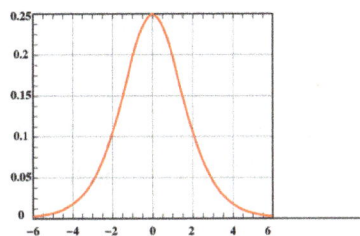

The standard Hubbert curve. For applications, the x and y scales are replaced by time and production scales.

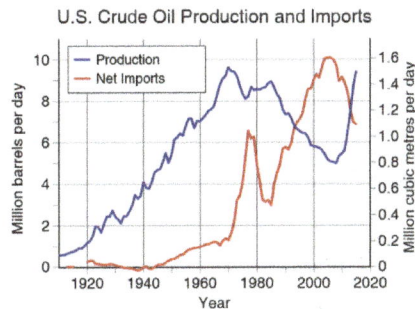

U.S. Oil Production and Imports 1910 to 2012

In 1956, Hubbert proposed that fossil fuel production in a given region over time would follow a roughly bell-shaped curve without giving a precise formula; he later used the Hubbert curve, the derivative of the logistic curve, for estimating future production using past observed discoveries.

Hubbert assumed that after fossil fuel reserves (oil reserves, coal reserves, and natural gas reserves) are discovered, production at first increases approximately exponentially, as more extraction commences and more efficient facilities are installed. At some point, a peak output is reached, and production begins declining until it approximates an exponential decline.

The Hubbert curve satisfies these constraints. Furthermore, it is roughly symmetrical, with the peak of production reached when about half of the fossil fuel that will ultimately be produced has been produced. It also has a single peak.

Given past oil discovery and production data, a Hubbert curve that attempts to approximate past discovery data may be constructed and used to provide estimates for future production. In particular, the date of peak oil production or the total amount of oil ultimately produced can be estimated that way. Cavallo defines the Hubbert curve used to predict the U.S. peak as the derivative of:

$$Q(t) = \frac{Q_{max}}{1 + ae^{-bt}}$$

where Q_{max} is the total resource available (ultimate recovery of crude oil), $Q(t)$ the cumulative production, and a and b are constants. The year of maximum annual production (peak) is:

$$t_{max} = \frac{1}{b}\ln(a).$$

so now the cumulative production $Q(t)$ reaches the half of the total available resource:

$$Q(t) = Q_{max}/2$$

Use of Multiple Curves

The sum of multiple Hubbert curves, a technique not developed by Hubbert himself, may be used in order to model more complicated real life scenarios.

Reliability

Crude Oil

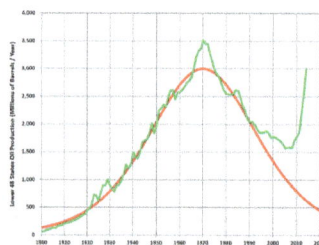

Hubbert's upper-bound prediction for US crude oil production (1956), and actual lower-48 states production through 2014

Hubbert, in his 1956 paper, presented two scenarios for US crude oil production:

- most likely estimate: a logistic curve with a logistic growth rate equal to 6%, an ultimate resource equal to 150 Giga-barrels (Gb) and a peak in 1965. The size of the ultimate resource was taken from a synthesis of estimates by well-known oil geologists and the US Geological Survey, which Hubbert judged to be the most likely case.

- upper-bound estimate: a logistic curve with a logistic growth rate equal to 6% and ultimate resource equal to 200 Giga-barrels and a peak in 1970.

Hubbert's upper-bound estimate, which he regarded as optimistic, accurately predicted that US oil production would peak in 1970, although the actual peak was 17% higher than Hubbert's curve. Production declined, as Hubbert had predicted, and stayed within 10 percent of Hubbert's predicted value from 1974 through 1994; since then, actual production has been significantly greater than the Hubbert curve. The development of new technologies has provided access to large quantities of unconventional resources, and the boost of production has largely discounted Huppert's prediction. In the future, pressure to limit the use of fossil fuels (and so reduce the release of greenhouse gasses) will curb production, not exhaustion of resources.

Hubbert's 1956 production curves depended on geological estimates of ultimate recoverable oil resources, but he was dissatisfied by the uncertainty this introduced, given the various estimates ranging from 110 billion to 590 billion barrels for the US. Starting in his 1962 publication, he made his calculations, including that of ultimate recovery, based only on mathematical analysis of production rates, proved reserves, and new discoveries, independent of any geological estimates of future discoveries. He concluded that the ultimate recoverable oil resource of the contiguous 48 states was 170 billion barrels, with a production peak in 1966 or 1967. He considered that because his model incorporated past technical advances, that any future advances would occur at the same rate, and were also incorporated. Hubbert continued to defend his calculation of 170 billion barrels in his publications of 1965 and 1967, although by 1967 he had moved the peak forward slightly, to 1968 or 1969.

A post-hoc analysis of peaked oil wells, fields, regions and nations found that Hubbert's model was the "most widely useful" (providing the best fit to the data), though many areas studied had a sharper "peak" than predicted.

A 2007 study of oil depletion by the UK Energy Research Centre pointed out that there is no theoretical and no robust practical reason to assume that oil production will follow a logistic curve. Neither is there any reason to assume that the peak will occur when half the ultimate recoverable resource has been produced; and in fact, empirical evidence appears to contradict this idea. An analysis of a 55 post-peak countries found that the average peak was at 25 percent of the ultimate recovery.

Natural Gas

Hubbert also predicted that natural gas production would follow a logistic curve similar to that of oil. At right is his gas production curve for the United States, published in 1962.

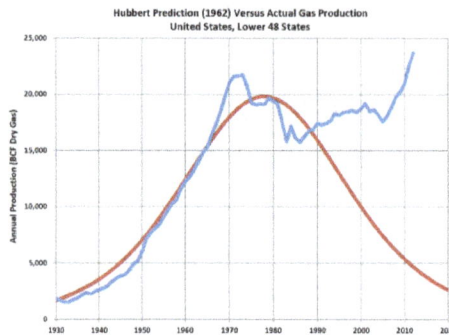

Hubbert's 1962 prediction of US lower 48-state gas production, versus actual production through 2012

Economics

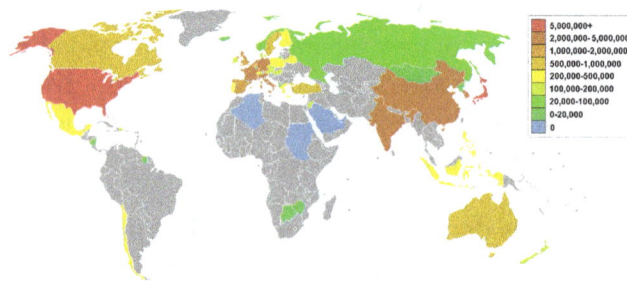

Oil imports by country Pre-2006

Energy Return On Energy Investment

The ratio of energy extracted to the energy expended in the process is often referred to as the Energy Return on Energy Investment (EROI or EROEI). As the EROEI drops to one, or equivalently the Net energy gain falls to zero, the oil production is no longer a net energy source. This happens long before the resource is physically exhausted.

Note that it is important to understand the distinction between a barrel of oil, which is a measure of oil, and a barrel of oil equivalent (BOE), which is a measure of energy. Many sources of energy, such as fission, solar, wind, and coal, are not subject to the same near-term supply restrictions that oil is. Accordingly, even an oil source with an EROEI of 0.5 can be usefully exploited if the energy required to produce that oil comes from a cheap and plentiful energy source. Availability of cheap, but hard to transport, natural gas in some oil fields has led to using natural gas to fuel enhanced oil recovery. Similarly, natural gas in huge amounts is used to power most Athabasca tar sands plants. Cheap natural gas has also led to ethanol fuel produced with a net EROEI of less than 1, although figures in this area are controversial because methods to measure EROEI are in debate.

Advances in technology or experience can lead to greater productivity. The US Energy Information Administration has reported that drilling for shale gas and light tight oil in the United States became much more efficient throughout the period 2007–2014. In terms of oil produced per day of rig drilling time, Bakken wells drilled in January 2014 produced 2.4 times as much oil as those drilled five years earlier, in January 2009. In the Marcellus Gas Trend, wells drilled in January 2014 produced more than nine times as much gas per day of drilling rig time as those drilled five years previously, in January 2009.

Growth-based Economic Models

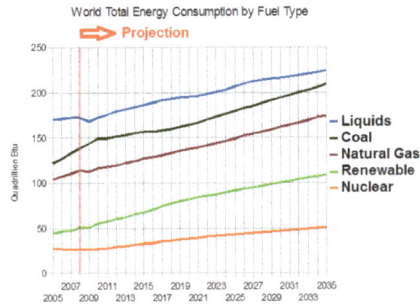

World energy consumption & predictions, 2005–2035. *Source: International Energy Outlook 2011.*

Insofar as economic growth is driven by oil consumption growth, post-peak societies must adapt. Hubbert believed:

> **"** Our principal constraints are cultural. During the last two centuries we have known nothing but exponential growth and in parallel we have evolved what amounts to an exponential-growth culture, a culture so heavily dependent upon the continuance of exponential growth for its stability that it is incapable of reckoning with problems of nongrowth. **"**

Some economists describe the problem as uneconomic growth or a false economy. At the political right, Fred Ikle has warned about "conservatives addicted to the Utopia of Perpetual Growth". Brief oil interruptions in 1973 and 1979 markedly slowed—but did not stop—the growth of world GDP.

Between 1950 and 1984, as the Green Revolution transformed agriculture around the globe, world grain production increased by 250%. The energy for the Green Revolution was provided by fossil fuels in the form of fertilizers (natural gas), pesticides (oil), and hydrocarbon fueled irrigation.

David Pimentel, professor of ecology and agriculture at Cornell University, and Mario Giampietro, senior researcher at the National Research Institute on Food and Nutrition (INRAN), place in their study *Food, Land, Population and the U.S. Economy* the maximum U.S. population for a sustainable economy at 200 million. To achieve a sustainable economy world population will have to be reduced by two-thirds, says the study. Without population reduction, this study predicts an agricultural crisis beginning in 2020, becoming critical c. 2050. The peaking of global oil along with the decline in regional natural gas production may precipitate this agricultural crisis sooner than generally expected. Dale Allen Pfeiffer claims that coming decades could see spiraling food prices without relief and massive starvation on a global level such as never experienced before.

Hubbert Peaks

Although Hubbert peak theory receives most attention in relation to peak oil production, it has also been applied to other natural resources.

Natural Gas

Doug Reynolds predicted in 2005 that the North American peak would occur in 2007. Bentley predicted a world "decline in conventional gas production from about 2020".

Coal

Although observers believe that peak coal is significantly further out than peak oil, Hubbert studied the specific example of anthracite in the USA, a high grade coal, whose production peaked in the 1920s. Hubbert found that anthracite matches a curve closely. Hubbert had recoverable coal reserves worldwide at 2.500×10^{12} metric tons and peaking around 2150 (depending on usage).

More recent estimates suggest an earlier peak. *Coal: Resources and Future Production* (PDF 630KB), published on April 5, 2007 by the Energy Watch Group (EWG), which reports to the German Parliament, found that global coal production could peak in as few as 15 years. Reporting on this, Richard Heinberg also notes that the date of peak annual energetic extraction from coal is likely to come earlier than the date of peak in quantity of coal (tons per year) extracted as the most energy-dense types of coal have been mined most extensively. A second study, *The Future of Coal* by B. Kavalov and S. D. Peteves of the Institute for Energy (IFE), prepared for European Commission Joint Research Centre, reaches similar conclusions and states that ""coal might not be so abundant, widely available and reliable as an energy source in the future".

Work by David Rutledge of Caltech predicts that the total of world coal production will amount to only about 450 gigatonnes. This implies that coal is running out faster than usually assumed.

Fissionable Materials

In a paper in 1956, after a review of US fissionable reserves, Hubbert notes of nuclear power:

> **"** There is promise, however, provided mankind can solve its international problems and not destroy itself with nuclear weapons, and provided world population (which is now expanding at such a rate as to double in less than a century) can somehow be brought under control, that we may at last have found an energy supply adequate for our needs for at least the next few centuries of the "foreseeable future." **"**

As of 2012, the identified resources of uranium are sufficient to provide more than 120 years of supply at the present rate of consumption. Technologies such as the thorium fuel cycle, reprocessing and fast breeders can, in theory, extend the life of uranium reserves from hundreds to thousands of years.

Caltech physics professor David Goodstein stated in 2004 that

> **"** ... you would have to build 10,000 of the largest power plants that are feasible by engineering standards in order to replace the 10 terawatts of fossil fuel we're burning today ... that's a staggering amount and if you did that, the known reserves of uranium would last for 10 to 20 years at that burn rate. So, it's at best a bridging technology ... You can use the rest of the uranium to breed plutonium 239 then we'd have at least 100 times as much fuel to use. But that means you're making plutonium, which is an extremely dangerous thing to do in the dangerous world that we live in. **"**

Helium

Almost all helium on Earth is a result of radioactive decay of uranium and thorium. Helium is extracted by fractional distillation from natural gas, which contains up to 7% helium. The world's largest helium-rich natural gas fields are found in the United States, especially in the Hugoton and nearby gas fields in Kansas, Oklahoma, and Texas. The extracted helium is stored underground in the National Helium Reserve near Amarillo, Texas, the self-proclaimed "Helium Capital of the World". Helium production is expected to decline along with natural gas production in these areas.

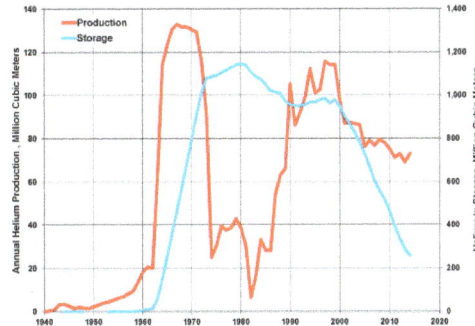

Helium production and storage in the United States, 1940–2014 (data from USGS)

Helium, which is the second-lightest chemical element, will rise to the upper layers of Earth's atmosphere, where it can forever break free from Earth's gravitational attraction. Approximately 1,600 tons of helium are lost per year as a result of atmospheric escape mechanisms.

Transition Metals

Hubbert applied his theory to "rock containing an abnormally high concentration of a given metal" and reasoned that the peak production for metals such as copper, tin, lead, zinc and others would occur in the time frame of decades and iron in the time frame of two centuries like coal. The price of copper rose 500% between 2003 and 2007 and was attributed by some to peak copper. Copper prices later fell, along with many other commodities and stock prices, as demand shrank from fear of a global recession.Lithium availability is a concern for a fleet of Li-ion battery using cars but a paper published in 1996 estimated that world reserves are adequate for at least 50 years. A similar prediction for platinum use in fuel cells notes that the metal could be easily recycled.

Precious Metals

In 2009, Aaron Regent president of the Canadian gold giant Barrick Gold said that global output has been falling by roughly one million ounces a year since the start of the decade. The total global mine supply has dropped by 10pc as ore quality erodes, implying that the roaring bull market of the last eight years may have further to run. "There is a strong case to be made that we are already at 'peak gold'," he told The Daily Telegraph at the RBC's annual gold conference in London. "Production peaked around 2000 and it has been in decline ever since, and we forecast that decline to continue. It is increasingly difficult to find ore," he said.

Ore grades have fallen from around 12 grams per tonne in 1950 to nearer 3 grams in the US, Canada, and Australia. South Africa's output has halved since peaking in 1970. Output fell a further 14 percent in South Africa in 2008 as companies were forced to dig ever deeper – at greater cost – to replace depleted reserves.

World mined gold production has peaked four times since 1900: in 1912, 1940, 1971, and 2001, each peak being higher than previous peaks. The latest peak was in 2001, when production reached

2,600 metric tons, then declined for several years. Production started to increase again in 2009, spurred by high gold prices, and achieved record new highs each year in 2012, 2013, and in 2014, when production reached 2,990 tonnes.

Phosphorus

Phosphorus supplies are essential to farming and depletion of reserves is estimated at somewhere from 60 to 130 years. According to a 2008 study, the total reserves of phosphorus are estimated to be approximately 3,200 MT, with a peak production at 28 MT/year in 2034. Individual countries' supplies vary widely; without a recycling initiative America's supply is estimated around 30 years. Phosphorus supplies affect agricultural output which in turn limits alternative fuels such as bio-diesel and ethanol. Its increasing price and scarcity (global price of rock phosphate rose 8-fold in the 2 years to mid 2008) could change global agricultural patterns. Lands, perceived as marginal because of remoteness, but with very high phosphorus content, such as the Gran Chaco may get more agricultural development, while other farming areas, where nutrients are a constraint, may drop below the line of profitability.

Peak Water

Hubbert's original analysis did not apply to renewable resources. However, over-exploitation often results in a Hubbert peak nonetheless. A modified Hubbert curve applies to any resource that can be harvested faster than it can be replaced.

For example, a reserve such as the Ogallala Aquifer can be mined at a rate that far exceeds replenishment. This turns much of the world's underground water and lakes into finite resources with peak usage debates similar to oil. These debates usually center around agriculture and suburban water usage but generation of electricity from nuclear energy or coal and tar sands mining mentioned above is also water resource intensive. The term fossil water is sometimes used to describe aquifers whose water is not being recharged.

Renewable Resources

- Fisheries: At least one researcher has attempted to perform Hubbert linearization (Hubbert curve) on the whaling industry, as well as charting the transparently dependent price of caviar on sturgeon depletion. The Atlantic northwest cod fishery was a renewable resource, but the numbers of fish taken exceeded the fish's rate of recovery. The end of the cod fishery does match the exponential drop of the Hubbert bell curve. Another example is the cod of the North Sea.

- Air/oxygen: Half the world's oxygen is produced by phytoplankton. The numbers of plankton have dropped by 40% since the 1950s.

Criticisms of Peak Oil

Economist Michael Lynch argues that the theory behind the Hubbert curve is too simplistic and relies on an overly Malthusian point of view. Lynch claims that Campbell's predictions for world oil production are strongly biased towards underestimates, and that Campbell has repeatedly pushed back the date.

Leonardo Maugeri, vice president of the Italian energy company Eni, argues that nearly all of peak estimates do not take into account unconventional oil even though the availability of these resources is significant and the costs of extraction and processing, while still very high, are falling because of improved technology. He also notes that the recovery rate from existing world oil fields has increased from about 22% in 1980 to 35% today because of new technology and predicts this trend will continue. The ratio between proven oil reserves and current production has constantly improved, passing from 20 years in 1948 to 35 years in 1972 and reaching about 40 years in 2003. These improvements occurred even with low investment in new exploration and upgrading technology because of the low oil prices during the last 20 years. However, Maugeri feels that encouraging more exploration will require relatively high oil prices.

Edward Luttwak, an economist and historian, claims that unrest in countries such as Russia, Iran and Iraq has led to a massive underestimate of oil reserves. The Association for the Study of Peak Oil and Gas (ASPO) responds by claiming neither Russia nor Iran are troubled by unrest currently, but Iraq is.

Cambridge Energy Research Associate[s] authored a report that is critical of Hubbert-influenced predictions:

> " Despite his valuable contribution, M. King Hubbert's methodology falls down because it does not consider likely resource growth, application of new technology, basic commercial factors, or the impact of geopolitics on production. His approach does not work in all cases-including on the United States itself-and cannot reliably model a global production outlook. Put more simply, the case for the imminent peak is flawed. As it is, production in 2005 in the Lower 48 in the United States was 66 percent higher than Hubbert projected. "

CERA does not believe there will be an endless abundance of oil, but instead believes that global production will eventually follow an "undulating plateau" for one or more decades before declining slowly, and that production will reach 40 Mb/d by 2015.

Alfred J. Cavallo, while predicting a conventional oil supply shortage by no later than 2015, does not think Hubbert's peak is the correct theory to apply to world production.

Criticisms of Peak Element Scenarios

Although M. King Hubbert himself made major distinctions between decline in petroleum production versus depletion (or relative lack of it) for elements such as fissionable uranium and thorium, some others have predicted peaks like peak uranium and peak phosphorus soon on the basis of published reserve figures compared to present and future production. According to some economists, though, the amount of proved reserves inventoried at a time may be considered "a poor indicator of the total future supply of a mineral resource."

As some illustrations, tin, copper, iron, lead, and zinc all had both production from 1950 to 2000 and reserves in 2000 much exceed world reserves in 1950, which would be impossible except for how "proved reserves are like an inventory of cars to an auto dealer" at a time, having little relationship to the actual total affordable to extract in the future. In the example of peak phosphorus, additional concentrations exist intermediate between 71,000 Mt of identified reserves (USGS) and the approximately 30,000,000,000 Mt of other phosphorus in Earth's crust, with the average rock being 0.1% phosphorus, so showing decline in human phosphorus production will occur soon would require far more than comparing the former figure to the 190 Mt/year of phosphorus extracted in mines (2011 figure).

Fractional Distillation

Fractional distillation is the separation of a mixture into its component parts, or fractions, separating chemical compounds by their boiling point by heating them to a temperature at which one or more fractions of the compound will vaporize. It uses distillation to fractionate. Generally the component parts have boiling points that differ by less than 25 °C from each other under a pressure of one atmosphere. If the difference in boiling points is greater than 25 °C, a simple distillation is typically used.

Laboratory Setup

Fractional distillation in a laboratory makes use of common laboratory glassware and apparatuses, typically including a Bunsen burner, a round-bottomed flask and a condenser, as well as the single-purpose fractionating column.

Apparatus

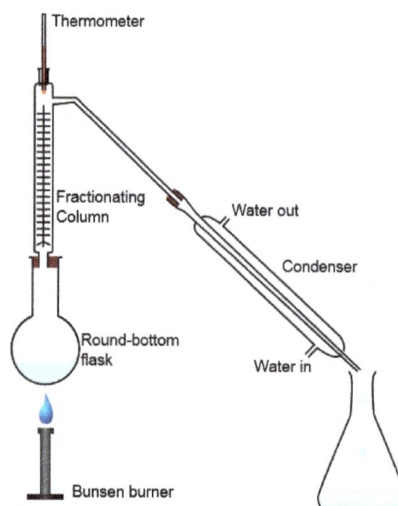

Fractional distillation
An Erlenmeyer flask is used as a receiving flask. Here the distillation head and fractionating column are combined in one piece.

- heat source, such as a hot plate with a bath, and ideally with a magnetic stirrer.

- distilling flask, typically a round-bottom flask

- receiving flask, often also a round-bottom flask

- fractionating column

- distillation head

- thermometer and adapter if needed

- condenser, such as a Liebig condenser or Allihn condenser

- vacuum adapter (not used in image to the right)

- boiling chips, also known as anti-bumping granules

- Standard laboratory glassware with ground glass joints, e.g. quickfit apparatus.

Discussion

As an example consider the distillation of a mixture of water and ethanol. Ethanol boils at 78.4 °C while water boils at 100 °C. So, by heating the mixture, the most volatile component (ethanol) will concentrate to a greater degree in the vapor leaving the liquid. Some mixtures form azeotropes, where the mixture boils at a lower temperature than either component. In this example, a mixture of 96% ethanol and 4% water boils at 78.2 °C; the mixture is more volatile than pure ethanol. For this reason, ethanol cannot be completely purified by direct fractional distillation of ethanol-water mixtures.

The apparatus is assembled as in the diagram. (The diagram represents a batch apparatus as opposed to a continuous apparatus.) The mixture is put into the round bottomed flask along with a few anti-bumping granules (or a Teflon coated magnetic stirrer bar if using magnetic stirring), and the fractionating column is fitted into the top. The fractional distillation column is set up with the heat source at the bottom on the still pot. As the distance from the stillpot increases, a temperature gradient is formed in the column; it is coolest at the top and hottest at the bottom. As the mixed vapor ascends the temperature gradient, some of the vapor condenses and revaporizes along the temperature gradient. Each time the vapor condenses and vaporizes, the composition of the more volatile component in the vapor increases. This distills the vapor along the length of the column, and eventually the vapor is composed solely of the more volatile component (or an azeotrope). The vapor condenses on the glass platforms, known as trays, inside the column, and runs back down into the liquid below, refluxing distillate. The efficiency in terms of the amount of heating and time required to get fractionation can be improved by insulating the outside of the column in an insulator such as wool, aluminium foil or preferably a vacuum jacket. The hottest tray is at the bottom and the coolest is at the top. At steady state conditions, the vapor and liquid on each tray are at equilibrium. The most volatile component of the mixture exits as a gas at the top of the column. The vapor at the top of the column then passes into the condenser, which cools it down until it liquefies. The separation is more pure with the addition of more trays (to a practical limitation of heat, flow, etc.) Initially, the condensate will be close to the azeotropic composition, but when much of the ethanol has been drawn off, the condensate becomes gradually richer in water.The process continues until all the ethanol boils out of the mixture. This point can be recognized by the sharp rise in temperature shown on the thermometer.

The above explanation reflects the theoretical way fractionation works. Normal laboratory fractionation columns will be simple glass tubes (often vacuum-jacketed, and sometimes internally silvered) filled with a packing, often small glass helices of 4 to 7 mm diameter. Such a column can be calibrated by the distillation of a known mixture system to quantify the column in terms of number of theoretical trays. To improve fractionation the apparatus is set up to return condensate to the column by the use of some sort of reflux splitter (reflux wire, gago, Magnetic swinging bucket, etc.) - a typical careful fractionation would employ a reflux ratio of around 4:1 (4 parts returned condensate to 1 part condensate take off).

In laboratory distillation, several types of condensers are commonly found. The Liebig condenser is simply a straight tube within a water jacket, and is the simplest (and relatively least expensive) form of condenser. The Graham condenser is a spiral tube within a water jacket, and the Allihn condenser has a series of large and small constrictions on the inside tube, each increasing the surface area upon which the vapor constituents may condense.

Alternate set-ups may use a multi−outlet distillation receiver flask (referred to as a "cow" or "pig") to connect three or four receiving flasks to the condenser. By turning the cow or pig, the distillates can be channeled into any chosen receiver. Because the receiver does not have to be removed and replaced during the distillation process, this type of apparatus is useful when distilling under an inert atmosphere for air-sensitive chemicals or at reduced pressure. A Perkin triangle is an alternative apparatus often used in these situations because it allows isolation of the receiver from the rest of the system, but does require removing and reattaching a single receiver for each fraction.

Vacuum distillation systems operate at reduced pressure, thereby lowering the boiling points of the materials. Anti-bumping granules, however, become ineffective at reduced pressures.

Industrial Distillation

Typical industrial fractional distillation columns

Fractional distillation is the most common form of separation technology used in petroleum refineries, petrochemical and chemical plants, natural gas processing and cryogenic air separation plants. In most cases, the distillation is operated at a continuoussteady state. New feed is always being added to the distillation column and products are always being removed. Unless the process is disturbed due to changes in feed, heat, ambient temperature, or condensing, the amount of feed being added and the amount of product being removed are normally equal. This is known as continuous, steady-state fractional distillation.

Industrial distillation is typically performed in large, vertical cylindrical columns known as "distillation or fractionation towers" or "distillation columns" with diameters ranging from about 65 centimeters to 6 meters and heights ranging from about 6 meters to 60 meters or more. The distillation towers have liquid outlets at intervals up the column which allow for the withdrawal of different fractions or products having different boiling points or boiling ranges. By increasing the temperature of the product inside the columns, the different hydrocarbons are separated. The "lightest" products (those with the lowest boiling point) exit from the top of the columns and the "heaviest" products (those with the highest boiling point) exit from the bottom of the column.

For example, fractional distillation is used in oil refineries to separate crude oil into useful substances (or fractions) having different hydrocarbons of different boiling points. The crude oil fractions with higher boiling points:

- have more carbon atoms

- have higher molecular weights

- are less branched chain alkanes

- are darker in color

- are more viscous

- are more difficult to ignite and to burn

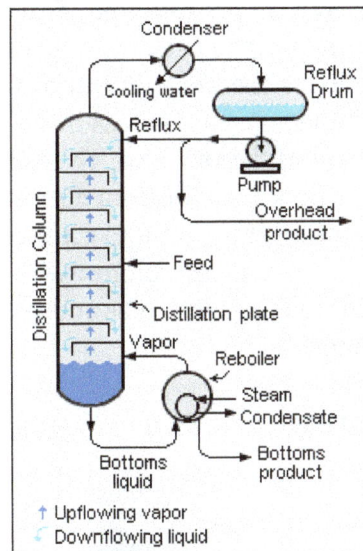

Diagram of a typical industrial distillation tower

Large-scale industrial towers use reflux to achieve a more complete separation of products. Reflux refers to the portion of the condensed overhead liquid product from a distillation or fractionation tower that is returned to the upper part of the tower as shown in the schematic diagram of a typical, large-scale industrial distillation tower. Inside the tower, the reflux liquid flowing downwards provides the cooling needed to condense the vapors flowing upwards, thereby increasing the effectiveness of the distillation tower. The more reflux is provided for a given number of theoretical plates, the better the tower's separation of lower boiling materials from higher boiling materials. Alternatively, the more reflux provided for a given desired separation, the fewer theoretical plates are required.

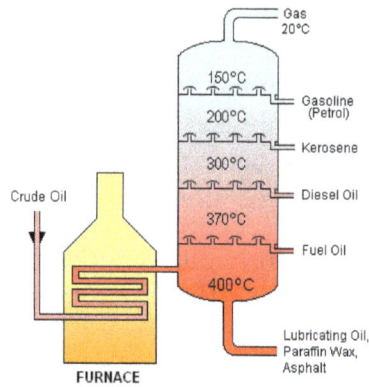

Crude oil is separated into fractions by fractional distillation. The fractions at the top of the fractionating column have lower boiling points than the fractions at the bottom. All of the fractions are processed further in other refining units.

Fractional distillation is also used in air separation, producing liquid oxygen, liquid nitrogen, and highly concentrated argon. Distillation of chlorosilanes also enable the production of high-purity silicon for use as a semiconductor.

In industrial uses, sometimes a packing material is used in the column instead of trays, especially when low pressure drops across the column are required, as when operating under vacuum. This packing material can either be random dumped packing (1-3" wide) such as Raschig rings or structured sheet metal. Typical manufacturers are Koch, Sulzer and other companies. Liquids tend to wet the surface of the packing and the vapors pass across this wetted surface, where mass transfer takes place. Unlike conventional tray distillation in which every tray represents a separate point of vapor liquid equilibrium the vapor liquid equilibrium curve in a packed column is continuous. However, when modeling packed columns it is useful to compute a number of "theoretical plates" to denote the separation efficiency of the packed column with respect to more traditional trays. Differently shaped packings have different surface areas and void space between packings. Both of these factors affect packing performance.

Design of Industrial Distillation Columns

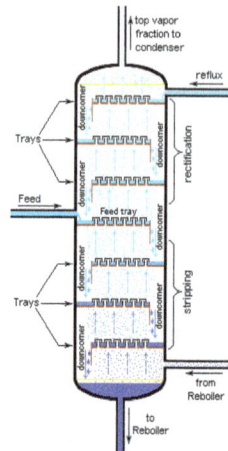

Chemical engineering schematic of typical bubble-cap trays in a distillation tower

Design and operation of a distillation column depends on the feed and desired products. Given a simple, binary component feed, analytical methods such as the McCabe–Thiele method or the Fenske equation can be used. For a multi-component feed, simulation models are used both for design and operation.

Moreover, the efficiencies of the vapor–liquid contact devices (referred to as *plates* or *trays*) used in distillation columns are typically lower than that of a theoretical 100% efficient equilibrium stage. Hence, a distillation column needs more plates than the number of theoretical vapor–liquid equilibrium stages.

Reflux refers to the portion of the condensed overhead product that is returned to the tower. The reflux flowing downwards provides the cooling required for condensing the vapours flowing upwards. The reflux ratio, which is the ratio of the (internal) reflux to the overhead product, is conversely related to the theoretical number of stages required for efficient separation of the distillation products. Fractional distillation towers or columns are designed to achieve the required separation efficiently. The design of fractionation columns is normally made in two steps; a process design, followed by a mechanical design. The purpose of the process design is to calculate the number of required theoretical stages and stream flows including the reflux ratio, heat reflux and other heat duties. The purpose of the mechanical design, on the other hand, is to select the tower internals, column diameter and height. In most cases, the mechanical design of fractionation towers is not straightforward. For the efficient selection of tower internals and the accurate calculation of column height and diameter, many factors must be taken into account. Some of the factors involved in design calculations include feed load size and properties and the type of distillation column used.

The two major types of distillation columns used are tray and packing columns. Packing columns are normally used for smaller towers and loads that are corrosive or temperature sensitive or for vacuum service where pressure drop is important. Tray columns, on the other hand, are used for larger columns with high liquid loads. They first appeared on the scene in the 1820s. In most oil refinery operations, tray columns are mainly used for the separation of petroleum fractions at different stages of oil refining.

In the oil refining industry, the design and operation of fractionation towers is still largely accomplished on an empirical basis. The calculations involved in the design of petroleum fractionation columns require in the usual practice the use of numerable charts, tables and complex empirical equations. In recent years, however, a considerable amount of work has been done to develop efficient and reliable computer-aided design procedures for fractional distillation.

Octane Rating

An octane rating, or octane number, is a standard measure of the performance of an engine or aviation fuel. The higher the octane number, the more compression the fuel can withstand before detonating (igniting). In broad terms, fuels with a higher octane rating are used in high performance gasoline engines that require higher compression ratios. In contrast, fuels with lower octane numbers (but higher cetane numbers) are ideal for diesel engines, because diesel engines (also referred to as compression-ignition engines) do not compress the fuel, but rather compress only air and then inject fuel into the air which was heated by compression. Gasoline engines rely on ignition of air and fuel compressed together as a mixture, which is ignited at the end of the compression stroke using spark plugs. Therefore, high compressibility of the fuel matters mainly for gasoline engines. Use of gasoline with lower octane numbers may lead to the problem of engine knocking.

Principles

The Problem: Pre-ignition and Knocking

In a normal spark-ignition engine, the air-fuel mixture is heated due to being compressed and is then triggered to burn rapidly by the spark plug. If it is heated (or compressed) too much, it will self-ignite before the ignition system sparks. This causes much higher pressures than engine components are designed for, and can cause a "knocking" or "pinging" sound. Knocking can cause major engine damage if severe.

The most typically used engine management systems found in automobiles today have a knock sensor that monitors if knock is being produced by the fuel being used. In modern computer-controlled engines, the ignition timing will be automatically altered by the engine management system to reduce the knock to an acceptable level.

Isooctane as a Reference Standard

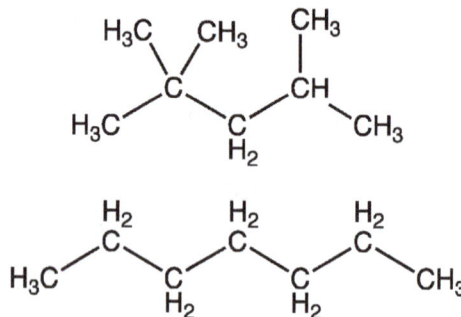

2,2,4-Trimethylpentane (iso-octane) (upper) has an octane rating of 100 whereas n-heptane has an octane rating of 0.

Octanes are a family of hydrocarbon that are typical components of gasoline. They are colorless liquids that boil around 125 °C (260 °F). One member of the octane family, isooctane, is used as a reference standard to benchmark the tendency of gasoline or LPG fuels to resist self-ignition.

The octane rating of gasoline is measured in a test engine and is defined by comparison with the mixture of 2,2,4-trimethylpentane (iso-octane) and heptane that would have the same anti-knocking capacity as the fuel under test: the percentage, by volume, of 2,2,4-trimethylpentane in that mixture is the octane number of the fuel. For example, gasoline with the same knocking characteristics as a mixture of 90% iso-octane and 10% heptane would have an octane rating of 90. A rating of 90 does not mean that the gasoline contains just iso-octane and heptane in these proportions but that it has the same detonation resistance properties (generally, gasoline sold for common use never consists solely of iso-octane and heptane; it is a mixture of many hydrocarbons and often other additives). Because some fuels are more knock-resistant than pure iso-octane, the definition has been extended to allow for octane numbers greater than 100.

Octane ratings are not indicators of the energy content of fuels. . They are only a measure of the fuel's tendency to burn in a controlled manner, rather than exploding in an uncontrolled manner. Where the octane number is raised by blending in ethanol, energy content per volume is reduced. Ethanol BTUs can be compared with gasoline BTUs in heat of combustion tables.

It is possible for a fuel to have a Research Octane Number (RON) more than 100, because iso-octane is not the most knock-resistant substance available. Racing fuels, avgas, LPG and alcohol fuels such as methanol may have octane ratings of 110 or significantly higher. Typical "octane booster" gasoline additives include MTBE, ETBE, isooctane and toluene. Lead in the form of tetraethyllead was once a common additive, but its use for fuels for road vehicles has been progressively phased-out worldwide, beginning in the 1970s.

Measurement Methods

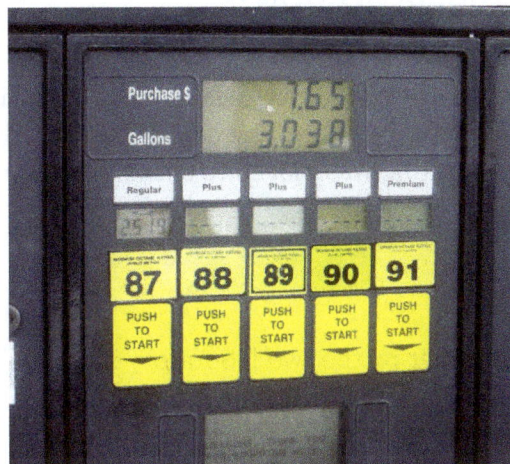

A US gas station pump offering five different (R+M)/2 octane ratings

Research Octane Number (RON)

The most common type of octane rating worldwide is the Research Octane Number (RON). RON is determined by running the fuel in a test engine with a variable compression ratio under controlled conditions, and comparing the results with those for mixtures of iso-octane and n-heptane.

Motor Octane Number (MON)

Another type of octane rating, called Motor Octane Number (MON), is determined at 900 rpm engine speed instead of the 600 rpm for RON. MON testing uses a similar test engine to that used in RON testing, but with a preheated fuel mixture, higher engine speed, and variable ignition timing to further stress the fuel's knock resistance. Depending on the composition of the fuel, the MON of a modern pump gasoline will be about 8 to 12 octane lower than the RON, but there is no direct link between RON and MON. Pump gasoline specifications typically require both a minimum RON and a minimum MON.

Anti-Knock Index (AKI) or (R+M)/2

In most countries, including Australia, New Zealand and all of those in Europe,the "headline" octane rating shown on the pump is the RON, but in Canada, the United States, Brazil, and some other countries, the headline number is the average of the RON and the MON, called the Anti-Knock Index (AKI), and often written on pumps as (R+M)/2. It may also sometimes be called the Posted Octane Number (PON).

Difference between RON, MON, and AKI

Because of the 8 to 12 octane number difference between RON and MON noted above, the AKI shown in Canada and the United States is 4 to 6 octane numbers lower than elsewhere in the world for the same fuel. This difference between RON and MON is known as the fuel's Sensitivity, and is not typically published for those countries that use the Anti-Knock Index labelling system.

Observed Road Octane Number (RdON)

Another type of octane rating, called Observed Road Octane Number (RdON), is derived from testing gasolines in real world multi-cylinder engines, normally at wide open throttle. It was developed in the 1920s and is still reliable today. The original testing was done in cars on the road but as technology developed the testing was moved to chassis dynamometers with environmental controls to improve consistency.

Octane Index

The evaluation of the octane number by the two laboratory methods requires a standard engine, and the test procedure can be both expensive and time-consuming. The standard engine required for the test may not always be available, especially in out-of-the-way places or in small or mobile laboratories. These and other considerations led to the search for a rapid method for the evaluation of the anti-knock quality of gasoline. Such methods include FTIR, near infrared on-line analyzers (ASTM D-2885) and others. Deriving an equation that can be used for calculating the octane quality would also serve the same purpose with added advantages. The term Octane Index is often used to refer to the calculated octane quality in contradistinction to the (measured) research or motor octane numbers. The octane index can be of great service in the blending of gasoline. Motor gasoline, as marketed, is usually a blend of several types of refinery grades that are derived from different processes such as straight-run gasoline, reformate, cracked gasoline etc. These different grades are considered as one group when blending to meet final product specifications. Most refiners produce and market more than one grade of motor gasoline, differing principally in their anti-knock quality. The ability to predict the octane quality of the blends prior to blending is essential, something for which the calculated octane index is specially suited.

Aviation Gasoline Octane Ratings

Aviation gasoline used in piston aircraft common in general aviation have slightly different methods of measuring the octane of the fuel. Similar to AKI, it has two different ratings, although it is referred to only by the lower of the two. One is referred to as the "aviation lean" rating and is the same as the MON of the fuel up to 100. The second is the "aviation rich" rating and corresponds to the octane rating of a test engine under forced induction operation common in high-performance and military piston aircraft. This utilizes a supercharger, and uses a significantly richer fuel/air ratio for improved detonation resistance.

The most commonly used current fuel, 100LL, has an aviation lean rating of 100 octane, and an aviation rich rating of 130.

Examples

The RON/MON values of n-heptane and iso-octane are exactly 0 and 100, respectively, by the definition of octane rating. The following table lists octane ratings for various other fuels.

Fuel	RON	MON	AKI or (R+M)/2
hexadecane	< −30		
n-octane	−20		
n-heptane (RON and MON 0 by definition)	0	0	0
diesel fuel	15–25		
2-methylheptane	23	23.8	23
n-hexane	25	26.0	26
1-pentene	34		
2-methylhexane	44	46.4	45.2
3-methylhexane		55.0	
1-heptene	60		
n-pentane	62	61.9	62
requirement for a typical two-strokeoutboard motor	69	65	67
Pertamina "Premium" in Indonesia	88	78	83
Pertamina "Pertalite" in Indonesia	90		
"Regular gasoline" in Japan (Japanese Industrial Standards)	90		
n-butanol	92	71	83
Neopentane (dimethylpropane)		80.2	
n-butane	94	90.1	92
Isopentane (methylbutane)		90.3	
"Regular Gasoline" in Australia, New Zealand and the US	91–92	82–83	87
Pertamina "Pertamax" in Indonesia	92	82	87
"Shell Super" in Indonesia, "Total Performance 92" in Indonesia	92		
2,2-dimethylbutane		93.4	
2,3-dimethylbutane		94.4	
"YPF Super" in Argentina	95	84	90
Pertamina "Pertamax Plus" gasoline in Indonesia, "Super/Premium" in New Zealand and Australia	95	85	90
"Aral Super 95" in Germany, "Aral Super 95 E10" (10% Ethanol) in Germany	95	85	90
"Shell V-Power" in Indonesia, "Total Performance 95" in Indonesia, "Shell FuelSave " in Malaysia	95		
"EuroSuper" or "EuroPremium" or "Regular unleaded" in Europe, "SP95" in France, "Super 95" in Belgium	95	85–86	90–91

Fuel	RON	MON	AKI or (R+M)/2
"Premium" or "Super unleaded" gasoline in US (10% ethanol blend)	97	87-88	92-93
"Shell V-Power 97" in Malaysia	97		
"IES 98 Plus" in Italy, "Aral SuperPlus 98" in Germany, Pertamina "Pertamax Turbo" in Indonesia	98		
"YPF Infinia" in Argentina	98	87	93
"Corriente (Regular)" in Colombia	91.5	70	81
"Extra (Super/Plus)" in Colombia	95	79	87
"SuperPlus" in Germany	98	88	93
"Shell V-Power 98", "Caltex Platinum 98 with Techron", "Esso Mobil Synergy 8000" and "SPC LEVO 98" in Singapore, "BP Ultimate 98/ Mobil Synergy 8000" in New Zealand, "SP98" in France, "Super 98" in Belgium, Great Britain, Slovenia and Spain	98	89–90	93–94
"Shell V-Power Nitro+ 99" "Tesco Momentum 99" In the United Kingdom	99	87	93
Pertamina "Pertamina Racing Fuel" (bioethanol blend) in Indonesia	100	86	93
"Premium" gasoline in Japan (Japanese Industrial Standards), "IP Plus 100" in Italy, "Tamoil WR 100" in Italy, "Shell V-Power Racing" in Australia - discontinued July 2008	100		
"Shell V-Power" in Italy and Germany	100	88	94
"Eni(or Agip) Blu Super +(or Tech)" in Italy	100	87	94
"isooctane" (RON and MON 100 by definition)	100	100	100
" Petron Blaze 100 Euro 4M " in Malaysia	100		
"San Marco Petroli F-101" in Italy (northern Italy only, just a few gas stations)	101		
benzene	101		
2,5-Dimethylfuran	101.3	88.1	94.7
Petro-Canada "Ultra 94" in Canada	101.5	88	94
Aral Ultimate 102 in Germany	102	88	95
ExxonMobil Avgas 100		99.5 (min)	
Petrobras Podium in Brazil	102	88	95
E85 gasoline	102-105	85-87	94-96
i-butane	102	97.6	100
"BP Ultimate 102 - now discontinued"	102	93–94	97–98
t-butanol	103	91	97
2,3,3-trimethylpentane	106.1	99.4	103
ethane	108		
ethanol	108.6	89.7	99.15
methanol	108.7	88.6	98.65
2,2,3-trimethylpentane	109.6	99.9	105
propane	112	97	105
2,2,3-trimethylbutane	112.1	101.3	106
xylene	118	115	116.5
isopropanol	118	98	108

Fuel	RON	MON	AKI or (R+M)/2
1-propanol	118	98	108
toluene	121	107	114
VP C16 Race Fuel	117	118	117.5
methane	120	120	120
hydrogen	> 130		

Effects

Higher octane ratings correlate to higher activation energies: the amount of applied energy required to initiate combustion. Since higher octane fuels have higher activation energy requirements, it is less likely that a given compression will cause uncontrolled ignition, otherwise known as autoignition or detonation.

The compression ratio is directly related to power and to thermodynamic efficiency of an internal combustion engine. Engines with higher compression ratios develop more area under the Otto-Cycle curve, thus they extract more energy from a given quantity of fuel.

During the compression stroke of an internal combustion engine, as the air-fuel mix is compressed its temperature rises.

A fuel with a higher octane rating is less prone to auto-ignition and can withstand a greater rise in temperature during the compression stroke of an internal combustion engine without auto-igniting, thus allowing more power to be extracted from the Otto-Cycle.

If during the compression stroke the air-fuel mix reaches a temperature greater than the auto-ignition temperature of the fuel, the fuel self or auto-ignites. When auto-ignition occurs (before the piston reaches the top of its travel) the up-rising piston is then attempting to squeeze the rapidly heating fuel charge. This will usually destroy an engine quickly if allowed to continue.

There are two types of induction systems on internal combustion engines: naturally aspirated engine (air is sucked in using the engine's pistons), or forced induction engines.

In the case of the normally aspirated engine, at the start of the compression stroke the cylinder air / fuel volume is very high, this translates into a low starting pressure. As the piston travels upward, abnormally high cylinder pressures may result in the mixture auto-igniting or detonating, which is why conservative compression ratios are used in consumer vehicles. In a forced induction engine where at the start of the compression stroke the cylinder pressure is already raised (having a greater volume of air/fuel) Exp. 202kPa (29.4Psi) the starting pressure or air / fuel volume would be 2 times that of the normally aspirated engine. This would translate into an effective compression ratio of 20:1 vs. 10:1 for the normally aspirated. This is why many forced induction engines have compression ratios in the 8:1 range.

Many high-performance engines are designed to operate with a high maximum compression, and thus demand fuels of higher octane. A common misconception is that power output or fuel efficiency can be improved by burning fuel of higher octane than that specified by the engine manufacturer.

The power output of an engine depends in part on the energy density of the fuel being burnt. Fuels of different octane ratings may have similar densities, but because switching to a higher octane fuel does not add more hydrocarbon content or oxygen, the engine cannot develop more power.

However, burning fuel with a lower octane rating than that for which the engine is designed often results in a reduction of power output and efficiency. Many modern engines are equipped with a knock sensor (a small piezoelectric microphone), which sends a signal to the engine control unit, which in turn retards the ignition timing when detonation is detected. Retarding the ignition timing reduces the tendency of the fuel-air mixture to detonate, but also reduces power output and fuel efficiency. Because of this, under conditions of high load and high temperature, a given engine may have a more consistent power output with a higher octane fuel, as such fuels are less prone to detonation. Some modern high performance engines are actually optimized for higher than pump premium (93 AKI in the US). The 2001 - 2007 BMW M3 with the S54 engine is one such car. Car and Driver magazine tested a car using a dynamometer, and found that the power output increased as the AKI was increased up to approximately 96 AKI.

Most fuel filling stations have two storage tanks (even those offering 3 or 4 octane levels): those motorists who purchase intermediate grade fuels are given a mixture of higher and lower octane fuels. "Premium" grade is fuel of higher octane, and the minimum grade sold is fuel of lower octane. Purchasing 91 octane fuel (where offered) simply means that more fuel of higher octane is blended with commensurately less fuel of lower octane, than when purchasing a lower grade. The detergents and other additives in the fuel are often, but not always, identical.

The octane rating was developed by chemist Russell Marker at the Ethyl Corporation in 1926. The selection of n-heptane as the zero point of the scale was due to its availability in high purity. Other isomers of heptane produced from crude oil have greatly different ratings.

Regional Variations

The selection of octane ratings available at the pump can vary greatly from region to region.

- Australia: "regular" unleaded fuel is 91 RON, "premium" unleaded with 95 RON is widely available, and 98 RON fuel is also reasonably common. Shell used to sell 100 RON fuel (5% ethanol content) from a small number of service stations, most of which are located in major cities (stopped in August 2008). United Petroleum used to sell 100 RON unleaded fuel (10% ethanol content) at a small number of its service stations (originally only two, but then expanded to 67 outlets nationwide) (stopped in September 2014). All fuel in Australia is unleaded except for some aviation fuels. E85 unleaded fuel is also available at several United service stations across the country.

- Bahrain: 91 and 95 (RON), standard in all gasoline stations in the country and advertised as (Jayyid) for Regular or 91 and (Mumtaz) for Premium or 95.

- China: 92 and 95 (RON) (previously 93 and 97) are commonly offered. In limited areas higher rating such as 98 RON is available. In some rural areas it can be difficult to find fuel with over 92 RON.

- Chile: 93, 95 and 97 RON are standard at almost all gas stations thorough Chile. The three types are unleaded.

- Colombia: "Ecopetrol", Colombia's monopoly of refining and distribution of gasoline establishes a minimum AKI of 81 octanes for "Corriente" gasoline and minimum AKI of 87 octanes for "Extra" gasoline. (91.5 RON corriente, and 91 RON for extra)

- Costa Rica: RECOPE, Costa Rica's distribution monopoly, establishes the following ratings: Plus 91 (at least 91 RON) and Super (at least 95 RON).

- Croatia: All fuel stations offer unleaded "Eurosuper BS" (abbreviation "BS" meaning "no sulfur content") 95 RON fuel, many also offer "Eurosuper Plus BS" 98 RON. Some companies offer 100 RON fuel instead of 98.

- Cyprus: All fuel stations offer unleaded 95 and 98 RON and a few offer 100 RON as well.

- Denmark: 92 and 95 RON are common choices. Most Shell stations offer Shell V-Power 99 RON. Some varieties of low percentage Ethanol mixtures are offered at larger gas stations.

- Ecuador: "Extra" with 87 and "Super" with 92 (RON) are available in all fuel stations. "Extra" is the most commonly used. All fuels are unleaded.

- Egypt: Egyptian Fuel Stations had 90 RON until July 2014 where the Government found no use for it. Leaving only 92 RON and 95 RON. 80 RON is found in a very limited amount of fuel stations as they are used only for extremely old cars that cannot cope with high octane fuel. 95 RON in used by limit due to its high price (more than twice the price of 92 RON).

- Estonia: 95 RON and 98 RON are widely available.

- Finland: 95 and 98 (RON), advertised as such, at almost all gas stations. Most cars run on 95, but 98 is available for vehicles that need higher octane fuel, or older models containing parts easily damaged by high ethanol content. Shell offers V-Power, advertised as "over 99 octane", instead of 98. In the beginning of 2011 95 RON was replaced by 95E10 containing 10% ethanol, and 98 RON by 98E5, containing 5% ethanol. ST1 also offers RE85 on some stations, which is 85% ethanol made from biodegradable waste (from which the advertised name "ReFuel" comes). RE85 is only suitable for flexifuel cars that can run on high-percentage ethanol.

- Germany: "Super E10" 95 RON and "Super Plus E5" 98 RON are available practically everywhere. Big suppliers such as Shell or Aral offer 100 RON gasoline (Shell V-Power, Aral Ultimate) at almost every fuel station. "Normal" 91 RON is only rarely offered because lower production amounts make it more expensive than "Super" 95 RON. Due to a new European Union law, gas stations are being required to offer a minimum rate of the new mixture of "Super" 95 RON with up to 10% Ethanol branded as "Super E10". Producers are discontinuing "Super E5" 95 RON with <5% Ethanol so cars that are unable to use E10 must use 98 RON gasoline automotive fuel instead. E85 is also available is most areas.

- Greece (Hellas): 95 RON (standard unleaded), 97+ & 100 RON unleaded offered by some companies (e.g. EKO, Shell, BP). Also available Super LRP 96 RON for older (non-catalytic) vehicles.

- Hong Kong: only 98 RON is available in the market. There have been calls to re-introduce 95 RON, but the calls have been rejected by all automotive fuel station chains, citing that 95 RON was phased out because of market forces.

- India: India's ordinary and premium gasolines are of 91 RON. The premium gasolines are generally ordinary fuels with additives, that do not really change the octane value. Two variants, "Speed 93" and "Speed 97", were launched, with RON values of 93 and 97. India's economy-class vehicles usually have compression ratios under 10:1, thus enabling them to use lower-octane gasoline without engine knocking.

- Indonesia: Indonesia's "Premium" gasoline rated at 88 RON and being subsidized it cost only about US$0.65/liter. Other options are newly introduced "Pertalite" rated at 90 RON (the price is not subsidized), "Pertamax" rated at 92 RON and the "Pertamax Plus" rated at 95 RON. "Pertamax Racing Fuel", a bioethanol flexfuel rated at RON 100 (sold in gallon can container only). Starting from August 2016 Pertamina has started selling a new fuel variant rated at 98 RON marketed by Pertamax Turbo. This fuel is said to replace Pertamax Plus which is still currently sold in almost every fuel station nation wide, making Pertamina to probably the only energy company in the world to currently provide six octane variants of gasoline, 88, 90, 92, 95, 98, and 100 RON. Total and Shell stations only sell RON 92 and 95 gasoline. Petronas has decided to shut down its retail business in Indonesia in 2012, after years of sluggish sales.

- Iran: 92 RON (marketed as regular) and 95 RON (marketed as Super) are widely available in gas stations.

- Ireland: 95 RON "unleaded" is the only gasoline type available through stations, although E5 (99 RON) is becoming more commonplace.

- Italy: 95 RON is the only compulsory gasoline offered (verde, "green"), only a few fuel stations (Agip, IP, IES, OMV) offer 98 RON as the premium type, many Shell and Tamoil stations close to the cities offer also V-Power Gasoline rated at 100 RON. Recently Agip introduced "Blu Super+", a 100 RON gasoline.

- Israel: 95 RON & 98 RON are normally available at most automotive fuel stations. 96 RON is also available at a large number of gas stations but 95 RON is more preferred because it's cheaper and performance differences aren't very wide and noticeable. "Regular" fuel is 95 RON. All variants are unleaded.

- Japan: Since 1986, "regular" is >=89 RON, and "high octane" is >=96 RON, lead free. Those values are defined in standard JIS K 2202. Sometimes "high octane" is sold under different names, such as "F-1".

- Latvia: 95 RON and 98 RON widely available.

- Lebanon: 95 RON and 98 RON are widely available.

- Lithuania: 95 RON and 98 RON widely available. In some gas stations E85 (bioethanol) gasoline, 98E15 (15% of ethanol), 98E25 (25% of ethanol) are available.

- Malaysia: 95 RON, 97 RON and 100 RON. "Regular" unleaded fuel is 95 RON; "Premium" fuel is rated at 97 RON (Shell's V-Power Racing is rated 97 RON). Petron sells 100 RON in selected outlets.

- México: Pemex Magna (87 AKI) is sold as a "regular" fuel and is available at every station. And Pemex Premium (92 AKI) is sold at almost all gas stations. Both variants are unleaded.

- Mongolia: 92 RON and 95 RON (advertised as A92 and A95 respectively) are available at nearly all stations while slightly fewer stations offer 80 RON (advertised as A80). 98 RON (advertised as A98) is available in select few stations.

- Montenegro: 95 RON is sold as a "regular" fuel. As a "premium" fuel, 98 RON is sold. Both variants are unleaded.

- Netherlands: 95 RON "Euro" is sold at every station, whereas 98 RON "Super Plus" is being phased out in favor of "premium" fuels, which are all 95 RON fuels with extra additives. Shell V-Power is a 97 RON (labelled as 95 due to the legalities of only using 95 or 98 labelling), some independent tests have shown that one year after introduction it was downgraded to 95 RON, whereas in neighboring Germany Shell V-Power consists of the regular 100 RON fuel.

- New Zealand: 91 RON "Regular" and 95 RON "Premium" are both widely available. 98 RON is available instead of 95 RON at some (BP, Mobil, Gull) service stations in larger urban areas.

- Norway: 95 RON are widely available, but you can also get 98 RON at Shell but it is about 10-20% more expensive compared 95 RON octane. Statoil has discontinued production and sale due to low demand.

- Philippines: A brand of Petron, Petron Blaze is rated at 100 RON (the only brand of gasoline in the Philippines without an ethanol blend). Other "super premium" brands like Petron XCS, Caltex Gold, Shell V-Power are rated at 95-97 RON, while Petron Xtra Unleaded, Caltex Silver, and Shell Super Unleaded are rated at 93 RON.

- Poland: Eurosuper 95 (RON 95) is sold in every gas station. Super Plus 98 (RON 98) is available in most stations, sometimes under brand (Orlen - Verva, BP - Ultimate, Shell - V-Power) and usually containing additives. Shell offers V-Power Racing fuel which is rated RON 100.

- Portugal: 95 RON "Euro" is sold in every station and 98 RON "Super" being offered in almost every station.

- Russia and CIS countries: 80 RON (76 MON) is the minimum available, the standard is 92 RON and 95 RON. 98 RON is available on some stations but it's usually quite expensive compared to the lower octane rating fuels.

- Saudi Arabia: Two types of fuel are available at all gas stations in Saudi Arabia. "Premium 91" (RON 91) where the pumps are coloured green, and "Super Premium 95" (RON 95) where the pumps are coloured red. While gas stations in Saudi Arabia are privatized, the prices are regulated by the authorities and have a fixed at S.A.R. 0.75 (U.S. $0.20) and S.A.R. 0.90 (U.S. $0.24) (as of Dec 28, 2015) per litre respectively (. Prior to 2006, only Super Premium RON 95 was available and the pumps weren't coloured in any specific order. The public didn't know what Octane rating was, therefore big educating campaigns were spread, telling the people to use the "red gas" only for high end cars, and save money on using the "green gas" for regular cars and trucks.

- Singapore: All four providers, Caltex, ExxonMobil, SPC and Shell have 3 grades of gasoline. Typically, these are 92, 95, and 98 RON. However, since 2009, Shell has removed 92 RON.

- South Africa: "regular" unleaded fuel is 95 RON in coastal areas. Inland (higher elevation) "regular" unleaded fuel is 93 RON; once again most fuel stations optionally offer 95 RON.

- Spain: 95 RON "Euro" is sold in every station with 98 RON "Super" being offered in most stations. Many stations around cities and highways offer other high-octane "premium" brands.

- Sri Lanka: In Ceypetco filling stations, 92 RON is the regular automotive fuel and 95 RON is called 'Super Petrol', which comes at a premium price. In LIOC filling stations, 90 RON remains as regular automotive fuel and 92 RON is available as 'Premium Petrol'. The cost of premium gasoline is lower than the cost of super gasoline. (Sri Lanka switched their regular gasoline from 90 RON to 92 RON on January 1, 2014)

- Sweden: 95 RON, 98 RON and E85 are widely available.

- Taiwan: 92 RON, 95 RON and 98 RON are widely available at gas stations in Taiwan.

- Thailand: 91 RON and 95 RON are widely available. 91 RON automotive fuel withdrawn on Jan 1st 2013 to increase uptake of gasohol fuels.

- Trinidad and Tobago: 92 RON (Super) and 95 RON (Premium) are widely available.

- Turkey: 95 RON and 95+ RON widely available in gas stations. 91 RON (Regular) has been dropped in 2006. 98 and 100 RON (Shell V-Power Racing) has been dropped in late 2009. The Gas which has been advertised 97 RON has been dropped in 2014 and renamed 95+.

- Ukraine: the standard gasoline is 95 RON, but 92 RON gasoline is also widely available and popular as a less expensive replacement for 95 RON gasoline. 80 RON gasoline is available for old cars and motorcycles.

- United Kingdom: 'regular' gasoline has an octane rating of 95 RON, with 97 RON fuel being widely available as the *Super Unleaded*. Tesco and Shell both offer 99 RON fuel. In April 2006, BP started a public trial of the super-high octanegasoline*BP Ultimate Unleaded 102*, which as the name suggests, has an octane rating of 102 RON. Although BP Ultimate Unleaded (with an octane rating of 97 RON) and BP Ultimate Diesel are both widely available throughout the UK, BP Ultimate Unleaded 102 was available throughout the UK in only 10 filling stations, and was priced at about two and half times more than their 97 RON fuel. In March 2010, BP stopped sales of Ultimate Unleaded 102, citing the closure of their specialty fuels manufacturing facility. Shell V-Power is also available, but in a 99 RON octane rating, and Tesco fuel stations also supply the Greenergy produced 99 RON "Momentum99".

- United States: in the US octane rating is displayed in AKI. In most areas, the standard grades are 87, 89-90 and 91-94 AKI. In the Rocky Mountain (high elevation) states, 85 AKI (90 RON) is the minimum octane, and 91 AKI (95 RON) is the maximum octane available in fuel. The reason for this is that in higher-elevation areas, a typical naturally aspirated engine draws in less air mass per cycle because of the reduced density of the atmosphere. This directly translates to less fuel and reduced absolute compression in the cylinder, therefore deterring knock. It is safe to fill a carbureted car that normally takes 87 AKI fuel at sea level with 85 AKI fuel in the mountains, but at sea level the fuel may cause damage to the engine.

However, since virtually all cars produced since the mid-1980s have fuel injection, 85 AKI fuel is not recommended for modern automobiles and may cause damage to the engine and decreased performance. Another disadvantage to this strategy is that most turbocharged vehicles are unable to produce full power, even when using the "premium" 91 AKI fuel. In some east coast states, up to 94 AKI (98 RON) is available. As of January, 2011, over 40 states and a total of over 2500 stations offer ethanol-based E-85 fuel with 105 AKI. Often, filling stations near US racing tracks will offer higher octane levels such as 100 AKI .

- Venezuela: 91 RON and 95 RON gasoline is available nationwide, in all PDV gas stations. 95 RON gasoline is the most widely used in the country, although most cars in Venezuela would work with 91 RON gasoline. This is because gasoline prices are heavily subsidized by the government (0.$083 per gallon 95 RON,vs 0.$061 per gallon 91 RON). All gasoline in Venezuela is unleaded.

- Vietnam: 92 is in every gas station and 95 is in the urban areas.

- Zimbabwe: 93 octane available with no other grades of fuels available, E10 which is an ethanol blend of fuel at 10% ethanol is available the octane rating however is still to be tested and confirmed but it is assumed that its around 95 Octane. E85 available from 3 outlets with an octane rating AKI index of between 102-105 depending on the base gasoline the ethanol is blended with.

References

- NEA, IAEA (2014). Uranium 2014 – Resources, Production and Demand (PDF). OECD Publishing. doi:10.1787/uranium-2014-en. ISBN 978-92-64-22351-6.

- Greenwood, N. N.; & Earnshaw, A. (1997). Chemistry of the Elements (2nd Edn.), Oxford:Butterworth-Heinemann. ISBN 0-7506-3365-4.

- Laurence M. Harwood; Christopher J. Moody (13 June 1989). Experimental organic chemistry: Principles and Practice (Illustrated ed.). pp. 145–147. ISBN 978-0-632-02017-1.

- Ibrahim, Hassan Al-Haj (2014). "Chapter 5". In Bennett, Kelly. Matlab: Applications for the Practical Engineer. Sciyo. pp. 139–171. ISBN 978-953-51-1719-3.

- Kemp, Kenneth W.; Brown, Theodore; Nelson, John D. (2003). Chemistry: the central science. Englewood Cliffs, N.J: Prentice Hall. p. 992. ISBN 0-13-066997-0.

- "The Oil Drum: Europe | Agriculture Meets Peak Oil: Soil Association Conference". Europe.theoildrum.com. Retrieved 2013-11-03.

- M. King Hubbert (June 1956). "Nuclear Energy And The Fossil Fuels" (PDF). Shell Development Company. Retrieved 2013-12-27.

- Jones, Tony (23 November 2004). "Professor Goodstein discusses lowering oil reserves". Australian Broadcasting Corporation. Retrieved 14 April 2013.

- Daniel L. Edelstein (January 2008). "Copper" (PDF). U.S. Geological Survey, Mineral Commodity Summaries. Retrieved 2013-12-27.

- Stephen M. Jasinski (January 2006). "Phosphate Rock" (PDF). U.S. Geological Survey, Mineral Commodity Summaries. Retrieved 2013-12-27.

- Ecological Sanitation Research Programme (May 2008). "Closing the Loop on Phosphorus" (PDF). Stockholm Environment Institute. Retrieved 2013-12-27.

Petroleum Extraction: An Integrated Study

The process by which petroleum is drawn out of the Earth is known as extraction of petroleum. Shale oil extraction is the industrial process for oil production. Some of the topics discussed herein are unconventional oil, oil well, oil and gas well completion, drilling rig and peak oil. The topics discussed in this section help the reader in understanding petroleum extraction.

Extraction of Petroleum

Pumpjack on an oil well in Texas

The extraction of petroleum is the process by which usable petroleum is extracted and removed from the earth.

Locating the Oil Field

Geologists use seismic surveys to search for geological structures that may form oil reservoirs. The "classic" method includes making an underground explosion nearby and observing the seismic response that provides information about the geological structures under the ground . However, "passive" methods that extract information from naturally-occurring seismic waves are also known. Other instruments such as gravimeters and magnetometers are also sometimes used in the search for petroleum. Extracting crude oil normally starts with drilling wells into the underground reservoir. When an oil well has been tapped, a geologist (known on the rig as the "mudlogger") will note its presence. Such a "mudlogger" is known to be sitting on the rig. Historically, in the United States, some oil fields existed where the oil rose naturally to the surface, but most of these fields have long since been used up, except in certain places in Alaska. Often many wells (called *multilateral wells*) are drilled into the same reservoir, to ensure that the extraction rate will be economically viable. Also, some wells (*secondary wells*) may be used to pump water, steam, acids or various gas mixtures into the reservoir to raise or maintain the reservoir pressure, and so maintain an economic extraction rate.

Drilling

The oil well is created by drilling a long hole into the earth with an oil rig. A steel pipe (casing) is placed in the hole, to provide structural integrity to the newly drilled well bore. Holes are then made in the base of the well to enable oil to pass into the bore. Finally a collection of valves called a "Christmas Tree" is fitted to the top, the valves regulate pressures and control flow.

Oil Extraction and Recovery

Primary Recovery

During the *primary recovery stage*, reservoir drive comes from a number of natural mechanisms. These include: natural water displacing oil downward into the well, expansion of the natural gas at the top of the reservoir, expansion of gas initially dissolved in the crude oil, and gravity drainage resulting from the movement of oil within the reservoir from the upper to the lower parts where the wells are located. Recovery factor during the primary recovery stage is typically 5-15%.

While the underground pressure in the oil reservoir is sufficient to force the oil to the surface, all that is necessary is to place a complex arrangement of valves (the Christmas tree) on the well head to connect the well to a pipeline network for storage and processing. Sometimes pumps, such as beam pumps and electrical submersible pumps (ESPs), are used to bring the oil to the surface; these are known as artificial lifting mechanisms.

Secondary Recovery

Over the lifetime of the well the pressure will fall, and at some point there will be insufficient underground pressure to force the oil to the surface. After natural reservoir drive diminishes, *secondary recovery* methods are applied. They rely on the supply of external energy into the reservoir in the form of injecting fluids to increase reservoir pressure, hence replacing or increasing the natural reservoir drive with an artificial drive. Secondary recovery techniques increase the reservoir's pressure by water injection, natural gas reinjection and gas lift, which injects air, carbon dioxide or some other gas into the bottom of an active well, reducing the overall density of fluid in the wellbore. Typical recovery factor from water-flood operations is about 30%, depending on the properties of oil and the characteristics of the reservoir rock. On average, the recovery factor after primary and secondary oil recovery operations is between 35 and 45%.

Enhanced Recovery

Enhanced, or Tertiary oil recovery methods increase the mobility of the oil in order to increase extraction.

Thermally enhanced oil recovery methods (TEOR) are tertiary recovery techniques that heat the oil, thus reducing its viscosity and making it easier to extract. Steam injection is the most common form of TEOR, and is often done with a cogeneration plant. In this type of cogeneration plant, a gas turbine is used to generate electricity and the waste heat is used to produce steam, which is then injected into the reservoir. This form of recovery is used extensively to increase oil extraction in the San Joaquin Valley, which has very heavy oil, yet accounts for 10% of the United States' oil extraction.Fire flooding (In-situ burning) is another form of TEOR, but instead of steam, some of the oil is burned to heat the surrounding oil.

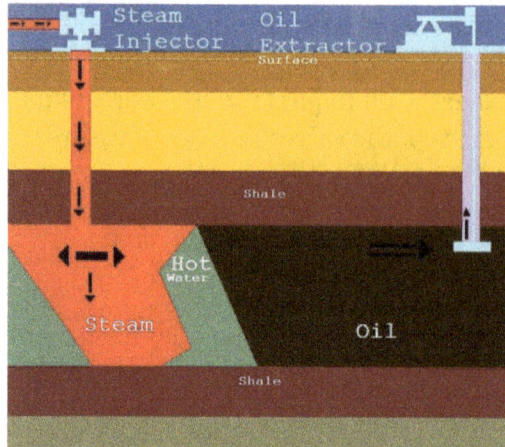

Steam is injected into many oil fields where the oil is thicker and heavier than normal crude oil

Occasionally, surfactants (detergents) are injected to alter the surface tension between the water and oil in the reservoir, mobilizing oil which would otherwise remain in the reservoir as residual oil.

Another method to reduce viscosity is carbon dioxide flooding.

Tertiary recovery allows another 5% to 15% of the reservoir's oil to be recovered. In some California heavy oil fields, steam injection has doubled or even tripled the oil reserves and ultimate oil recovery. For example, Midway-Sunset Oil Field, California's largest oilfield.

Tertiary recovery begins when secondary oil recovery isn't enough to continue adequate extraction, but only when the oil can still be extracted profitably. This depends on the cost of the extraction method and the current price of crude oil. When prices are high, previously unprofitable wells are brought back into use and when they are low, extraction is curtailed.

Microbial treatments is another tertiary recovery method. Special blends of the microbes are used to treat and break down the hydrocarbon chain in oil thus making the oil easy to recover as well as being more economic versus other conventional methods. In some states, such as Texas, there are tax incentives for using these microbes in what is called a secondary tertiary recovery. Very few companies supply these, however companies like Bio Tech, Inc. have proven very successful in waterfloods across Texas.

Recovery Rates and Factors

The amount of oil that is recoverable is determined by a number of factors including the permeability of the rocks, the strength of natural drives (the gas present, pressure from adjacent water or gravity), and the viscosity of the oil. When the reservoir rocks are "tight" such as shale, oil generally cannot flow through but when they are permeable such as in sandstone, oil flows freely.

Estimated Ultimate Recovery

Although recovery of a well cannot be known with certainty until the well ceases production, petroleum engineers will often estimate an estimated ultimate recovery (EUR) based on decline rate projections years into the future. Various models, mathematical techniques and approximations are used.

Shale gas EUR is difficult to predict and it is possible to choose recovery methods that tend to underestimate decline of the well beyond that which is reasonable.

Shale Oil Extraction

Shale oil extraction is an industrial process for unconventional oil production. This process converts kerogen in oil shale into shale oil by pyrolysis, hydrogenation, or thermal dissolution. The resultant shale oil is used as fuel oil or upgraded to meet refinery feedstock specifications by adding hydrogen and removing sulfur and nitrogen impurities.

Shale oil extraction is usually performed above ground (*ex situ* processing) by mining the oil shale and then treating it in processing facilities. Other modern technologies perform the processing underground (on-site or *in situ* processing) by applying heat and extracting the oil via oil wells.

The earliest description of the process dates to the 10th century. In 1684, Great Britain granted the first formal extraction process patent. Extraction industries and innovations became widespread during the 19th century. The industry shrank in the mid-20th century following the discovery of large reserves of conventional oil, but high petroleum prices at the beginning of the 21st century have led to renewed interest, accompanied by the development and testing of newer technologies.

As of 2010, major long-standing extraction industries are operating in Estonia, Brazil, and China. Its economic viability usually requires a lack of locally available crude oil. National energy security issues have also played a role in its development. Critics of shale oil extraction pose questions about environmental management issues, such as waste disposal, extensive water use, waste water management, and air pollution.

History

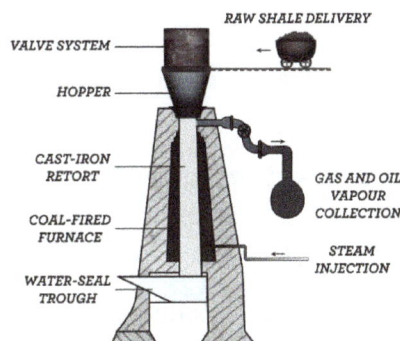

Alexander C. Kirk's retort, used in the mid-to-late 19th century, was one of the first vertical oil shale retorts. Its design is typical of retorts used in the end of 19th and beginning of 20th century.

In the 10th century, the Arabian physician Masawaih al-Mardini (Mesue the Younger) wrote of his experiments in extracting oil from "some kind of bituminous shale". The first shale oil extraction patent was granted by the British Crown in 1684 to three people who had "found a way to extract and make great quantities of pitch, tarr, and oyle out of a sort of stone". Modern industrial extraction of shale oil originated in France with the implementation of a process invented by Alexander Selligue in 1838, improved upon a decade later in Scotland using a process invented by James Young. During the late 19th century, plants were built in Australia, Brazil, Canada, and the United States. The 1894 invention of the Pumpherston retort, which was much less reliant on coal heat than its predecessors, marked the separation of the oil shale industry from the coal industry.

China (Manchuria), Estonia, New Zealand, South Africa, Spain, Sweden, and Switzerland began extracting shale oil in the early 20th century. However, crude oil discoveries in Texas during the 1920s and in the Middle East in the mid 20th century brought most oil shale industries to a halt. In 1944, the US recommenced shale oil extraction as part of its Synthetic Liquid Fuels Program. These industries continued until oil prices fell sharply in the 1980s. The last oil shale retort in the US, operated by Unocal Corporation, closed in 1991. The US program was restarted in 2003, followed by a commercial leasing program in 2005 permitting the extraction of oil shale and oil sands on federal lands in accordance with the Energy Policy Act of 2005.

As of 2010, shale oil extraction is in operation in Estonia, Brazil, and China. In 2008, their industries produced about 930,000 metric tonnes (17,700 barrels per day) of shale oil. Australia, the US, and Canada have tested shale oil extraction techniques via demonstration projects and are planning commercial implementation; Morocco and Jordan have announced their intent to do the same. Only four processes are in commercial use: Kiviter, Galoter, Fushun, and Petrosix.

Processing Principles

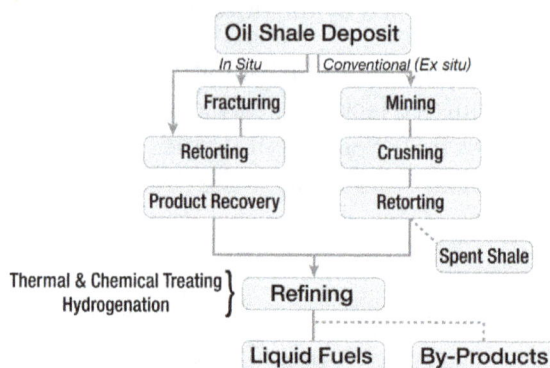

Overview of shale oil extraction

Shale oil extraction process decomposes oil shale and converts its kerogen into shale oil—a petroleum-like synthetic crude oil. The process is conducted by pyrolysis, hydrogenation, or thermal dissolution. The efficiencies of extraction processes are often evaluated by comparing their yields to the results of a Fischer Assay performed on a sample of the shale.

The oldest and the most common extraction method involves pyrolysis (also known as *retorting* or destructive distillation). In this process, oil shale is heated in the absence of oxygen until its kerogen decomposes into condensable shale oil vapors and non-condensable combustibleoil shale gas. Oil vapors and oil shale gas are then collected and cooled, causing the shale oil to condense. In addition, oil shale processing produces spent oil shale, which is a solid residue. Spent shale consists of inorganic compounds (minerals) and char—a carbonaceous residue formed from kerogen. Burning the char off the spent shale produces oil shale ash. Spent shale and shale ash can be used as ingredients in cement or brick manufacture. The composition of the oil shale may lend added value to the extraction process through the recovery of by-products, including ammonia, sulfur, aromatic compounds, pitch, asphalt, and waxes.

Heating the oil shale to pyrolysis temperature and completing the endothermic kerogen decomposition reactions require a source of energy. Some technologies burn other fossil fuels such as natural gas, oil, or coal to generate this heat and experimental methods have used electricity, radio waves, microwaves, or reactive fluids for this purpose. Two strategies are used to reduce, and even eliminate, external heat energy requirements: the oil shale gas and char by-products generated by pyrolysis may be burned as a source of energy, and the heat contained in hot spent oil shale and oil shale ash may be used to pre-heat the raw oil shale.

For *ex situ* processing, oil shale is crushed into smaller pieces, increasing surface area for better extraction. The temperature at which decomposition of oil shale occurs depends on the time-scale of the process. In *ex situ* retorting processes, it begins at 300 °C (570 °F) and proceeds more rapidly and completely at higher temperatures. The amount of oil produced is the highest when the temperature ranges between 480 and 520 °C (900 and 970 °F). The ratio of oil shale gas to shale oil generally increases along with retorting temperatures. For a modern *in situ* process, which might take several months of heating, decomposition may be conducted at temperatures as low as 250 °C (480 °F). Temperatures below 600 °C (1,110 °F) are preferable, as this prevents the decomposition of limestone and dolomite in the rock and thereby limits carbon dioxide emissions and energy consumption.

Hydrogenation and thermal dissolution (reactive fluid processes) extract the oil using hydrogen donors, solvents, or a combination of these. Thermal dissolution involves the application of solvents at elevated temperatures and pressures, increasing oil output by cracking the dissolved organic matter. Different methods produce shale oil with different properties.

Classification of Extraction Technologies

Industry analysts have created several classifications of the technologies used to extract shale oil from oil shale.

By process principles: Based on the treatment of raw oil shale by heat and solvents the methods are classified as pyrolysis, hydrogenation, or thermal dissolution.

By location: A frequently used distinction considers whether processing is done above or below ground, and classifies the technologies broadly as *ex situ* (displaced) or *in situ* (in place). In *ex situ* processing, also known as above-ground retorting, the oil shale is mined either underground or at the surface and then transported to a processing facility. In contrast, *in situ* processing converts the kerogen while it is still in the form of an oil shale deposit, following which it is then extracted via oil wells, where it rises in the same way as conventional crude oil. Unlike *ex situ* processing, it does not involve mining or spent oil shale disposal aboveground as spent oil shale stays underground.

By heating method: The method of transferring heat from combustion products to the oil shale may be classified as direct or indirect. While methods that allow combustion products to contact the oil shale within the retort are classified as *direct*, methods that burn materials external to the retort to heat another material that contacts the oil shale are described as *indirect*

By heat carrier: Based on the material used to deliver heat energy to the oil shale, processing technologies have been classified into gas heat carrier, solid heat carrier, wall conduction, reactive fluid, and volumetric heating methods. Heat carrier methods can be sub-classified as direct or indirect.

The following table shows extraction technologies classified by heating method, heat carrier and location (*in situ* or *ex situ*).

	Classification of processing technologies by heating method and location (according to Alan Burnham)	
Heating Method	**Above ground (*ex situ*)**	**Underground (*in situ*)**
Internal combustion	Gas combustion, NTU, Kiviter, Fushun, Union A, Paraho Direct, Superior Direct	Occidental Petroleum MIS, LLNL RISE, Geokinetics Horizontal, Rio Blanco
Hot recycled solids (inert or burned shale)	Alberta Taciuk, Galoter, Enefit, Lurgi-Ruhrgas, TOSCO II, Chevron STB, LLNL HRS, Shell Spher, KENTORT II	–
Conduction through a wall (various fuels)	Pumpherston, Fischer Assay, Oil-Tech, EcoShale In-Capsule, Combustion Resources	Shell ICP (primary method), American Shale Oil CCR, IEP Geothermic Fuel Cell
Externally generated hot gas	PetroSIX, Union B, Paraho Indirect, Superior Indirect, Syntec (Smith process)	Chevron CRUSH, Omnishale, MWE IGE
Reactive fluids	IGT Hytort (high-pressure H_2), donor solvent processes Rendall ProcessChattanooga fluidized bed reactor	Shell ICP (some embodiments)
Volumetric heating	–	Radio wave, microwave, and electric current processes

By raw oil shale particle size: The various *ex situ* processing technologies may be differentiated by the size of the oil shale particles that are fed into the retorts. As a rule, gas heat carrier technologies process oil shale lumps varying in diameter from 10 to 100 millimeters (0.4 to 3.9 in), while solid heat carrier and wall conduction technologies process fines which are particles less than 10 millimeters (0.4 in) in diameter.

By retort orientation: "Ex-situ" technologies are sometimes classified as vertical or horizontal. Vertical retorts are usually shaft kilns where a bed of shale moves from top to bottom by gravity. Horizontal retorts are usually horizontal rotating drums or screws where shale moves from one end to the other. As a general rule, vertical retorts process lumps using a gas heat carrier, while horizontal retorts process fines using solid heat carrier.

By complexity of technology: *In situ* technologies are usually classified either as *true in situ* processes or *modified in situ* processes. *True in situ* processes do not involve mining or crushing the oil shale. *Modified in situ* processes involve drilling and fracturing the target oil shale deposit to create voids in the deposit. The voids enable a better flow of gases and fluids through the deposit, thereby increasing the volume and quality of the shale oil produced.

Ex Situ Technologies

Internal Combustion

Internal combustion technologies burn materials (typically char and oil shale gas) within a vertical shaft retort to supply heat for pyrolysis. Typically raw oil shale particles between 12 millimetres (0.5 in) and 75 millimetres (3.0 in) in size are fed into the top of the retort and are heated by the rising hot gases, which pass through the descending oil shale, thereby causing decomposition of the kerogen at about 500 °C (932 °F) . Shale oil mist, evolved gases and cooled combustion gases are removed from the top of the retort then moved to separation equipment. Condensed shale oil is collected, while non-condensable gas is recycled and used to carry heat up the retort. In the lower part of the retort, air is injected for the combustion which heats the spent oil shale and gases to between 700 °C (1,292 °F) and 900 °C (1,650 °F). Cold recycled gas may enter the bottom of the retort to cool the shale ash. The Union A and Superior Direct processes depart from this pattern. In the Union A process, oil shale is fed through the bottom of the retort and a pump moves it upward. In the Superior Direct process, oil shale is processed in a horizontal, segmented, doughnut-shaped traveling-grate retort.

Internal combustion technologies such as the Paraho Direct are thermally efficient, since combustion of char on the spent shale and heat recovered from the shale ash and evolved gases can provide all the heat requirements of the retort. These technologies can achieve 80-90% of Fischer assay yield. Two well-established shale oil industries use internal combustion technologies: Kiviter process facilities have been operated continuously in Estonia since the 1920s, and a number of Chinese companies operate Fushun process facilities.

Common drawbacks of internal combustion technologies are that the combustible oil shale gas is diluted by combustion gases and particles smaller than 10 millimeters (0.4 in) can not be processed. Uneven distribution of gas across the retort can result in blockages when hot spots cause particles to fuse or disintegrate.

Hot Recycled Solids

Hot recycled solids technologies deliver heat to the oil shale by recycling hot solid particles—typically oil shale ash. These technologies usually employ rotating kiln or fluidized bed retorts, fed by fine oil shale particles generally having a diameter of less than 10 millimeters (0.4 in); some technologies use particles even smaller than 2.5 millimeters (0.10 in). The recycled particles are heated in a separate chamber or vessel to about 800 °C (1,470 °F) and then mixed with the raw oil shale to cause the shale to decompose at about 500 °C (932 °F). Oil vapour and shale oil gas are separated from the solids and cooled to condense and collect the oil. Heat recovered from the combustion gases and shale ash may be used to dry and preheat the raw oil shale before it is mixed with the hot recycle solids.

In the Galoter and Enefit processes, the spent oil shale is burnt in a separate furnace and the resulting hot ash is separated from the combustion gas and mixed with oil shale particles in a rotating kiln. Combustion gases from the furnace are used to dry the oil shale in a dryer before mixing with hot ash. The TOSCO II process uses ceramic balls instead of shale ash as the hot recycled solids. The distinguishing feature of the Alberta Taciuk Process (ATP) is that the entire process occurs in a single rotating multi–chamber horizontal vessel.

Because the hot recycle solids are heated in a separate furnace, the oil shale gas from these technologies is not diluted with combustion exhaust gas. Another advantage is that there is no limit on the smallest particles that the retort can process, thus allowing all the crushed feed to be used. One disadvantge is that more water is used to handle the resulting finer shale ash.

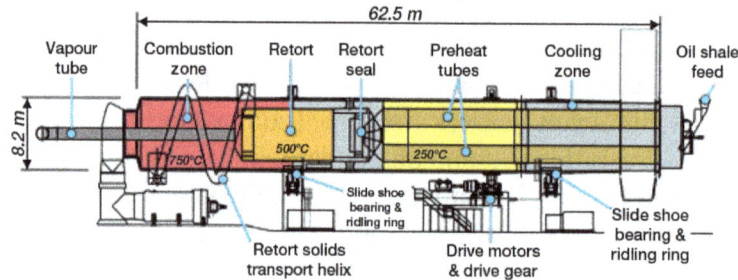

Alberta Taciuk Processor retort

Conduction through a Wall

These technologies transfer heat to the oil shale by conducting it through the retort wall. The shale feed usually consists of fine particles. Their advantage lies in the fact that retort vapors are not combined with combustion exhaust. The Combustion Resources process uses a hydrogen–fired rotating kiln, where hot gas is circulated through an outer annulus. The Oil-Tech staged electrically heated retort consists of individual inter-connected heating chambers, stacked atop each other. Its principal advantage lies in its modular design, which enhances its portability and adaptability. The Red Leaf Resources EcoShale In-Capsule Process combines surface mining with a lower-temperature heating method similar to *in situ* processes by operating within the confines of an earthen structure. A hot gas circulated through parallel pipes heats the oil shale rubble. An installation within the empty space created by mining would permit rapid reclamation of the topography. A general drawback of conduction through a wall technologies is that the retorts are more costly when scaled-up due to the resulting large amount of heat conducting walls made of high-temperature alloys.

Externally Generated Hot Gas

In general, externally generated hot gas technologies are similar to internal combustion technologies in that they also process oil shale lumps in vertical shaft kilns. Significantly, though, the heat in these technologies is delivered by gases heated outside the retort vessel, and therefore the retort vapors are not diluted with combustion exhaust. The Petrosix and Paraho Indirect employ this technology. In addition to not accepting fine particles as feed, these technologies do not utilize the potential heat of combusting the char on the spent shale and thus must burn more valuable fuels. However, due to the lack of combustion of the spent shale, the oil shale does not exceed 500 °C (932 °F) and significant carbonate mineral decomposition and subsequent CO_2 generation can be avoided for some oil shales. Also, these technologies tend to be the more stable and easier to control than internal combustion or hot solid recycle technologies.

Reactive Fluids

Kerogen is tightly bound to the shale and resists dissolution by most solvents. Despite this constraint, extraction using especially reactive fluids has been tested, including those in a supercritical state. Reactive fluid technologies are suitable for processing oil shales with a low hydrogen content. In these technologies, hydrogen gas (H_2) or hydrogen donors (chemicals that donate hydrogen during chemical reactions) react with coke precursors (chemical structures in the oil shale that are prone to form char during retorting but have not yet done so). Reactive fluid technologies include the IGT Hytort (high-pressure H_2) process, donor solvent processes, and the Chattanooga fluidized bed reactor. In the IGT Hytort oil shale is processed in a high-pressure hydrogen environment. The Chattanooga process uses a fluidized bed reactor and an associated hydrogen-fired heater for oil shale thermal cracking and hydrogenation. Laboratory results indicate that these technologies can often obtain significantly higher oil yields than pyrolysis processes. Drawbacks are the additional cost and complexity of hydrogen production and high-pressure retort vessels.

Plasma Gasification

Several experimental tests have been conducted for the oil-shale gasification by using plasma technologies. In these technologies, oil shale is bombarded by radicals (ions). The radicals crack kerogen molecules forming synthetic gas and oil. Air, hydrogen or nitrogen are used as plasma gas and processes may operate in an arc, plasma arc, or plasma electrolysis mode. The main benefit of these technologies is processing without using water.

In Situ Technologies

In situ technologies heat oil shale underground by injecting hot fluids into the rock formation, or by using linear or planar heating sources followed by thermal conduction and convection to distribute heat through the target area. Shale oil is then recovered through vertical wells drilled into the formation. These technologies are potentially able to extract more shale oil from a given area of land than conventional *ex situ* processing technologies, as the wells can reach greater depths than surface mines. They present an opportunity to recover shale oil from low-grade deposits that traditional mining techniques could not extract.

During World War II a modified *in situ* extraction process was implemented without significant success in Germany. One of the earliest successful *in situ* processes was underground gasification by electrical energy (Ljungström method)—a process exploited between 1940 and 1966 for shale oil extraction at Kvarntorp in Sweden. Prior to the 1980s, many variations of the *in situ* process were explored in the United States. The first modified *in situ* oil shale experiment in the United States was conducted by Occidental Petroleum in 1972 at Logan Wash, Colorado. Newer technologies are being explored that use a variety of heat sources and heat delivery systems.

Wall Conduction

Wall conduction *in situ* technologies use heating elements or heating pipes placed within the oil shale formation. The Shell in situ conversion process (Shell ICP) uses electrical heating elements for heating the oil shale layer to between 650 and 700 °F (340 and 370 °C) over a period of approximately four years. The processing area is isolated from surrounding groundwater by a freeze wall consisting of wells filled with a circulating super-chilled fluid. Disadvantages of this process are large electrical power consumption, extensive water use, and the risk of groundwater pollution. The process was tested since the early 1980s at the Mahogany test site in the Piceance Basin. 1,700 barrels (270 m³) of oil were extracted in 2004 at a 30-by-40-foot (9.1 by 12.2 m) testing area.

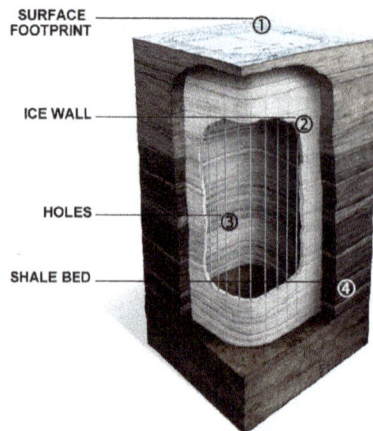

Shell's freeze wall for *in situ* shale oil production separates the process from its surroundings

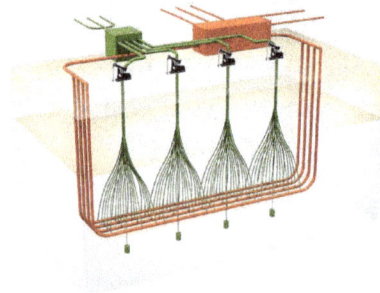

American Shale Oil CCR Process

In the CCR Process proposed by American Shale Oil, superheated steam or another heat transfer medium is circulated through a series of pipes placed below the oil shale layer to be extracted. The system combines horizontal wells, through which steam is passed, and vertical wells, which provide both vertical heat transfer through refluxing of converted shale oil and a means to collect the produced hydrocarbons. Heat is supplied by combustion of natural gas or propane in the initial phase and by oil shale gas at a later stage.

The Geothermic Fuels Cells Process (IEP GFC) proposed by Independent Energy Partners extracts shale oil by exploiting a high-temperature stack of fuel cells. The cells, placed in the oil shale formation, are fueled by natural gas during a warm-up period and afterward by oil shale gas generated by its own waste heat.

Externally Generated Hot Gas

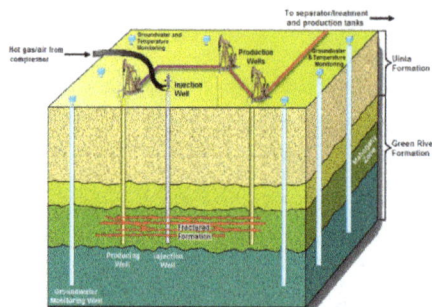

Chevron CRUSH process

Externally generated hot gas *in situ* technologies use hot gases heated above-ground and then injected into the oil shale formation. The Chevron CRUSH process, which was researched by Chevron Corporation in partnership with Los Alamos National Laboratory, injects heated carbon dioxide into the formation via drilled wells and to heat the formation through a series of horizontal fractures through which the gas is circulated. General Synfuels International has proposed the Omnishale process involving injection of super-heated air into the oil shale formation.Mountain West Energy's In Situ Vapor Extraction process uses similar principles of injection of high-temperature gas.

ExxonMobil Electrofrac

ExxonMobil's *in situ* technology (ExxonMobil Electrofrac) uses electrical heating with elements of both wall conduction and volumetric heating methods. It injects an electrically conductive material such as calcined petroleum coke into the hydraulic fractures created in the oil shale formation which then forms a heating element. Heating wells are placed in a parallel row with a second horizontal well intersecting them at their toe. This allows opposing electrical charges to be applied at either end.

Volumetric Heating

Artist's rendition of a radio wave-based extraction facility

The Illinois Institute of Technology developed the concept of oil shale volumetric heating using radio waves (radio frequency processing) during the late 1970s. This technology was further developed by Lawrence Livermore National Laboratory. Oil shale is heated by vertical electrode arrays. Deeper volumes could be processed at slower heating rates by installations spaced at tens of meters. The concept presumes a radio frequency at which the skin depth is many tens of meters, thereby overcoming the thermal diffusion times needed for conductive heating. Its drawbacks include intensive electrical demand and the possibility that groundwater or char would absorb undue amounts of the energy. Radio frequency processing in conjunction with critical fluids is being developed by Raytheon together with CF Technologies and tested by Schlumberger.

Microwave heating technologies are based on the same principles as radio wave heating, although it is believed that radio wave heating is an improvement over microwave heating because its energy can penetrate farther into the oil shale formation. The microwave heating process was tested by Global Resource Corporation. Electro-Petroleum proposes electrically enhanced oil recovery by the passage of direct current between cathodes in producing wells and anodes located either at the surface or at depth in other wells. The passage of the current through the oil shale formation results in resistive Joule heating.

Economics

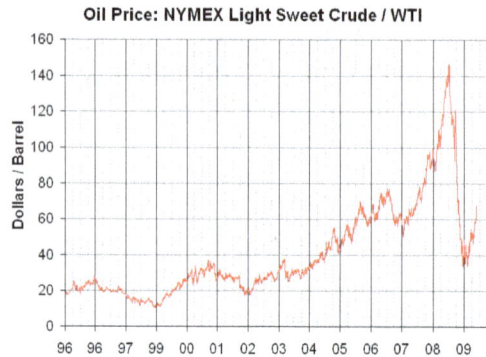

NYMEXlight-sweet crude oil prices 1996–2009 (not adjusted for inflation)

The dominant question for shale oil production is under what conditions shale oil is economically viable. According to the United States Department of Energy, the capital costs of a 100,000 barrels per day (16,000 m³/d) *ex-situ* processing complex are $3–10 billion. The various attempts to develop oil shale deposits have succeeded only when the shale-oil production cost in a given region is lower than the price of petroleum or its other substitutes. According to a survey conducted by the RAND Corporation, the cost of producing shale oil at a hypothetical surface retorting complex in the United States (comprising a mine, retorting plant, upgrading plant, supporting utilities, and spent oil shale reclamation), would be in a range of $70–95 per barrel ($440–600/m³), adjusted to 2005 values. Assuming a gradual increase in output after the start of commercial production, the analysis projects a gradual reduction in processing costs to $30–40 per barrel ($190–250/m³) after achieving the milestone of 1 billion barrels (160×10⁶ m³). The United States Department of Energy estimates that the *ex-situ* processing would be economic at sustained average world oil prices above $54 per barrel and *in-situ* processing would be economic at prices above $35 per barrel. These estimates assume a return rate of 15%.Royal Dutch Shell announced in 2006 that its Shell ICP technology would realize a profit when crude oil prices are higher than $30 per barrel ($190/m³), while some technologies at full-scale production assert profitability at oil prices even lower than $20 per barrel ($130/m³).

To increase the efficiency of oil shale retorting and by this the viability of the shale oil production, researchers have proposed and tested several co-pyrolysis processes, in which other materials such as biomass, peat, waste bitumen, or rubber and plastic wastes are retorted along with the oil shale. Some modified technologies propose combining a fluidized bed retort with a circulated fluidized bed furnace for burning the by-products of pyrolysis (char and oil shale gas) and thereby improving oil yield, increasing throughput, and decreasing retorting time.

Other ways of improving the economics of shale oil extraction could be to increase the size of the operation to achieve economies of scale, use oil shale that is a by-product of coal mining such as at Fushun China, produce specialty chemicals as by Viru Keemia Grupp in Estonia, co-generate electricity from the waste heat and process high grade oil shale that yields more oil per shale processed.

A possible measure of the viability of oil shale as an energy source lies in the ratio of the energy in the extracted oil to the energy used in its mining and processing (Energy Returned on Energy Invested, or EROEI). A 1984 study estimated the EROEI of the various known oil shale deposits as varying between 0.7–13.3; Some companies and newer technologies assert an EROEI between 3 and 10. According to the World Energy Outlook 2010, the EROEI of *ex-situ* processing is typically 4 to 5 while of *in-situ* processing it may be even as low as 2.

To increase the EROEI, several combined technologies were proposed. These include the usage of process waste heat, e.g. gasification or combustion of the residual carbon (char), and the usage of waste heat from other industrial processes, such as coal gasification and nuclear power generation.

The water requirements of extraction processes are an additional economic consideration in regions where water is a scarce resource.

Environmental Considerations

Objections to its potential environmental impact have stalled governmental support for extraction of shale oil in some countries, such as Australia. Shale oil extraction may involve a number of different environmental impacts that vary with process technologies. Depending on the geological conditions and mining techniques, mining impacts may include acid drainage induced by the sudden rapid exposure and subsequent oxidation of formerly buried materials, the introduction of metals into surface water and groundwater, increased erosion, sulfur gas emissions, and air pollution caused by the production of particulates during processing, transport, and support activities. Surface mining for *ex situ* processing, as with *in situ* processing, requires extensive land use and *ex situ* thermal processing generates wastes that require disposal. Mining, processing, spent oil shale disposal, and waste treatment require land to be withdrawn from traditional uses. Depending on the processing technology, the waste material may contain pollutants including sulfates, heavy metals, and polycyclic aromatic hydrocarbons, some of which are toxic and carcinogenic. Experimental *in situ* conversion processes may reduce some of these impacts, but may instead cause other problems, such as groundwater pollution.

Spent shale often presents a disposal problem

The production and usage of oil shale usually generates more greenhouse gas emissions, including carbon dioxide, than conventional fossil fuels. Depending on the technology and the oil shale composition, shale oil extraction processes may also emit sulfur dioxide, hydrogen sulfide, carbonyl sulfide, and nitrogen oxides. Developing carbon capture and storage technologies may reduce the processes' carbon footprint.

Concerns have been raised over the oil shale industry's use of water, particularly in arid regions where water consumption is a sensitive issue. Above-ground retorting typically consumes between one and five barrels of water per barrel of produced shale oil, depending on technology. Water is usually used for spent oil shale cooling and oil shale ash disposal. *In situ* processing, according to one estimate, uses about one-tenth as much water. In other areas, water must be pumped out of oil shale mines. The resulting fall in the water table may have negative effects on nearby arable land and forests.

A 2008 programmatic environmental impact statement issued by the United States Bureau of Land Management stated that surface mining and retort operations produce 2 to 10 U.S. gallons (7.6 to 37.9 l; 1.7 to 8.3 imp gal) of waste water per 1 short ton (0.91 t) of processed oil shale.

Unconventional Oil

Unconventional oil is petroleum produced or extracted using techniques other than the conventional (oil well) method. Oil industries and governments across the globe are investing in unconventional oil sources due to the increasing scarcity of conventional oil reserves.

Sources of Unconventional Oil

According to the International Energy Agency's (IEA) *World Energy Outlook 2001* unconventional oil included "oil shales, oil sands-based synthetic crudes and derivative products, (heavy oil, Orimulsion®), coal-based liquid supplies, biomass-based liquid supplies, gas to liquid (GTL) - liquids arising from chemical processing of gas."

In the IEA's *World Energy Outlook 2011* report, "[u]nconventional oil include[d] extra-heavy oil, natural bitumen (oil sands), kerogen oil, liquids and gases arising from chemical processing of natural gas (GTL), coal-to-liquids (CTL) and additives."

Defining Unconventional Oil

In their 2013 webpage jointly published with the Organisation for Economic Co-operation and Development (OECD), the IEA observed that as technologies and economies change, definitions for unconventional and conventional oils also change.

Conventional oil is a category that includes crude oil - and natural gas and its condensates. Crude oil production in 2011 stood at approximately 70 million barrels per day. Unconventional oil consists of a wider variety of liquid sources including oil sands, extra heavy oil, gas to liquids and other liquids. In general conventional oil is easier and cheaper to produce than unconventional oil. However, the categories "conventional" and "unconventional" do not remain fixed, and over time, as economic and technological conditions evolve, resources hitherto considered unconventional can migrate into the conventional category.

—IEA 2013

According to the US Department of Energy (DOE), "unconventional oils have yet to be strictly defined."

In a communication to the UK entitled *Oil Sands Crude* in the series *The Global Range of Crude Oils*, it was argued that commonly used definitions of unconventional oil based on production techniques are imprecise and time-dependent. They noted that the International Energy Agency does not recognize any universally accepted definition for "conventional" or "unconventional" oil. Extraction techniques that are categorized as "conventional" use "unconventional means" such as gas re-injection or the use of heat" not traditional oil extraction methods. As the use of newer technologies increase, "unconventional" oil recovery has become the norm not the exception. They noted that the Canadian oil sands production "pre-dates oil production from areas such as the North Sea (the source of a benchmark crude oil known as "Brent").

Under revised definitions, petroleum *products*, such as Western Canadian Select, a heavy crude benchmark blend produced in Hardisty, Alberta may migrate from its categorization as unconventional oil to conventional oil because of its density, even though the oil sands are an unconventional resource.

Oil Sands

Oil sands generally consist of extra heavy crude oil or crude bitumen trapped in unconsolidated sandstone. These hydrocarbons are forms of crude oil that are extremely dense and viscous, with a consistency ranging from that of molasses for some extra-heavy oil to as solid as peanut butter for some bitumen at room temperature, making extraction difficult. These heavy crude oils have a density (specific gravity) approaching or even exceeding that of water. As a result of their high viscosity, they cannot be produced by conventional methods, transported without heating or dilution with lighter hydrocarbons, or refined by older oil refineries without major modifications. Such heavy crude oils often contain high concentrations of sulfur and heavy metals, particularly nickel and vanadium, which interfere with refining processes, although lighter crude oils can also suffer from sulfur and heavy metal contamination, too. These properties present significant environmental challenges to the growth of heavy oil production and use. Canada's Athabasca oil sands and Venezuela's Orinoco heavy oil belt are the best known example of this kind of unconventional reserve. In 2003 the estimated reserves were 1.2 trillion barrels (1.9×10^{11} m³).

Heavy oil sands and bituminous sands occur world-wide. The two most important deposits are the Athabasca Oil Sands in Alberta, Canada and the Orinoco heavy oil belt in Venezuela. The hydrocarbon content of these deposits is either crude bitumen or extra-heavy crude oil, the former of which is often upgraded to synthetic crude (syncrude) and the latter of which the Venezuelan fuel Orimulsion is based. The Venezuelan extra heavy oil deposits differ from the Canadian bituminous sands in that they flow more readily at Venezuela's higher reservoir temperatures and could be produced by conventional techniques, but the recovery rates would be less than the unconventional Canadian techniques (about 8% versus up to 90% for surface mining and 60% for steam assisted gravity drainage).

In 2011, Alberta's total proven oil reserves were 170.2 billion barrels representing 11 percent of the total global oil reserves (1,523 billion barrels) and 99% of Canada's oil reserves. By 2011 Alberta was supplying 15% of the United States crude oil imports, exporting about 1.3 million barrels per day (210,000 m³/d) of crude oil. The 2006 projections for 2015, were about 3 million barrels per day (480,000 m³/d). At that rate the Athabasca oil sands reserves would last less than 160 years. About 80 percent of Alberta's bituminous deposits can be extracted using in-situ methods such as steam assisted gravity drainage and 20 percent by surface mining methods. The Northern Alberta oil sands in Athabasca, Cold Lake and Peace River areas contain an estimated 2 trillion barrels (initial volume in place) of crude bitumen and extra-heavy oil of which 9 percent was considered recoverable using technology available in 2013.

It is estimated by oil companies that the Athabasca and Orinoco sites (both of similar size) have as much as two-thirds of total global oil deposits. They have only recently been considered proven reserves of oil. This is because oil prices have risen since 2003 and costs to extract oil from these mines have fallen. Between 2003 and 2008, world oil prices rose to over $140, and costs to extract the oil fell to less than $15 per barrel at the Suncor and Syncrude mines.

In 2013 crude oil from the Canadian oil sands was expensive oil to produce, although new US tight oil production was similarly expensive. Supply costs for Athabasca oil sands projects were approximately US$50 to US$90 per barrel. However, costs for Bakken, Eagle Ford and Niobrara were higher at approximately US$70 to US$90 according to 135 global oil and gas companies surveyed reported by the Financial Post.

Extracting a significant percentage of world oil production from these deposits will be difficult since the extraction process takes a great deal of capital, manpower and land. Another constraint is energy for project heat and electricity generation, currently coming from natural gas, which in recent years has seen a surge in production and a corresponding drop in price in North America. With the new supply of shale gas in North America, the need for alternatives to natural gas has been greatly diminished.

A 2009 study by CERA estimated that production from Canada's oil sands emits "about 5–15% more carbon dioxide, over the "well-to-wheels" lifetime analysis of the fuel, than average crude oil." Author and investigative journalist David Strahan that same year stated that IEA figures show that carbon dioxide emissions from the tar sands are 20% higher than average emissions from oil.

Tight Oil

Tight oil, including light tight oil (sometimes confusingly the term 'shale oil' is used instead of 'light tight oil') is crude oil contained in petroleum-bearing formations of low permeability, often shale or tight sandstone. Economic production from tight oil formations requires the same hydraulic fracturing and often uses the same horizontal well technology used in the production of shale gas. It should not be confused with oil shale, which is shale rich in kerogen, or shale oil, which is synthetic oil produced from oil shales. Therefore, the International Energy Agency recommends to use the term "light tight oil" for oil produced from shales or other very low permeability formations, while World Energy Resources 2013 report by the World Energy Council uses the term "tight oil".

Oil Shale

Oil shale is an organic-rich fine-grainedsedimentary rock containing significant amounts of kerogen (a solid mixture of organic chemical compounds) from which technology can extract liquid hydrocarbons (shale oil) and combustible oil shale gas. The kerogen in oil shale can be converted to shale oil through the chemical processes of pyrolysis, hydrogenation, or thermal dissolution. The temperature when perceptible decomposition of oil shale occurs depends on the time-scale of the pyrolysis; in the above ground retorting process the perceptible decomposition occurs at 300 °C

(570 °F), but proceeds more rapidly and completely at higher temperatures. The rate of decomposition is the highest at a temperature of 480 °C (900 °F) to 520 °C (970 °F). The ratio of shale gas to shale oil depends on the retorting temperature and as a rule increases with the rise of temperature. For the modern *in-situ* process, which might take several months of heating, decomposition may be conducted as low as 250 °C (480 °F). Depending on the exact properties of oil shale and the exact processing technology, the retorting process may be water and energy extensives. Oil shale has also been burnt directly as a low-grade fuel.

Estimates of global deposits range from 2.8 to 3.3 trillion barrels (450×10^9 to 520×10^9 m³) of recoverable oil. There are around 600 known oil shale deposits around the world, including major deposits in the United States of America. Although oil shale deposits occur in many countries, only 33 countries possess known deposits of possible economic value. The largest deposits in the world occur in the United States in the Green River Formation, which covers portions of Colorado, Utah, and Wyoming; about 70% of this resource lies on land owned or managed by the United States federal government. Deposits in the United States constitute 62% of world resources; together, the United States, Russia and Brazil account for 86% of the world's resources in terms of shale-oil content. These figures remain tentative, with exploration or analysis of several deposits still outstanding. Well-explored deposits, potentially possessing economic value, include the Green River deposits in the western United States, the Tertiary deposits in Queensland, Australia, deposits in Sweden and Estonia, the El-Lajjun deposit in Jordan, and deposits in France, Germany, Brazil, Morocco, China, southern Mongolia and Russia. These deposits have given rise to expectations of yielding at least 40 litres (0.25 bbl) of shale oil per tonne of shale, using the Fischer Assay method.

According to a survey conducted by the RAND Corporation, the cost of producing a barrel of oil at a surface retorting complex in the United States (comprising a mine, retorting plant, upgrading plant, supporting utilities, and spent shale reclamation), would range between US$70–95 ($440–600/m³, adjusted to 2005 values). As of 2008, industry uses oil shale for shale oil production in Brazil, China and Estonia. Several additional countries started assessing their reserves or had built experimental production plants. In the USA, if oil shale could be used to meet a quarter of the current 20 million barrels per day (3,200,000 m³/d) demand, 800 billion barrels (1.3×10^{11} m³) of recoverable resources would last for more than 400 years.

Thermal Depolymerization

Thermal depolymerization (TDP) has the potential to recover energy from existing sources of waste such as petroleum coke as well as pre-existing waste deposits. This process, which imitates those that occur in nature, uses heat and pressure to break down organic and inorganic compounds through a method known as hydrous pyrolysis. Because energy output varies greatly based on feedstock, it is difficult to estimate potential energy production. According to Changing World Technologies, Inc., this process even has the ability to break down several types of materials, many of which are poisonous to both humans and the environment.

Coal and Gas Conversion

Using synthetic fuel processes, the conversion of coal and natural gas has the potential to yield great quantities of unconventional oil and/or refined products, albeit at much lower net energy output than the historic average for conventional oil extraction.

In its day - prior to the drilling of oilwells to tap reservoirs of crude oil- the pyrolysis of mined solid organic-rich deposits was the conventional method of producing mineral oils. Historically, petroleum was already being produced on an industrial scale in the United Kingdom and the United States by dry distillation of cannel coal or oil shale in the first half of the 19th Century. Yields of oil from simple pyrolysis, however, are limited by the composition of the material being pyrolysed, and modern 'oil-from-coal' processes aim for a much higher yield of organic liquids, brought about by chemical reaction with the solid feedstuff.

The four primary conversion technologies used for the production of unconventional oil and refined products from coal and gas are the indirect conversion processes of the Fischer-Tropsch process and the Mobil Process (also known as Methanol to Gasoline), and the direct conversion processes of the Bergius process and the Karrick process.

Sasol has run a 150,000 barrels per day (24,000 m³/d) coal-to-liquids plant based on Fischer Tropsch conversion in South Africa since the 1970s.

Because of the high cost of transporting natural gas, many known but remote fields were not being developed. On-site conversion to liquid fuels are making this energy available under present market conditions. Fischer Tropsch fuels plants converting natural gas to fuel, a process broadly known as gas-to-liquids are operating in Malaysia, South Africa, and Qatar. Large direct conversion coal to liquids plants are currently under construction, or undergoing start-up in China.

Total global synthetic fuel production capacity exceeds 240,000 barrels per day (38,000 m³/d), and is expected to grow rapidly in coming years, with multiple new plants currently under construction.

Environmental Concerns

As with all forms of mining, there are hazardous tailings and waste generated from the varied processes of oil extraction and production.

Environmental concerns with heavy oils are similar to those with lighter oils. However, they provide additional concerns, such as the need to heat heavy oils to pump them out of the ground. Extraction also requires large volumes of water.

The environmental impacts of oil shale differ depending on the type of extraction; however, there are some common trends. The mining process releases carbon dioxide, in addition to other oxides and pollutants, as the shale is heated. Furthermore, there is some concern about some of the chemicals mixing with ground water (either as runoff or through seeping). There are processes either in use or under development to help mitigate some of these environmental concerns.

The conversion of coal or natural gas into oil generates large amounts of carbon dioxide in addition to all the impacts of gaining these resources to begin with. However, placing plants in key areas can reduce the effective emissions due to pumping the carbon dioxide into oil beds or coal beds to enhance the recovery of oil and methane.

Carbon dioxide is a greenhouse gas, so the increased carbon dioxide produced from both the more involved extraction process with unconventional oil, as well as burning the oil itself of course, has led to deep concerns about unconventional oil worsening the impacts climate change.

Economics

Sources of unconventional oil will be increasingly relied upon when conventional oil becomes more expensive due to depletion. Conventional oil sources are currently preferred because they are less expensive than unconventional sources. New technologies, such as steam injection for oil sands deposits, are being developed to reduce unconventional oil production costs.

In May 2013 the IEA in its *Medium-Term Oil Market Report* (MTOMR) said that the North American oil production surge led by unconventional oils - US light, tight oil (LTO) and Canadian oil sands - had produced a global supply shock that would reshape the way oil is transported, stored, refined and marketed.

Oil Well

The pumpjack, such as this one located south of Midland, Texas, is a common sight in West Texas

An oil well is a boring in the Earth that is designed to bring petroleum oil hydrocarbons to the surface. Usually some natural gas is produced along with the oil. A well that is designed to produce mainly or only gas may be termed a gas well.

History

Bottom Part of an Oil Drilling Derrick in Brazoria County, Texas (Harry Walker Photograph, circa 1940)

The earliest known oil wells were drilled in China in 347 CE. These wells had depths of up to about 240 metres (790 ft) and were drilled using bits attached to bamboo poles. The oil was burned to evaporate brine and produce salt. By the 10th century, extensive bamboo pipelines connected oil wells with salt springs. The ancient records of China and Japan are said to contain many allusions to the use of natural gas for lighting and heating. Petroleum was known as *Burning water* in Japan in the 7th century.

According to Kasem Ajram, petroleum was distilled by the Persian alchemist Muhammad ibn Zakarīya Rāzi (Rhazes) in the 9th century, producing chemicals such as kerosene in the alembic (*al-ambiq*), and which was mainly used for kerosene lamps.Arab and Persian chemists also distilled crude oil in order to produce flammable products for military purposes. Through Islamic Spain, distillation became available in Western Europe by the 12th century.

Some sources claim that from the 9th century, oil fields were exploited in the area around modern Baku, Azerbaijan, to produce naphtha for the petroleum industry. These fields were described by Marco Polo in the 13th century, who described the output of those oil wells as hundreds of shiploads. When Marco Polo in 1264 visited the Azerbaijani city of Baku, on the shores of the Caspian Sea, he saw oil being collected from seeps. He wrote that "on the confines toward Geirgine there is a fountain from which oil springs in great abundance, in as much as a hundred shiploads might be taken from it at one time."

A FOUNTAIN AT BIBI-EIBAT IN FLAMES, BAKU

1904 oil well fire at Bibi-Eibat

In North America, the first commercial oil well entered operation in Oil Springs, Ontario in 1858, while the first offshore oil well was drilled in 1896 at the Summerland Oil Field on the California Coast.

The earliest oil wells in modern times were drilled percussively, by repeatedly raising and dropping a cable tool into the earth. In the 20th century, cable tools were largely replaced with rotary drilling, which could drill boreholes to much greater depths and in less time. The record-depth Kola Borehole used non-rotary mud motor drilling to achieve a depth of over 12,000 metres (39,000 ft).

Until the 1970s, most oil wells were vertical, although lithological and mechanical imperfections cause most wells to deviate at least slightly from true vertical. However, modern directional drilling technologies allow for strongly deviated wells which can, given sufficient depth and with the proper tools, actually become horizontal. This is of great value as the reservoir rocks which contain hydrocarbons are usually horizontal or nearly horizontal; a horizontal wellbore placed in a production zone has more surface area in the production zone than a vertical well, resulting in a higher production rate. The use of deviated and horizontal drilling has also made it possible to reach reservoirs several kilometers or miles away from the drilling location (extended reach drilling), allowing for the production of hydrocarbons located below locations that are either difficult to place a drilling rig on, environmentally sensitive, or populated.

Life of a Well

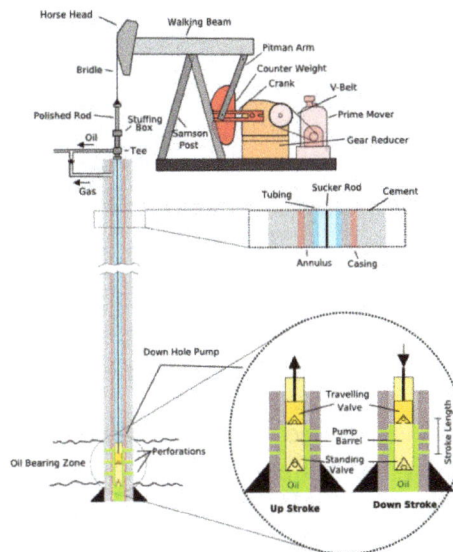

A schematic of a typical oil well being produced by a pumpjack, which is used to produce the remaining recoverable oil after natural pressure is no longer sufficient to raise oil to the surface

The creation and life of a well can be divided up into five segments:

- Planning
- Drilling
- Completion
- Production
- Abandonment

Drilling

The well is created by drilling a hole 12 cm to 1 meter (5 in to 40 in) in diameter into the earth with a drilling rig that rotates a drill string with a bit attached. After the hole is drilled, sections of steel pipe (casing), slightly smaller in diameter than the borehole, are placed in the hole. Cement may be placed between the outside of the casing and the borehole known as the annulus. The casing provides structural integrity to the newly drilled wellbore, in addition to isolating potentially dangerous high pressure zones from each other and from the surface.

With these zones safely isolated and the formation protected by the casing, the well can be drilled deeper (into potentially more-unstable and violent formations) with a smaller bit, and also cased with a smaller size casing. Modern wells often have two to five sets of subsequently smaller hole sizes drilled inside one another, each cemented with casing.

To drill the well

Well Casing

- The drill bit, aided by the weight of thick walled pipes called "drill collars" above it, cuts into the rock. There are different types of drill bit; some cause the rock to disintegrate by compressive failure, while others shear slices off the rock as the bit turns.

- Drilling fluid, a.k.a. "mud", is pumped down the inside of the drill pipe and exits at the drill bit. The principal components of drilling fluid are usually water and clay, but it also typically contains a complex mixture of fluids, solids and chemicals that must be carefully tailored to provide the correct physical and chemical characteristics required to safely drill the well. Particular functions of the drilling mud include cooling the bit, lifting rock cuttings to the surface, preventing destabilisation of the rock in the wellbore walls and overcoming the pressure of fluids inside the rock so that these fluids do not enter the wellbore. Some oil wells are drilled with air or foam as the drilling fluid.

Mud log in process, a common way to study the lithology when drilling oil wells

- The generated rock "cuttings" are swept up by the drilling fluid as it circulates back to surface outside the drill pipe. The fluid then goes through "shakers" which strain the cuttings from the good fluid which is returned to the pit. Watching for abnormalities in the returning cuttings and monitoring pit volume or rate of returning fluid are imperative to catch "kicks" early. A "kick" is when the formation pressure at the depth of the bit is more than the hydrostatic head of the mud above, which if not controlled temporarily by closing the blowout preventers and ultimately by increasing the density of the drilling fluid would allow formation fluids and mud to come up through the annulus uncontrollably.

- The pipe or drill string to which the bit is attached is gradually lengthened as the well gets deeper by screwing in additional 9 m (30 ft) sections or "joints" of pipe under the kelly or topdrive at the surface. This process is called making a connection, or "tripping". Joints can be combined for more efficient tripping when pulling out of the hole by creating stands of multiple joints. A conventional triple, for example, would pull pipe out of the hole three joints at a time and stack them in the derrick. Many modern rigs, called "super singles", trip pipe one at a time, laying it out on racks as they go.

This process is all facilitated by a drilling rig which contains all necessary equipment to circulate the drilling fluid, hoist and turn the pipe, control downhole, remove cuttings from the drilling fluid, and generate on-site power for these operations.

Completion

Modern drilling rig in Argentina

After drilling and casing the well, it must be 'completed'. Completion is the process in which the well is enabled to produce oil or gas.

In a cased-hole completion, small holes called perforations are made in the portion of the casing which passed through the production zone, to provide a path for the oil to flow from the surrounding rock into the production tubing. In open hole completion, often 'sand screens' or a 'gravel pack' is installed in the last drilled, uncased reservoir section. These maintain structural integrity of the wellbore in the absence of casing, while still allowing flow from the reservoir into the wellbore. Screens also control the migration of formation sands into production tubulars and surface equipment, which can cause washouts and other problems, particularly from unconsolidated sand formations of offshore fields.

After a flow path is made, acids and fracturing fluids may be pumped into the well to fracture, clean, or otherwise prepare and stimulate the reservoir rock to optimally produce hydrocarbons into the wellbore. Finally, the area above the reservoir section of the well is packed off inside the casing, and connected to the surface via a smaller diameter pipe called tubing. This arrangement provides a redundant barrier to leaks of hydrocarbons as well as allowing damaged sections to be replaced. Also, the smaller cross-sectional area of the tubing produces reservoir fluids at an increased velocity in order to minimize liquid fallback that would create additional back pressure, and shields the casing from corrosive well fluids.

In many wells, the natural pressure of the subsurface reservoir is high enough for the oil or gas to flow to the surface. However, this is not always the case, especially in depleted fields where the pressures have been lowered by other producing wells, or in low permeability oil reservoirs. Installing a smaller diameter tubing may be enough to help the production, but artificial lift methods may also be needed. Common solutions include downhole pumps, gas lift, or surface pump jacks. Many new systems in the last ten years have been introduced for well completion. Multiple packer systems with frac ports or port collars in an all in one system have cut completion costs and improved production, especially in the case of horizontal wells. These new systems allow casings to run into the lateral zone with proper packer/frac port placement for optimal hydrocarbon recovery.

Production

The production stage is the most important stage of a well's life; when the oil and gas are produced. By this time, the oil rigs and workover rigs used to drill and complete the well have moved off the wellbore, and the top is usually outfitted with a collection of valves called a Christmas tree or production tree. These valves regulate pressures, control flows, and allow access to the wellbore in case further completion work is needed. From the outlet valve of the production tree, the flow can be connected to a distribution network of pipelines and tanks to supply the product to refineries, natural gas compressor stations, or oil export terminals.

As long as the pressure in the reservoir remains high enough, the production tree is all that is required to produce the well. If the pressure depletes and it is considered economically viable, an artificial lift method mentioned in the completions section can be employed.

Workovers are often necessary in older wells, which may need smaller diameter tubing, scale or paraffin removal, acid matrix jobs, or completing new zones of interest in a shallower reservoir. Such remedial work can be performed using workover rigs – also known as *pulling units*, *completion rigs* or "service rigs" – to pull and replace tubing, or by the use of well intervention techniques utilizing coiled tubing. Depending on the type of lift system and wellhead a rod rig or flushby can be used to change a pump without pulling the tubing.

Enhanced recovery methods such as water flooding, steam flooding, or CO_2 flooding may be used to increase reservoir pressure and provide a "sweep" effect to push hydrocarbons out of the reservoir. Such methods require the use of injection wells (often chosen from old production wells in a carefully determined pattern), and are used when facing problems with reservoir pressure depletion, high oil viscosity, or can even be employed early in a field's life. In certain cases – depending on the reservoir's geomechanics – reservoir engineers may determine that ultimate recoverable oil may be increased by applying a waterflooding strategy early in the field's development rather than later. Such enhanced recovery techniques are often called "tertiary recovery".

Abandonment

A well is said to reach an "economic limit" when its most efficient production rate does not cover the operating expenses, including taxes.

The economic limit for oil and gas wells can be expressed using these formulae:

$$\text{Oil fields: } EL_{oil} = \frac{WI \times LOE}{NRI[P_o + (P_g \times GOR)/1,000] \times (1-T)}$$

$$\text{Gas fields: } EL_{gas} = \frac{WI \times LOE}{NRI[(P_o \times Y) + P_g] \times (1-T)}$$

Where: EL_{oil} is an oil well's economic limit in oil barrels per month (bbls/month). EL_{gas} is a gas

well's economic limit in thousand standard cubic feet per

month (MSCF/month). P_o, P_g are the current prices of oil and gas in dollars per barrels and dollars per MSCF respectively. LOE is the lease operating expenses in dollars per well per month. WI working interest, as a fraction. NRI net revenue interest, as a fraction. GOR gas/oil ratio as SCF/bbl. Y condensate yield as barrel/million standard cubic feet. T production and severance taxes, as a fraction.

When the economic limit is raised, the life of the well is shortened and proven oil reserves are lost. Conversely, when the economic limit is lowered, the life of the well is lengthened.

When the economic limit is reached, the well becomes a liability and is abandoned. In this process, tubing is removed from the well and sections of well bore are filled with concrete to isolate the flow path between gas and water zones from each other, as well as the surface. Completely filling the well bore with concrete is costly and unnecessary. The surface around the wellhead is then excavated, and the wellhead and casing are cut off, a cap is welded in place and then buried.

At the economic limit there often is still a significant amount of unrecoverable oil left in the reservoir. It might be tempting to defer physical abandonment for an extended period of time, hoping that the oil price will go up or that new supplemental recovery techniques will be perfected. In these cases, temporary plugs will be placed downhole and locks attached to the wellhead to prevent tampering. There are thousands of "abandoned" wells throughout North America, waiting to see what the market will do before permanent abandonment. Often, lease provisions and governmental regulations usually require quick abandonment; liability and tax concerns also may favor abandonment.

In theory an abandoned well can be reentered and restored to production (or converted to injection service for supplemental recovery or for downhole hydrocarbons storage), but reentry often proves to be difficult mechanically and not cost effective.

Types of Well

Burning of natural gases at an oil drilling site, presumably at Pangkalan Brandan, East Coast of Sumatra - circa 1905

Fossil-fuel wells come in many varieties. By produced fluid, there can be wells that produce oil, wells that produce oil*and*natural gas, or wells that *only* produce natural gas. Natural gas is almost always a byproduct of producing oil, since the small, light gas carbon chains come out of solution as they undergo pressure reduction from the reservoir to the surface, similar to uncapping a bottle of soda pop where the carbon dioxide effervesces. Unwanted natural gas can be a disposal problem at the well site. If there is not a market for natural gas near the wellhead it is virtually valueless since it must be piped to the end user. Until recently, such unwanted gas was burned off at the wellsite, but due to environmental concerns this practice is becoming less common. Often, unwanted (or 'stranded' gas without a market) gas is pumped back into the reservoir with an 'injection' well for disposal or repressurizing the producing formation. Another solution is to export the natural gas as a liquid. Gas to liquid, (GTL) is a developing technology that converts stranded natural gas into synthetic gasoline, diesel or jet fuel through the Fischer-Tropsch process developed in World War II Germany. Such fuels can be transported through conventional pipelines and tankers to users. Proponents claim GTL fuels burn cleaner than comparable petroleum fuels. Most major international oil companies are in advanced development stages of GTL production, e.g. the 140,000 bbl/d (22,000 m³/d) Pearl GTL plant in Qatar, scheduled to come online in 2011. In locations such as the United States with a high natural gas demand, pipelines are constructed to take the gas from the wellsite to the end consumer.

A natural gas well in the southeast Lost Hills Field, California, US.

Another obvious way to classify oil wells is by land or offshore wells. There is very little difference in the well itself. An offshore well targets a reservoir that happens to be underneath an ocean. Due to logistics, drilling an offshore well is far more costly than an onshore well. By far the most common type is the onshore well. These wells dot the Southern and Central Great Plains, Southwestern United States, and are the most common wells in the Middle East.

Raising the derrick

Oil extraction in Boryslav in 1909

Another way to classify oil wells is by their purpose in contributing to the development of a resource. They can be characterized as:

- *wildcat wells* are drilled where little or no known geological information is available. The site may have been selected because of wells drilled some distance from the proposed location but on a terrain that appeared similar to the proposed site.

- *exploration wells* are drilled purely for exploratory (information gathering) purposes in a new area, the site selection is usually based on seismic data, satellite surveys etc. Details gathered in this well includes the presence of Hydrocarbon in the drilled location, the amount of fluid present and the depth at which oil or/and gas occurs.

- *appraisal wells* are used to assess characteristics (such as flow rate, reserve quantity) of a proven hydrocarbon accumulation. The purpose of this well is to reduce uncertainty about the characteristics and properties of the hydrocarbon present in the field.

- *production wells* are drilled primarily for producing oil or gas, once the producing structure and characteristics are determined.

- *development wells* are wells drilled for the production of oil or gas already proven by appraisal drilling to be suitable for exploitation.

- *Abandoned well* are wells permanently plugged in the drilling phase for technical reasons.

At a producing well site, active wells may be further categorised as:

- *oil producers* producing predominantly liquid hydrocarbons, but mostly with some associated gas.

- *gas producers* producing almost entirely gaseous hydrocarbons.

- *water injectors* injecting water into the formation to maintain reservoir pressure, or simply to dispose of water produced with the hydrocarbons because even after treatment, it would be too oily and too saline to be considered clean for dumping overboard offshore, let alone into a fresh water resource in the case of onshore wells. Water injection into the producing zone frequently has an element of reservoir management; however, often produced water disposal is into shallower zones safely beneath any fresh water zones.

- *aquifer producers* intentionally producing water for re-injection to manage pressure. If possible this water will come from the reservoir itself. Using aquifer produced water rather than water from other sources is to preclude chemical incompatibility that might lead to reservoir-plugging precipitates. These wells will generally be needed only if produced water from the oil or gas producers is insufficient for reservoir management purposes.

- *gas injectors* injecting gas into the reservoir often as a means of disposal or sequestering for later production, but also to maintain reservoir pressure.

Lahee Classification

- *New Field Wildcat* (NFW) – far from other producing fields and on a structure that has not previously produced.

- *New Pool Wildcat* (NPW) – new pools on already producing structure.

- *Deeper Pool Test* (DPT) – on already producing structure and pool, but on a deeper pay zone.

- *Shallower Pool Test* (SPT) – on already producing structure and pool, but on a shallower pay zone.

- *Outpost* (OUT) – usually two or more locations from nearest productive area.

- *Development Well* (DEV) – can be on the extension of a pay zone, or between existing wells (*Infill*).

Cost

The cost of a well depends mainly on the daily rate of the drilling rig, the extra services required to drill the well, the duration of the well program (including downtime and weather time), and the remoteness of the location (logistic supply costs).

The daily rates of offshore drilling rigs vary by their capability, and the market availability. Rig rates reported by industry web service show that the deepwater water floating drilling rigs are over twice that of the shallow water fleet, and rates for jackup fleet can vary by factor of 3 depending upon capability.

With deepwater drilling rig rates in 2015 of around \$520,000/day, and similar additional spread costs, a deep water well of duration of 100 days can cost around US\$100 million.

With high performance jackup rig rates in 2015 of around \$177,000, and similar service costs, a high pressure, high temperature well of duration 100 days can cost about US\$30 million.

Onshore wells can be considerably cheaper, particularly if the field is at a shallow depth, where costs range from less than \$1 million to \$15 million for deep and difficult wells.

The total cost of an oil well mentioned does not include the costs associated with the risk of explosion and leakage of oil. Those costs include the cost of protecting against such disasters, the cost of the cleanup effort, and the hard-to-calculate cost of damage to the company's image.

Reefs

Offshore platforms (the structure supporting the wells) often provide habitat for marine life. After the wells have been abandoned, sometimes the platforms can be toppled in place or moved elsewhere to be dropped to the ocean floor to produce artificial reefs.

Completion (Oil and Gas Wells)

Completion, in petroleum production, is the process of making a well ready for production (or injection). This principally involves preparing the bottom of the hole to the required specifications, running in the production tubing and its associated down hole tools as well as perforating and stimulating as required. Sometimes, the process of running in and cementing the casing is also included.

Perforated Shoe

Lower Completion

This refers to the portion of the well across the production or injection zone. The well designer has many tools and options available to design the lower completion according to the conditions of the reservoir. Typically, the lower completion is set across the productive zone using a liner hanger system, which anchors the lower completion to the production casing string. The broad categories of lower completion are listed below.

Barefoot Completion

This type is the most basic, but can be a good choice for hard rock, multi-laterals and underbalance drilling. It involves leaving the productive reservoir section without any tubulars. This effectively removes control of flow of fluids from the formation; it is not suitable for weaker formations which might require sand control, nor for formations requiring selective isolation of oil, gas and water intervals. However, advances in interventions such as coiled tubing and tractors means that barefoot wells can be successfully produced.

Open Hole

The production casing is set above the zone of interest before drilling the zone. The zone is open to the well bore. In this case little expense is generated with perforations log interpretation is not critical. The well can be deepened easily and it is easily converted to screen and liner. However, excessive gas and water production is difficult to control, and may require frequent clean outs. Also the interval can not be selectively stimulated.

Open Hole Completion

This designation refers to a range of completions where no casing or liner is cemented in place across the production zone. In competent formations, the zone might be left entirely bare, but some sort of sand-control and/or flow-control means are usually incorporated.

Openhole completions have seen significant uptake in recent years, and there are many configurations, often developed to address specific reservoir challenges. There have been many recent developments that have boosted the success of openhole completions, and they also tend to be popular in horizontal wells, where cemented installations are more expensive and technically more difficult. The common options for openhole completions are:

Pre-holed Liner

Also often called pre-drilled liner. The liner is prepared with multiple small drilled holes, then set across the production zone to provide wellbore stability and an intervention conduit. Pre-holed liner is often combined with openhole packers, such as swelling elastomers, mechanical packers or external casing packers, to provide zonal segregation and isolation. It is now quite common to see a combination of pre-holed liner, solid liner and swelling elastomer packers to provide an initial isolation of unwanted water or gas zones. Multiple sliding sleeves can also be used in conjunction with openhole packers to provide considerable flexibility in zonal flow control for the life of the wellbore.

This type of completion is also being adopted in some water injection wells, although these require a much greater performance envelope for openhole packers, due to the considerable pressure and temperature changes that occur in water injectors.

Openhole completions (in comparison with cemented pipe) require better understanding of formation damage, wellbore clean-up and fluid loss control. A key difference is that perforating penetrates through the first 6-18 inches (15–45 cm) of formation around the wellbore, whilst openhole completions require the reservoir fluids to flow through all of the filtrate-invaded zone around the wellbore and lift-off of the mud filter cake.

Many openhole completions will incorporate fluid loss valves at the top of the liner to provide well control whilst the upper completion is run.

There are an increasing number of ideas coming into the market place to extend the options for openhole completions; for example, electronics can be used to actuate a self-opening or self-closing liner valve. This might be used in an openhole completion to improve clean-up, by bringing the well onto production from the toe-end for 100 days, then self-opening the heel-end. Inflow control devices and intelligent completions are also installed as openhole completions.

Pre-holed liner may provide some basic control of solids production, where the wellbore is thought to fail in aggregated chunks of rubble, but it is not typically regarded as a sand control completion.

Slotted Liner

Slotted liners can be selected as an alternative to pre-holed liner, sometimes as a personal preference or from established practice on a field. It can also be selected to provide a low cost control of sand/solids production. The slotted liner is machined with multiple longitudinal slots, for example 2 mm x 50mm, spread across the length and circumference of each joint. Recent advances in laser cutting means that slotting can now be done much cheaper to much smaller slot widths and in some situation slotted liner is now used for the same functionality as sand control screens.

Openhole Sand Control

This is selected where the liner is required to mechanically hold back the movement of formation sand. There are many variants of openhole sand control, the three popular choices being stand-alone screens, openhole gravel packs (also known as external gravel packs, where a sized sand 'gravel' is placed as an annulus around the sand control screen) and expandable screens. Screen designs are mainly wire-wrap or premium; wire-wrap screens use spiral-welded corrosion-resistant wire wrapped around a drilled basepipe to provide a consistent small helical gap (such as 0.012-inch (0.30 mm), termed 12 gauge). Premium screens use a woven metal cloth wrapped around a basepipe. Expandable screens are run to depth before being mechanically swaged to a larger diameter. Ideally, expandable screens will be swaged until they contact the wellbore wall.

Horizontal Open Hole Completions

This is the most common open hole completion used today. It is basically the same described on the vertical open hole completion but on a horizontal well it enlarges significantly the contact with the reservoir, increasing the production or injection rates of your well. Sand control on a horizontal well is completely different from a vertical well. We can no longer rely on the gravity for the gravel placement. Most service companies uses an alpha and beta wave design to cover the total length of the horizontal well with gravel. It's known that very long wells (around 6000 ft) were successfully gravel packed in many occasions, including deepwater reservoirs in Brazil.

Liner Completions

In this case the casing is set above the primary zone. An un-cemented screen and liner assembly is installed across the pay section. This technique minimizes formation damage and gives the ability to control sand. It also makes cleanout easy. Perforating expense is also low to non-existent. However gas and water build up is difficult to control and selective stimulation not possible the well can't be easily deepened and additional rig time may be needed.

Perforated Liner

Casing is set above the producing zone, the zone is drilled and the liner casing is cemented in place. The liner is then perforated for production. This time additional expense in perforating the casing is incurred, also log interpretation is critical and it may be difficult to obtain good quality cement jobs.

Perforated Casing

Production casing is cemented through the zone and the pay section is selectively perforated. Gas and water are easily controlled as is sand. The formation can be selectively stimulated and the well can be deepened. This selection is adaptable to other completion configurations and logs are available to assist casing decisions. Much better primary casing. It can however cause damage to zones and needs good log interpretation. The perforating cost can be very high.

Cased Hole Completion

This involves running casing or a liner down through the production zone, and cementing it in place. Connection between the well bore and the formation is made by perforating. Because perforation intervals can be precisely positioned, this type of completion affords good control of fluid flow, although it relies on the quality of the cement to prevent fluid flow behind the liner. As such it is the most common form of completion...

Conventional Completions

- *Casing flow*: means that the producing fluid flow has only one path to the surface through the casing.

- *Casing and tubing flow*: means that there is tubing within the casing that allows fluid to reach the surface. This tubing can be used as a kill string for chemical injection. The tubing may have a "no-go" nipple at the end as a means of pressure testing.

- *Pumping flow*: the tubing and pump are run to a depth beneath the working fluid. The pump and rod string are installed concentrically within the tubing. A tubing anchor prevents tubing movement while pumping.

- *Tubing flow*: a tubing string and a production packer are installed. The packer means that all the flow goes through the tubing. Within the tubing you can mount a combination of tools that will help to control fluid flow through the tubing.

- Gas lift *well*: gas is fed into valves installed in mandrels in the tubing strip. The hydrostatic head is lowered and the fluid is gas lifted to the surface.

- *Single-well alternate completions*: in this instance there is a well with two zones. In order to produce from both the zones are isolated with packers. Blast joints may be used on the tubing within the region of the perforations. These are thick walled subs that can withstand the fluid abrasion from the producing zone. This arrangement can also work if you have to produce from a higher zone given the depletion of a lower zone. The tubing may also have flow control mechanism.

- *Single-well concentric kill string*: within the well a small diameter concentric kill string is used to circulate kill fluids when needed.

- *Single-well 2-tubing completion*: in this instance 2 tubing strings are inserted down 1 well. They are connected at the lower end by a circulating head. Chemicals can be circulated down one tube and production can continue up the other.

Completion Components

The upper completion refers to all components from the bottom of the production tubing upwards. Proper design of this "completion string" is essential to ensure the well can flow properly given the reservoir conditions and to permit any operations as are deemed necessary for enhancing production and safety.

Wellhead

This is the pressure containing equipment at the surface of the well where casing strings are suspended and the blowout preventer or Christmas tree is connected.

Christmas Tree

This is the main assembly of valves that controls flow from the well to the process plant (or the other way round for injection wells) and allows access for chemical squeezes and well interventions.

Tubing Hanger

This is the component, which sits on top of the wellhead and serves as the main support for the production tubing.

Production Tubing

Production tubing is the main conduit for transporting hydrocarbons from the reservoir to surface (or injection material the other way). It runs from the tubing hanger at the top of the wellhead down to a point generally just above the top of the production zone.

Downhole Safety Valve (DHSV)

This component is intended as a last-resort method of protecting the surface from the uncontrolled release of hydrocarbons. It is a cylindrical valve with either a ball or flapper closing mechanism. It is installed in the production tubing and is held in the open position by a high-pressure hydraulic line from surface contained in a 6.35 mm (1/4") control line that is attached to the DHSV's hydraulic chamber and terminated at surface to an hydraulic actuator. The high pressure is needed to overcome the production pressure in the tubing upstream of the choke on the tree. The valve will operate if the umbilical HP line is cut or the wellhead/tree is destroyed.

This valve allows fluids to pass up or be pumped down the production tubing. When closed the DHSV forms a barrier in the direction of hydrocarbon flow, but fluids can still be pumped down for well kill operations. It is placed as far below the surface as is deemed safe from any possible surface disturbance including cratering caused by the wipeout of the platform. Where hydrates are likely to form (most production is at risk of this), the depth of the SCSSV (surface-controlled, sub-surface safety valve) below the seabed may be as much as 1 km: this will allow for the geothermal temperature to be high enough to prevent hydrates from blocking the valve.

Annular Safety Valve

On wells with gas lift capability, many operators consider it prudent to install a valve, which will isolate the *A* annulus for the same reasons a DHSV may be needed to isolate the production tubing in order to prevent the inventory of natural gas downhole from becoming a hazard as it became on Piper Alpha.

Side Pocket Mandrel

This is a welded/machined product which contains a "side pocket" alongside the main tubular conduit. The side pocket, typically 1" or 1½" diameter is designed to contain gas lift valve, which allows flow of High pressure gas into the tubing there by reducing the tubing pressure and allowing the hydrocarbons to move upwards.

Electrical Submersible Pump

This device is used for artificial lift to help provide energy to drive hydrocarbons to surface if reservoir pressure is insufficient.

Landing Nipple

A completion component fabricated as a short section of heavy wall tubular with a machined internal surface that provides a seal area and a locking profile. Landing nipples are included in most completions at predetermined intervals to enable the installation of flow-control devices, such as plugs and chokes. Three basic types of landing nipple are commonly used: no-go nipples, selective-landing nipples and ported or safety-valve nipples.

Sliding Sleeve

The sliding sleeve is hydraulically or mechanically actuated to allow communication between the tubing and the 'A' annulus. They are often used in multiple reservoir wells to regulate flow to and from the zones.

Production Packer

The packer isolates the annulus between the tubing and the inner casing and the foot of the well. This is to stop reservoir fluids from flowing up the full length of the casing and damaging it. It is generally placed close to the foot of the tubing, shortly above the production zone.

Downhole Gauges

This is an electronic or fiberoptic sensor to provide continuous monitoring of downhole pressure and temperature. Gauges either use a 1/4" control line clamped onto the outside of the tubing string to provide an electrical or fiberoptic communication to surface, or transmit measured data to surface by acoustic signal in the tubing wall. The information obtained from these monitoring devices can be used to model reservoirs or predict the life or problems in a specific wellbore.

Perforated Joint

This is a length of tubing with holes punched into it. If used, it will normally be positioned below the packer and will offer an alternative entry path for reservoir fluids into the tubing in case the shoe becomes blocked, for example, by a stuck perforation gun.

Formation Isolation Valve

This component, placed towards the foot of the completion string, is used to provide two way isolation from the formation for completion operations without the need for kill weight fluids. Their use is sporadic as they do not enjoy the best reputation for reliability when it comes to opening them at the end of the completion process.

Centralizer

In highly deviated wells, this component may be included towards the foot of the completion. It consists of a large collar, which keeps the completion string centralised within the hole.

Wireline Entry Guide

This component is often installed at the end of the tubing, or "the shoe". It is intended to make pulling out wireline tools easier by offering a guiding surface for the toolstring to re-enter the tubing without getting caught on the side of the shoe.

Perforating and Stimulating

In cased hole completions (the majority of wells), once the completion string is in place, the final stage is to make a connection between the wellbore and the formation. This is done by running perforation guns to blast holes in the casing or liner to make a connection. Modern perforations are made using shaped explosive charges, similar to the armor-penetrating charge used on antitank rockets (bazookas).

Sometimes once the well is fully completed, further stimulation is necessary to achieve the planned productivity. There are a number of stimulation techniques.

Acidizing

This involves the injection of chemicals to eat away at any skin damage, "cleaning up" the formation, thereby improving the flow of reservoir fluids. A strong acid (usually hydrochloric acid) is used to dissolve rock formations, but this acid does not react with the Hydrocarbons. As a result, the Hydrocarbons are more accessible. Acid can also be used to clean the wellbore of some scales that form from mineral laden produced water.

Fracturing

This means creating and extending fractures from the perforation tunnels deeper into the formation, increasing the surface area for formation fluids to flow into the well, as well as extending past any possible damage near the wellbore. This may be done by injecting fluids at high pressure (hydraulic fracturing), injecting fluids laced with round granular material (proppant fracturing), or using explosives to generate a high pressure and high speed gas flow (TNT or PETN up to 1,900,000 psi (13,000,000 kPa)) and (propellant stimulation up to 4,000 psi (28,000 kPa)).

Acidizing and Fracturing (Combined Method)

This involves use of explosives and injection of chemicals to increase acid-rock contact.

Nitrogen Circulation

Sometimes, productivity may be hampered due to the residue of completion fluids, heavy brines, in the wellbore. This is particularly a problem in gaswells. In these cases, coiled tubing may be used to pump nitrogen at high pressure into the bottom of the borehole to circulate out the brine.

Drilling Rig

Drilling the Bakken formation in the Williston Basin

A drilling rig is a machine that creates holes in the earth sub-surface. Drilling rigs can be massive structures housing equipment used to drill water wells, oil wells, or natural gas extraction wells, or they can be small enough to be moved manually by one person and are called augers. Drilling rigs can sample sub-surface mineral deposits, test rock, soil and groundwater physical properties, and also can be used to install sub-surface fabrications, such as underground utilities, instrumentation, tunnels or wells. Drilling rigs can be mobile equipment mounted on trucks, tracks or trailers, or more permanent land or marine-based structures (such as oil platforms, commonly called 'offshore oil rigs' even if they don't contain a drilling rig). The term "rig" therefore generally refers to the complex of equipment that is used to penetrate the surface of the Earth's crust.

Small to medium-sized drilling rigs are mobile, such as those used in mineral exploration drilling, blast-hole, water wells and environmental investigations. Larger rigs are capable of drilling through thousands of metres of the Earth's crust, using large "mud pumps" to circulate drilling mud (slurry) through the drill bit and up the casing annulus, for cooling and removing the "cuttings" while a well is drilled. Hoists in the rig can lift hundreds of tons of pipe. Other equipment can force acid or sand into reservoirs to facilitate extraction of the oil or natural gas; and in remote locations there can be permanent living accommodation and catering for crews (which may be more than a hundred). Marine rigs may operate thousands of miles distant from the supply base with infrequent crew rotation or cycle.

Petroleum Drilling Industry

Oil and natural gas drilling rigs are used not only to identify geologic reservoirs but also to create holes that allow the extraction of oil or natural gas from those reservoirs. Primarily in onshore oil and gas fields once a well has been drilled, the drilling rig will be moved off of the well and a service rig (a smaller rig) that is purpose-built for completions will be moved on to the well to get the well on line. This frees up the drilling rig to drill another hole and streamlines the operation as well as allowing for specialization of certain services, i.e., completions vs. drilling.

Water Well Drilling

New portable drillcat technology uses smaller portable trailer mounted rigs with shorter 3-metre (10 ft) drill pipe. The shorter drill pipe also allows a much smaller mast. Portable trailer mounted drilling rigs have drill ratings from 90 to 200 metres (300 to 800 ft) depending on mud pump flow and pressure ratings and drill pipe sizes.

Other, heavier, truck rigs are more complicated, thus requiring more skill to run. They're also more difficult to handle safely due to the longer 6-to-9-metre (20 to 30 ft) drill pipe. Large truck rigs also require a much higher overhead clearance to operate. Large truck drills can use over 570 litres (150 US gal) of fuel per day, while the smaller Deeprock Style portable drills use a mere 20 to 75 litres (5 to 20 US gal) of fuel per day. This makes smaller, more portable rigs preferable in remote or hard-to-reach places, and they are more cost effective when fuel prices are high.

Mining Drilling Industry

Mining drilling rigs are used for two main purposes, exploration drilling which aims to identify the location and quality of a mineral, and production drilling, used in the production-cycle for mining. Drilling rigs used for rock blasting for surface mines vary in size dependent on the size of the hole desired, and is typically classified into smaller pre-split and larger production holes. Underground mining (hard rock) uses a variety of drill rigs dependent on the desired purpose, such as production, bolting, cabling, and tunnelling.

History

Antique drilling rig now on display at Western History Museum in Lingle, Wyoming. It was used to drill many water wells in that area—many of those wells are still in use.

Antique drilling Rigs in Zigong, China

Until internal combustion engines were developed in the late 19th century, the main method for drilling rock was muscle power of man or animal. The drilling of wells for the manufacture of salt began by the Song Dynasty in China. The well had a particularly small mouth, "as small as a small bowl." Archaeological evidence of the drilling tools used in deep-well dwelling are kept and displayed in the Zigong Salt Industry Museum. According to *Salt: A World History*, a Qing Dynasty well, also located in Zigong, "continued down to 3,300 feet [1,000 m] making it at the time the deepest drilled well in the world." Mechanised versions of this system persisted until about 1970, using a cam to rapidly raise and drop what, by then, was a steel cable up to 3 mm

In the 1970s, outside of the oil and gas industry, roller bits using mud circulation were replaced by the first pneumatic reciprocating piston Reverse Circulation (RC) drills, and became essentially obsolete for most shallow drilling, and are now only used in certain situations where rocks preclude other methods. RC drilling proved much faster and more efficient, and continues to improve with better metallurgy, deriving harder, more durable bits, and compressors delivering higher air pressures at higher volumes, enabling deeper and faster penetration. Diamond drilling has remained essentially unchanged since its inception.

Mobile Drilling Rigs

Mobile drilling rig mounted on a truck

In early oil exploration, drilling rigs were semi-permanent in nature and the derricks were often built on site and left in place after the completion of the well. In more recent times drilling rigs are expensive custom-built machines that can be moved from well to well. Some light duty drilling rigs are like a mobile crane and are more usually used to drill water wells. Larger land rigs must be broken apart into sections and loads to move to a new place, a process which can often take weeks.

Small mobile drilling rigs are also used to drill or bore piles. Rigs can range from 100 toncontinuous flight auger (CFA) rigs to small air powered rigs used to drill holes in quarries, etc. These rigs use the same technology and equipment as the oil drilling rigs, just on a smaller scale.

The drilling mechanisms outlined below differ mechanically in terms of the machinery used, but also in terms of the method by which drill cuttings are removed from the cutting face of the drill and returned to surface.

Drilling Rig Classification

There are many types and designs of drilling rigs, with many drilling rigs capable of switching or combining different drilling technologies as needed. Drilling rigs can be described using any of the following attributes:

By Power Used

- Mechanical — the rig uses torque converters, clutches, and transmissions powered by its own engines, often diesel
- Electric — the major items of machinery are driven by electric motors, usually with power generated on-site using internal combustion engines
- Hydraulic — the rig primarily uses hydraulic power
- Pneumatic — the rig is primarily powered by pressurized air
- Steam — the rig uses steam-powered engines and pumps (obsolete after middle of 20th Century.)

By Pipe Used

- Cable — a cable is used to raise and drop the drill bit
- Conventional — uses metal or plastic drill pipe of varying types
- Coil tubing — uses a giant coil of tube and a downhole drilling motor

By Height

(Rigs are differentiated by height based on how many connected pipe they are able to "stand" in the derrick when needing to temporarily remove the drill pipe from the hole. Typically this is done when changing a drill bit or when "logging" the well.)

- Single — can pull only single drill pipes. The presence or absence of vertical pipe racking "fingers" varies from rig to rig.
- Double — can hold a stand of pipe in the derrick consisting of two connected drill pipes, called a "double stand".

- Triple — can hold a stand of pipe in the derrick consisting of three connected drill pipes, called a "triple stand".

- Quadri — can store stand of pipe in the derrick composed of four connected drill pipes, called a "quadri stand".

By Method of Rotation or Drilling Method

- No-rotation includes direct push rigs and most service rigs

- Rotary table — rotation is achieved by turning a square or hexagonal pipe (the «Kelly») at drill floor level.

- Top drive — rotation and circulation is done at the top of the drill string, on a motor that moves in a track along the derrick.

- Sonic — uses primarily vibratory energy to advance the drill string

- Hammer — uses rotation and percussive force

By Position of Derrick

- Conventional — derrick is vertical

- Slant — derrick is slanted at a 45 degree angle to facilitate horizontal drilling

Drill Types

There are a variety of drill mechanisms which can be used to sink a borehole into the ground. Each has its advantages and disadvantages, in terms of the depth to which it can drill, the type of sample returned, the costs involved and penetration rates achieved. There are two basic types of drills: drills which produce rock chips, and drills which produce core samples.

Auger Drilling

Auger drilling is done with a helical screw which is driven into the ground with rotation; the earth is lifted up the borehole by the blade of the screw. Hollow stem auger drilling is used for softer ground such as swamps where the hole will not stay open by itself for environmental drilling, geotechnical drilling, soilengineering and geochemistry reconnaissance work in exploration for mineral deposits. Solid flight augers/bucket augers are used in harder ground construction drilling. In some cases, mine shafts are dug with auger drills. Small augers can be mounted on the back of a utility truck, with large augers used for sinking piles for bridge foundations.

Auger drilling is restricted to generally soft unconsolidated material or weak weathered rock. It is cheap and fast.

Percussion Rotary Air Blast Drilling (RAB)

RAB drilling is used most frequently in the mineral exploration industry. (This tool is also known as a Down-the-hole drill.) The drill uses a pneumatic reciprocating piston-driven "hammer" to energetically drive a heavy drill bit into the rock. The drill bit is hollow, solid steel and has ~20 mm thick tungsten rods protruding from the steel matrix as "buttons". The tungsten buttons are the cutting face of the bit.

The cuttings are blown up the outside of the rods and collected at surface. Air or a combination of air and foam lift the cuttings.

RAB drilling is used primarily for mineralexploration, water bore drilling and blast-hole drilling in mines, as well as for other applications such as engineering, etc. RAB produces lower quality samples because the cuttings are blown up the outside of the rods and can be contaminated from contact with other rocks. RAB drilling at extreme depth, if it encounters water, may rapidly clog the outside of the hole with debris, precluding removal of drill cuttings from the hole. This can be counteracted, however, with the use of "stabilizers" also known as "reamers", which are large cylindrical pieces of steel attached to the drill string, and made to perfectly fit the size of the hole being drilled. These have sets of rollers on the side, usually with tungsten buttons, that constantly break down cuttings being pushed upwards.

The use of high-powered air compressors, which push 900-1150 cfm of air at 300-350 psi down the hole also ensures drilling of a deeper hole up to ~1250 m due to higher air pressure which pushes all rock cuttings and any water to the surface. This, of course, is all dependent on the density and weight of the rock being drilled, and on how worn the drill bit is.

Air Core Drilling

Air core drilling and related methods use hardened steel or tungsten blades to bore a hole into unconsolidated ground. The drill bit has three blades arranged around the bit head, which cut the unconsolidated ground. The rods are hollow and contain an inner tube which sits inside the hollow outer rod barrel. The drill cuttings are removed by injection of compressed air into the hole via the annular area between the innertube and the drill rod. The cuttings are then blown back to surface up the inner tube where they pass through the sample separating system and are collected if needed. Drilling continues with the addition of rods to the top of the drill string. Air core drilling can occasionally produce small chunks of cored rock.

This method of drilling is used to drill the weathered regolith, as the drill rig and steel or tungsten blades cannot penetrate fresh rock. Where possible, air core drilling is preferred over RAB drilling as it provides a more representative sample. Air core drilling can achieve depths approaching 300 metres in good conditions. As the cuttings are removed inside the rods and are less prone to contamination compared to conventional drilling where the cuttings pass to the surface via outside return between the outside of the drill rod and the walls of the hole. This method is more costly and slower than RAB.

Cable Tool Drilling

Cable tool rigs are a traditional way of drilling water wells. The majority of large diameter water supply wells, especially deep wells completed in bedrockaquifers, were completed using this drilling method. Although this drilling method has largely been supplanted in recent years by other, faster drilling techniques, it is still the most practicable drilling method for large diameter, deep bedrock wells, and in widespread use for small rural water supply wells. The impact of the drill bit fractures the rock and in many shale rock situations increases the water flow into a well over rotary.

Cable tool water well drilling rig in West Virginia. These slow rigs have mostly been replaced by rotary drilling rigs in the U.S.

Also known as ballistic well drilling and sometimes called "spudders", these rigs raise and drop a drill string with a heavy carbide tipped drilling bit that chisels through the rock by finely pulverizing the subsurface materials. The drill string is composed of the upper drill rods, a set of "jars" (inter-locking "sliders" that help transmit additional energy to the drill bit and assist in removing the bit if it is stuck) and the drill bit. During the drilling process, the drill string is periodically removed from the borehole and a bailer is lowered to collect the drill cuttings (rock fragments, soil, etc.). The bailer is a bucket-like tool with a trapdoor in the base. If the borehole is dry, water is added so that the drill cuttings will flow into the bailer. When lifted, the trapdoor closes and the cuttings are then raised and removed. Since the drill string must be raised and lowered to advance the boring, the casing (larger diameter outer piping) is typically used to hold back upper soil materials and stabilize the borehole.

Cable tool rigs are simpler and cheaper than similarly sized rotary rigs, although loud and very slow to operate. The world record cable tool well was drilled in New York to a depth of almost 12,000 feet (3,700 m). The common Bucyrus-Erie 22 can drill down to about 1,100 feet (340 m). Since cable tool drilling does not use air to eject the drilling chips like a rotary, instead using a cable strung bailer, technically there is no limitation on depth.

Cable tool rigs now are nearly obsolete in the United States. They are mostly used in Africa or Third-World countries. Being slow, cable tool rig drilling means increased wages for drillers. In the United States drilling wages would average around US$200 per day per man, while in Africa it is only US$6 per day per man, so a slow drilling machine can still be used in undeveloped countries with depressed wages. A cable tool rig can drill 25 feet (7.6 m) to 60 feet (18 m) of hard rock a day. A newer rotary drillcat top head rig equipped with down-the-hole (DTH) hammer can drill 500 feet (150 m) or more per day, depending on size and formation hardness.

Reverse Circulation (RC) Drilling

RC drilling is similar to air core drilling, in that the drill cuttings are returned to surface inside the rods. The drilling mechanism is a pneumatic reciprocating piston known as a "hammer" driving a tungsten-steel drill bit. RC drilling utilises much larger rigs and machinery and depths of up to 500 metres are routinely achieved. RC drilling ideally produces dry rock chips, as large air compressors dry the rock out ahead of the advancing drill bit. RC drilling is slower and costlier but achieves better penetration than RAB or air core drilling; it is cheaper than diamond coring and is thus preferred for most mineral exploration work.

Track mounted Reverse Circulation rig (side view).

Reverse circulation is achieved by blowing air down the rods, the differential pressure creating air lift of the water and cuttings up the "inner tube", which is inside each rod. It reaches the "divertor" at the top of the hole, then moves through a sample hose which is attached to the top of the "cyclone". The drill cuttings travel around the inside of the cyclone until they fall through an opening at the bottom and are collected in a sample bag.

Reverse Circulation Drilling set-up on Vertical Travel Leads at the Port of La Rochelle, France

The most commonly used RC drill bits are 5-8 inches (13–20 cm) in diameter and have round tungsten 'buttons' that protrude from the bit, which are required to drill through shale and abrasive rock. As the buttons wear down, drilling becomes slower and the rod string can potentially become bogged in the hole. This is a problem as trying to recover the rods may take hours and in some cases weeks. The rods and drill bits themselves are very expensive, often resulting in great cost to drilling companies when equipment is lost down the bore hole. Most companies will regularly re-grind the buttons on their drill bits in order to prevent this, and to speed up progress. Usually, when something is lost (breaks off) in the hole, it is not the drill string, but rather from the bit, hammer, or stabilizer to the bottom of the drill string (bit). This is usually caused by operator error, over-stressed metal, or adverse drilling conditions causing downhole equipment to get stuck in a part of the hole.

Although RC drilling is air-powered, water is also used to reduce dust, keep the drill bit cool, and assist in pushing cutting back upwards, but also when "collaring" a new hole. A mud called "Liqui-Pol" is mixed with water and pumped into the rod string, down the hole. This helps to bring up the sample to the surface by making the sand stick together. Occasionally, "Super-Foam" (a.k.a. "Quik-Foam") is also used, to bring all the very fine cuttings to the surface, and to clean the hole. When the drill reaches hard rock, a "collar" is put down the hole around the rods, which is normally PVC piping. Occasionally the collar may be made from metal casing. Collaring a hole is needed to stop the walls from caving in and bogging the rod string at the top of the hole. Collars may be up to 60 metres deep, depending on the ground, although if drilling through hard rock a collar may not be necessary.

Reverse circulation rig setups usually consist of a support vehicle, an auxiliary vehicle, as well as the rig itself. The support vehicle, normally a truck, holds diesel and water tanks for resupplying the rig. It also holds other supplies needed for maintenance on the rig. The auxiliary is a vehicle, carrying an auxiliary engine and a booster engine. These engines are connected to the rig by high pressure air hoses. Although RC rigs have their own booster and compressor to generate air pressure, extra power is needed which usually isn't supplied by the rig due to lack of space for these large engines. Instead, the engines are mounted on the auxiliary vehicle. Compressors on an RC rig have an output of around 1000 cfm at 500 psi (500 L·s⁻¹ at 3.4 MPa). Alternatively, stand-alone air compressors which have an output of 900-1150cfm at 300-350 psi each are used in sets of 2, 3, or 4, which are all routed to the rig through a multi-valve manifold.

Diamond Core Drilling

Multi-combination drilling rig (capable of both diamond and reverse circulation drilling). Rig is currently set up for diamond drilling.

Diamond core drilling (exploration diamond drilling) utilizes an annular diamond-impregnated drill bit attached to the end of hollow drill rods to cut a cylindrical core of solid rock. The diamonds used to make diamond core bits are a variety of sizes, fine to microfine industrial grade diamonds, and the ratio of diamonds to metal used in the matrix affects the performance of the bits cutting ability in different types of rock formations . The diamonds are set within a matrix of varying hardness, from brass to high-grade steel. Matrix hardness, diamond size and dosing can be varied according to the rock which must be cut. The bits made with hard steel with a low diamond count are ideal for softer highly fractured rock while others made of softer steels and high diamond ratio are good for coring in hard solid rock. Holes within the bit allow water to be delivered to the cutting face. This provides three essential functions — lubrication, cooling, and removal of drill cuttings from the hole.

Diamond drilling is much slower than reverse circulation (RC) drilling due to the hardness of the ground being drilled. Drilling of 1200 to 1800 metres is common and at these depths, ground is mainly hard rock. Techniques vary among drill operators and what the rig they are using is capable of, some diamond rigs need to drill slowly to lengthen the life of drill bits and rods, which are very expensive and time consuming to replace at extremely deep depths. As a diamond drill rig cores deeper and deeper the time consuming part of the process is not cutting 5 to 10 more feet of rock core but the retrieval of the core with the wire line & overshot tool. Core samples are retrieved via the use of a core tube, a hollow tube placed inside the rod string and pumped with water until it locks into the core barrel. As the core is drilled, the core barrel slides over the core as it is cut. An "overshot" attached to the end of the winch cable is lowered inside the rod string and locks on to the backend (aka head assembly), located on the top end of the core barrel. The winch is retracted, pulling the core tube to the surface. The core does not drop out of the inside of the core tube when lifted because either a split ring core lifter or basket retainer allow the core to move into, but not back out of the tube.

Diamond core drill bits

Once the core tube is removed from the hole, the core sample is then removed from the core tube and catalogued. The Driller's assistant unscrews the backend off the core tube using tube wrenches, then each part of the tube is taken and the core is shaken out into core trays. The core is washed, measured and broken into smaller pieces using a hammer or sawn through to make it fit into the sample trays. Once catalogued, the core trays are retrieved by geologists who then analyse the core and determine if the drill site is a good location to expand future mining operations.

Diamond rigs can also be part of a multi-combination rig. Multi-combination rigs are a dual setup rig capable of operating in either a reverse circulation (RC) and diamond drilling role (though not at the same time). This is a common scenario where exploration drilling is being performed in a very isolated location. The rig is first set up to drill as an RC rig and once the desired metres are drilled, the rig is set up for diamond drilling. This way the deeper metres of the hole can be drilled without moving the rig and waiting for a diamond rig to set up on the pad.

Direct Push Rigs

Direct push technology includes several types of drilling rigs and drilling equipment which advances a drill string by pushing or hammering without rotating the drill string. While this does not meet the proper definition of drilling, it does achieve the same result — a borehole. Direct push rigs include both cone penetration testing (CPT) rigs and direct push sampling rigs such as a PowerProbe or Geoprobe. Direct push rigs typically are limited to drilling in unconsolidated soil materials and very soft rock.

CPT rigs advance specialized testing equipment (such as electronic cones), and soil samplers using large hydraulic rams. Most CPT rigs are heavily ballasted (20 metric tons is typical) as a counter force against the pushing force of the hydraulic rams which are often rated up to 20 kN. Alternatively, small, light CPT rigs and offshore CPT rigs will use anchors such as screwed-in ground anchors to create the reactive force. In ideal conditions, CPT rigs can achieve production rates of up to 250–300 meters per day.

Direct push drilling rigs use hydraulic cylinders and a hydraulic hammer in advancing a hollow core sampler to gather soil and groundwater samples. The speed and depth of penetration is largely dependent on the soil type, the size of the sampler, and the weight and power of the rig. Direct push techniques are generally limited to shallow soil sample recovery in unconsolidated soil materials. The advantage of direct push technology is that in the right soil type it can produce a large number of high quality samples quickly and cheaply, generally from 50 to 75 meters per day. Rather than hammering, direct push can also be combined with sonic (vibratory) methods to increase drill efficiency.

Hydraulic Rotary Drilling

Oil well drilling utilises tri-cone roller, carbide embedded, fixed-cutter diamond, or diamond-impregnated drill bits to wear away at the cutting face. This is preferred because there is no need to return intact samples to surface for assay as the objective is to reach a formation containing oil or natural gas. Sizable machinery is used, enabling depths of several kilometres to be penetrated. Rotating hollow drill pipes carry down bentonite and barite infused drilling muds to lubricate, cool, and clean the drilling bit, control downhole pressures, stabilize the wall of the borehole and remove drill cuttings. The mud travels back to the surface around the outside of the drill pipe, called the annulus. Examining rock chips extracted from the mud is known as mud logging. Another form of well logging is electronic and is frequently employed to evaluate the existence of possible oil and gas deposits in the borehole. This can take place while the well is being drilled, using Measurement While Drilling tools, or after drilling, by lowering measurement tools into the newly drilled hole.

The rotary system of drilling was in general use in Texas in the early 1900s. It is a modification of one invented by Fauvelle in 1845, and used in the early years of the oil industry in some of the oil-producing countries in Europe. Originally pressurized water was used instead of mud, and was almost useless in hard rock before the diamond cutting bit. The main breakthrough for rotary drilling came in 1901, when Anthony Francis Lucas combined the use of a steam-driven rig and of mud instead of water in the Spindletop discovery well.

The drilling and production of oil and gas can pose a safety risk and a hazard to the environment from the ignition of the entrained gas causing dangerous fires and also from the risk of oil leakage polluting water, land and groundwater. For these reasons, redundant safety systems and highly trained personnel are required by law in all countries with significant production.

Sonic (Vibratory) Drilling

A sonic drill head works by sending high frequency resonant vibrations down the drill string to the drill bit, while the operator controls these frequencies to suit the specific conditions of the soil/ rock geology. Vibrations may also be generated within the drill head. The frequency is generally between 50 and 150 hertz (cycles per second) and can be varied by the operator.

Resonance magnifies the amplitude of the drill bit, which fluidizes the soil particles at the bit face, allowing for fast and easy penetration through most geological formations. An internal spring system isolates these vibrational forces from the rest of the drill rig.

Automated Drill Rig

Automated Drill Rig (ADR) is a state-of-the-art automated full-sized walking land-based drill rig that drills long lateral sections in horizontal wells for the oil and gas industry. ADRs are agile rigs that can move from pad to pad to new well sites faster than other full-sized drilling rigs. Each rig costs about $25 million. ADR is used extensively in the Athabasca oil sands. According to the "Oil Patch Daily News", "Each rig will generate 50,000 man-hours of work during the construction phase and upon completion, each operating rig will directly and indirectly employ more than 100 workers." Compared to conventional drilling rigs", Ensign, an international oilfield services contractor based in Calgary, Alberta, that makes ADRs claims that they are "safer to operate, have "enhanced controls intelligence," "reduced environmental footprint, quick mobility and advanced communications between field and office." In June 2005 the first specifically designed slant automated drilling rig (ADR), Ensign Rig No. 118, for steam assisted gravity drainage (SAGD) applications was mobilized by Deer Creek Energy Limited, a Calgary-based oilsands company.

Limits of the Technology

Drill technology has advanced steadily since the 19th century. However, there are several basic limiting factors which will determine the depth to which a bore hole can be sunk.

All holes must maintain outer diameter; the diameter of the hole must remain wider than the diameter of the rods or the rods cannot turn in the hole and progress cannot continue. Friction caused by the drilling operation will tend to reduce the outside diameter of the drill bit. This applies to all drilling methods, except that in diamond core drilling the use of thinner rods and casing may permit the hole to continue. Casing is simply a hollow sheath which protects the hole against collapse during drilling, and is made of metal or PVC. Often diamond holes will start off at a large diameter and when outside diameter is lost, thinner rods put down inside casing to continue, until finally the hole becomes too narrow. Alternatively, the hole can be reamed; this is the usual practice in oil well drilling where the hole size is maintained down to the next casing point.

For percussion techniques, the main limitation is air pressure. Air must be delivered to the piston at sufficient pressure to activate the reciprocating action, and in turn drive the head into the rock with sufficient strength to fracture and pulverise it. With depth, volume is added to the in-rod string, requiring larger compressors to achieve operational pressures. Secondly, groundwater is ubiquitous, and increases in pressure with depth in the ground. The air inside the rod string must be pressurised enough to overcome this water pressure at the bit face. Then, the air must be able to carry the rock fragments to surface. This is why depths in excess of 500 m for reverse circulation drilling are rarely achieved, because the cost is prohibitive and approaches the threshold at which diamond core drilling is more economic.

Diamond drilling can routinely achieve depths in excess of 1200 m. In cases where money is no issue, extreme depths have been achieved, because there is no requirement to overcome water pressure. However, water circulation must be maintained to return the drill cuttings to surface, and more importantly to maintain cooling and lubrication of the cutting surface of the bit; while at the same time reduce friction on the steel walls of the rods turning against the rock walls of the hole. When water return is lost the rods will vibrate, this is called "rod chatter", and that will damage the drill rods, and crack the joints.

Without sufficient lubrication and cooling, the matrix of the drill bit will soften. While diamond is the hardest substance known, at 10 on the Mohs hardness scale, it must remain firmly in the matrix to achieve cutting. Weight on bit, the force exerted on the cutting face of the bit by the drill rods in the hole above the bit, must also be monitored.

A unique drilling operation in deep ocean water was named Project Mohole.

New Oilfield Technologies

Research includes technologies based on the utilization of water jet, chemical plasma, hydrothermal spallation or laser.

Causes of Deviation

Most drill holes deviate slightly from their planned trajectory. This is because of the torque of the turning bit working against the cutting face, because of the flexibility of the steel rods and especially the screw joints, because of reaction to foliation and structure within the rock, and because of refraction as the bit moves into different rock layers of varying resistance. Additionally, inclined holes will tend to deviate upwards because the drill rods will lie against the bottom of the bore, causing the drill bit to be slightly inclined from true. It is because of deviation that drill holes must be surveyed if deviation will impact the usefulness of the information returned. Sometimes the surface location can be offset laterally to take advantage of the expected deviation tendency, so the bottom of the hole will end up near the desired location. Oil well drilling commonly uses a process of controlled deviation called directional drilling (e.g., when several wells are drilled from one surface location).

Rig Equipment

Drilling rigs typically include at least some of the following items: See Drilling rig (petroleum) for a more detailed description.

- Blowout preventers: (BOPs)

The equipment associated with a rig is to some extent dependent on the type of rig but (#23 & #24) are devices installed at the wellhead to prevent fluids and gases from unintentionally escaping from the borehole. #23 is the annular (often referred to as the "Hydril", which is one manufacturer) and #24 is the pipe rams and blind rams. In the place of #24 Variable bore rams or VBRs can be used. These offer the same pressure and sealing capacity found in standard pipe rams, while offering the versatility of sealing on various sizes of drill pipe, production tubing and casing without changing standard pipe rams. Normally VBRs are used when utilizing a tapered drill string (when different size drill pipe is used in the complete drill string).

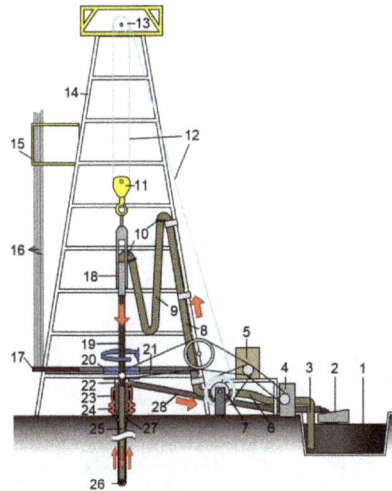

Simple diagram of a drilling rig and its basic operation

- Centrifuge: an industrial version of the device that separates fine silt and sand from the drilling fluid.

- Solids control: solids control equipment is for preparing drilling mud for the drilling rig.

- Chain tongs: wrench with a section of chain, that wraps around whatever is being tightened or loosened. Similar to a pipe wrench.

- Degasser: a device that separates air and/or gas from the drilling fluid.

- Desander / desilter: contains a set of hydrocyclones that separate sand and silt from the drilling fluid.

- Drawworks: (#7) is the mechanical section that contains the spool, whose main function is to reel in/out the drill line to raise/lower the traveling block (#11).

- Drill bit: (#26) is a device attached to the end of the drill string that breaks apart the rock being drilled. It contains jets through which the drilling fluid exits.

- Drill pipe: (#16) joints of hollow tubing used to connect the surface equipment to the bottom hole assembly (BHA) and acts as a conduit for the drilling fluid. In the diagram, these are "stands" of drill pipe which are 2 or 3 joints of drill pipe connected together and "stood" in the derrick vertically, usually to save time while tripping pipe.

- Elevators: a gripping device that is used to latch to the drill pipe or casing to facilitate the lowering or lifting (of pipe or casing) into or out of the borehole.

- Mud motor: a hydraulically powered device positioned just above the drill bit used to spin the bit independently from the rest of the drill string.

- Mud pump: (#4) reciprocal type of pump used to circulate drilling fluid through the system.

- Mud tanks: (#1) often called mud pits, provides a reserve store of drilling fluid until it is required down the wellbore.

- Rotary table: (#20) rotates the drill string along with the attached tools and bit.

- Shale shaker: (#2) separates drill cuttings from the drilling fluid before it is pumped back down the borehole.

Occupational Safety

Rig move safety for roughnecks

Drilling rigs create some safety challenges for those who work on them. One safety concern is the use of seatbelts for workers driving between two locations. Motor vehicle fatalities on the job for these workers is 8.5 times the rate of the rest of the US working population, which can be attributed to the low rate of seatbelt use.

Rig move safety for truckers

Peak Oil

Peak oil, an event based on M. King Hubbert's theory, is the point in time when the maximum rate of extraction of petroleum is reached, after which it is expected to enter terminal decline. Peak oil theory is based on the observed rise, peak, fall, and depletion of aggregate production rate in oil fields over time. It is often confused with oil depletion; however, peak oil is the point of maximum production, while depletion refers to a period of falling reserves and supply.

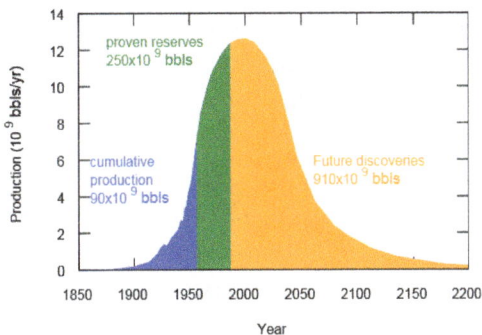

A 1956 world oil production distribution, showing historical data and future production, proposed by M. King Hubbert – it has a peak of 12.5 billion barrels per year in about the year 2000

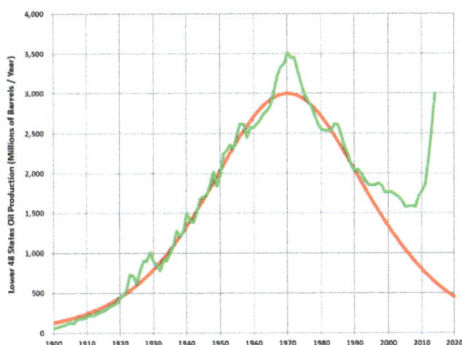

Hubbert's upper-bound prediction for US crude oil production (1956), and actual lower-48 states production through 2014

Some observers, such as petroleum industry experts Kenneth S. Deffeyes and Matthew Simmons, predict negative global economy implications following a post-peak production decline and subsequent oil price increase because of the high dependence of most modern industrial transport, agricultural, and industrial systems on the low cost and high availability of oil. Predictions vary greatly as to what exactly these negative effects would be.

Oil production forecasts on which predictions of peak oil are based are often made within a range which includes optimistic (higher production) and pessimistic (lower production) scenarios. Optimistic estimations of peak production forecast the global decline will begin after 2020, and assume major investments in alternatives will occur before a crisis, without requiring major changes in the lifestyle of heavily oil-consuming nations. Pessimistic predictions of future oil production made after 2007 stated either that the peak had already occurred, that oil production was on the cusp of the peak, or that it would occur shortly.

Hubbert's original prediction that US peak oil would be in about 1970 seemed accurate for a time, as US average annual production peaked in 1970 at 9.6 million barrels per day. However, the successful application of massive hydraulic fracturing to additional tight reservoirs caused US production to rebound, challenging the inevitability of post-peak decline for the US oil production. In addition, Hubbert's original predictions for world peak oil production proved premature.

Modeling Global Oil Production

The idea that the rate of oil production would peak and irreversibly decline is an old one. In 1919, David White, chief geologist of the United States Geological Survey, wrote of US petroleum: "… the peak of production will soon be passed, possibly within 3 years." In 1953, Eugene Ayers, a researcher for Gulf Oil, projected that if US ultimate recoverable oil reserves were 100 billion barrels, then production in the US would peak no later than 1960. If ultimate recoverable were to be as high as 200 billion barrels, which he warned was wishful thinking, US peak production would come no later than 1970. Likewise for the world, he projected a peak somewhere between 1985 (one trillion barrels ultimate recoverable) and 2000 (two trillion barrels recoverable). Ayers made his projections without a mathematical model. He wrote: "But if the curve is made to look reasonable, it is quite possible to adapt mathematical expressions to it and to determine, in this way, the peak dates corresponding to various ultimate recoverable reserve numbers"

By observing past discoveries and production levels, and predicting future discovery trends, the geoscientist M. King Hubbert used statistical modelling in 1956 to accurately predict that United States oil production would peak between 1965 and 1971. Hubbert used a semi-logistical curved model (sometimes incorrectly compared to a normal distribution). He assumed the production rate of a limited resource would follow a roughly symmetrical distribution. Depending on the limits of exploitability and market pressures, the rise or decline of resource production over time might be sharper or more stable, appear more linear or curved. That model and its variants are now called Hubbert peak theory; they have been used to describe and predict the peak and decline of production from regions, countries, and multinational areas. The same theory has also been applied to other limited-resource production.

In a 2006 analysis of Hubbert theory, it was noted that uncertainty in real world oil production amounts and confusion in definitions increases the uncertainty in general of production predictions. By comparing the fit of various other models, it was found that Hubbert's methods yielded the closest fit over all, but that none of the models were very accurate. In 1956 Hubbert himself recommended using "a family of possible production curves" when predicting a production peak and decline curve.

More recently, the term "peak oil" was popularized by Colin Campbell and Kjell Aleklett in 2002 when they helped form the Association for the Study of Peak Oil and Gas (ASPO). In his publications, Hubbert used the term "peak production rate" and "peak in the rate of discoveries".

Demand

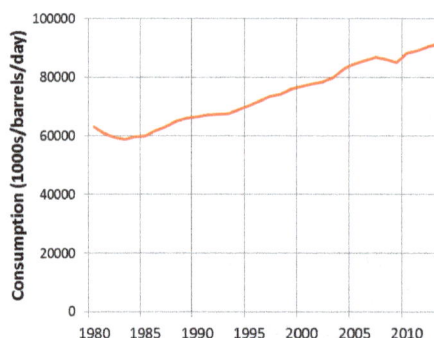

Global consumption of oil 1980–2013 (Energy Information Administration)

The demand side of peak oil over time is concerned with the total quantity of oil that the global market would choose to consume at various possible market prices and how this entire listing of quantities at various prices would evolve over time. Global demand for crude oil grew an average of 1.76% per year from 1994 to 2006, with a high growth of 3.4% in 2003–2004. After reaching a high of 85.6 million barrels (13,610,000 m³) per day in 2007, world consumption decreased in both 2008 and 2009 by a total of 1.8%, despite fuel costs plummeting in 2008. Despite this lull, world quantity-demanded for oil is projected to increase 21% over 2007 levels by 2030 (104 million barrels per day (16.5×10^6 m³/d) from 86 million barrels (13.7×10^6 m³)), or about 0.8% average annual growth, due in large part to increases in demand from the transportation sector. According to projections by the International Energy Agency (IEA) in 2013, growth in global oil demand will be significantly outpaced by growth in production capacity over the next 5 years. Developments in late 2014–2015 have seen an oversupply of global markets leading to a significant drop in the price of oil.

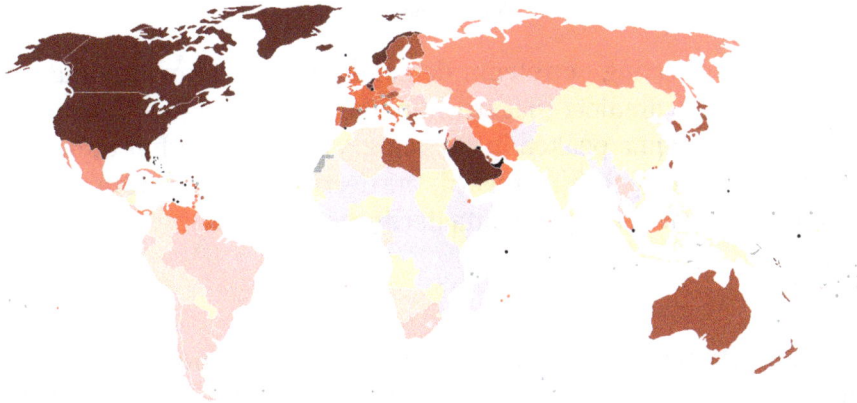

Oil consumption per capita (darker colors represent more consumption, gray represents no data)

Energy demand is distributed amongst four broad sectors: transportation, residential, commercial, and industrial. In terms of oil use, transportation is the largest sector and the one that has seen the largest growth in demand in recent decades. This growth has largely come from new demand for personal-use vehicles powered by internal combustion engines. This sector also has the highest consumption rates, accounting for approximately 71% of the oil used in the United States in 2013. and 55% of oil use worldwide as documented in the Hirsch report. Transportation is therefore of particular interest to those seeking to mitigate the effects of peak oil.

Although demand growth is highest in the developing world, the United States is the world's largest consumer of petroleum. Between 1995 and 2005, US consumption grew from 17,700,000 barrels per day (2,810,000 m³/d) to 20,700,000 barrels per day (3,290,000 m³/d), a 3,000,000 barrels per day (480,000 m³/d) increase. China, by comparison, increased consumption from 3,400,000 barrels per day (540,000 m³/d) to 7,000,000 barrels per day (1,100,000 m³/d), an increase of 3,600,000 barrels per day (570,000 m³/d), in the same time frame. The Energy Information Administration (EIA) stated that gasoline usage in the United States may have peaked in 2007, in part because of increasing interest in and mandates for use of biofuels and energy efficiency.

As countries develop, industry and higher living standards drive up energy use, oil usage being a major component. Thriving economies, such as China and India, are quickly becoming large oil consumers. For example, China surpassed the United States as the world's largest crude oil importer in 2015. Oil consumption growth is expected to continue; however, not at previous rates, as China's economic growth is predicted to decrease from the high rates of the early part of the 21st century. India's oil imports are expected to more than triple from 2005 levels by 2020, rising to 5 million barrels per day (790×103 m³/d).

Population

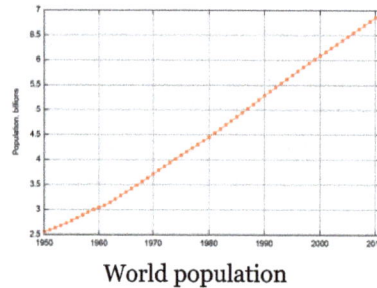

World population

Another significant factor affecting petroleum demand has been human population growth. The United States Census Bureau predicts that world population in 2030 will be almost double that of 1980. Oil production per capita peaked in 1979 at 5.5 barrels/year but then declined to fluctuate around 4.5 barrels/year since. In this regard, the decreasing population growth rate since the 1970s has somewhat ameliorated the per capita decline.

Economic Growth

Some analysts argue that the cost of oil has a profound effect on economic growth due to its pivotal role in the extraction of resources and the processing, manufacturing, and transportation of goods. As the industrial effort to extract new unconventional oil sources increases, this has a compounding negative effect on all sectors of the economy, leading to economic stagnation or even eventual contraction. Such a scenario would result in an inability for national economies to pay high oil prices, leading to declining demand and a price collapse.

Supply

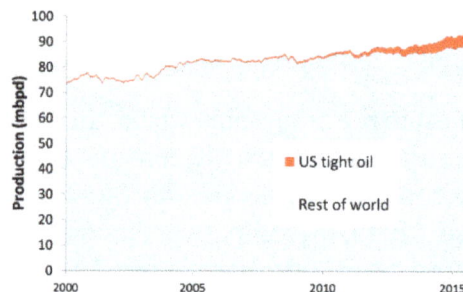

Global liquids production 2000-2015, indicating the component of US tight oil (Energy Information Administration)

Our analysis suggests there are ample physical oil and liquid fuel resources for the foreseeable future. However, the rate at which new supplies can be developed and the break-even prices for those new supplies are changing.

—International Energy Agency

Defining Sources of Oil

Oil may come from conventional or unconventional sources. The terms are not strictly defined, and vary within the literature as definitions based on new technologies tend to change over time. As a result, different oil forecasting studies have included different classes of liquid fuels. Some use the terms "conventional" oil for what is included in the model, and "unconventional" oil for classes excluded.

In 1956, Hubbert confined his peak oil prediction to that crude oil "producible by methods now in use." By 1962, however, his analyses included future improvements in exploration and production. All of Hubbert's analyses of peak oil specifically excluded oil manufactured from oil shale or mined from oil sands. A 2013 study predicting an early peak excluded deepwater oil, tight oil, oil with API gravity less than 17.5, and oil close to the poles, such as that on the North Slope of Alaska, all of which it defined as non-conventional. Some commonly used definitions for conventional and unconventional oil are detailed below.

Conventional Sources

Conventional oil is extracted on land and offshore using standard techniques, and can be categorized as light, medium, heavy, or extra heavy in grade. The exact definitions of these grades vary depending on the region from which the oil came. Light oil flows naturally to the surface or can be extracted by simply pumping it out of the ground. Heavy refers to oil that has higher density and therefore lower API gravity. It does not flow easily, and its consistency is similar to that of molasses. While some of it can be produced using conventional techniques, recovery rates are better using unconventional methods.

Unconventional Sources

Oil currently considered unconventional is derived from multiple sources.

- Tight oil is extracted from deposits of low-permeability rock, sometimes shale deposits but often other rock types, using hydraulic fracturing, or "fracking." It is often confused with shale oil, which is oil manufactured from the kerogen contained in an oil shale, Production of tight oil has led to a resurgence of US production in recent years. However, tight oil production peaked in 2015 and is not expected to increase again until there is a significant oil price recovery.

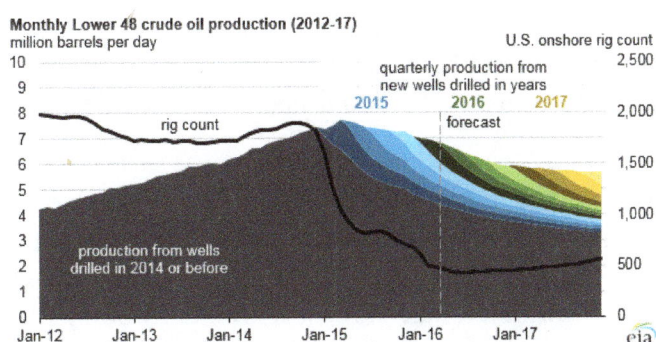

US Lower 48 oil production from 2012 and anticipated decline in production to the end of 2017, with rig count (Energy Information Administration)

- Oil shale is a common term for sedimentary rock such as shale or marl, containing kerogen, a waxy oil precursor that has not yet been transformed into crude oil by the high pressures and temperatures caused by deep burial. The term "oil shale" is somewhat confusing, because what is referred to in the U.S. as "oil shale" is not really oil and the rock it is found in is generally not shale. Since it is close to the surface rather than buried deep in the earth, the shale or marl is typically mined, crushed, and retorted, producing synthetic oil from the kerogen. Its net energy yield is much lower than conventional oil, so much so that estimates of the net energy yield of shale discoveries are considered extremely unreliable.

- Oil sands are unconsolidated sandstone deposits containing large amounts of very viscous crude bitumen or extra-heavy crude oil that can be recovered by surface mining or by in-situ oil wells using steam injection or other techniques. It can be liquefied by upgrading, blending with diluent, or by heating; and then processed by a conventional oil refinery. The recovery process requires advanced technology but is more efficient than that of oil shale. The reason is that, unlike U.S. "oil shale", Canadian oil sands actually contain oil, and the sandstones they are found in are much easier to produce oil from than shale or marl. In the U.S. dialect of English, these formations are often called "tar sands", but the material found in them is not tar but an extra-heavy and viscous form of oil technically known as bitumen.

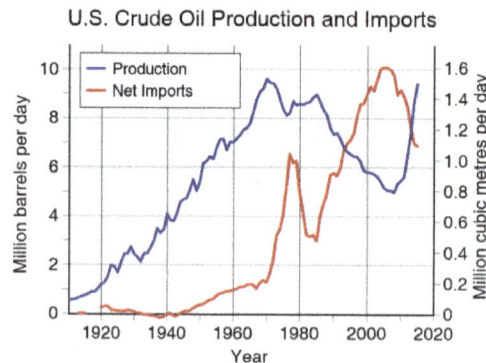

U.S. Crude Oil Production and Imports

United States crude oil production exceeds imports for the first time since the early 1990s

- Coal liquefaction or gas to liquids product are liquid hydrocarbons that are synthesised from the conversion of coal or natural gas by the Fischer-Tropsch process, Bergius process, or Karrick process. Currently, two companies SASOL and Shell, have synthetic oil technology proven to work on a commercial scale. Sasol's primary business is based on CTL (coal-to-liquid) and GTL (natural gas-to-liquid) technology, producing US$4.40 billion in revenues (FY2009). Shell has used these processes to recycle waste flare gas (usually burnt off at oil wells and refineries) into usable synthetic oil. However, for CTL there may be insufficient coal reserves to supply global needs for both liquid fuels and electric power generation.

- Minor sources include thermal depolymerization, as discussed in a 2003 article in Discover magazine, that could be used to manufacture oil indefinitely, out of garbage, sewage, and agricultural waste. The article claimed that the cost of the process was $15 per barrel. A follow-up article in 2006 stated that the cost was actually $80 per barrel, because the feedstock that had previously been considered as hazardous waste now had market value. A 2008 news bulletin published by Los Alamos Laboratory proposed that hydrogen (possibly produced using hot fluid from nuclear reactors to split water into hydrogen and oxygen) in combination with sequestered CO_2 could be used to produce methanol (CH_3OH), which could then be converted into gasoline.

Discoveries

All the easy oil and gas in the world has pretty much been found. Now comes the harder work in finding and producing oil from more challenging environments and work areas.

— William J. Cummings, Exxon-Mobil company spokesman, December 2005

It is pretty clear that there is not much chance of finding any significant quantity of new cheap oil. Any new or unconventional oil is going to be expensive.

— Lord Ron Oxburgh, a former chairman of Shell, October 2008

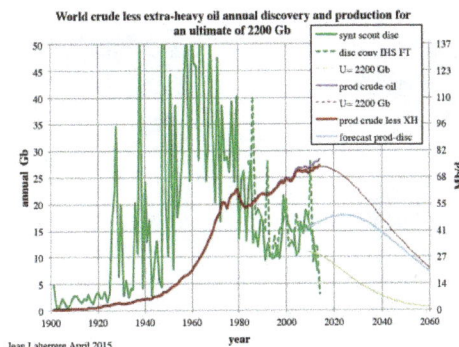

World oil discoveries peaked in the 1960s

The peak of world oilfield discoveries occurred in the 1960s at around 55 billion barrels (8.7×10^9 m³) (Gb)/year. According to the Association for the Study of Peak Oil and Gas (ASPO), the rate of discovery has been falling steadily since. Less than 10 Gb/yr of oil were discovered each year between 2002 and 2007. According to a 2010 Reuters article, the annual rate of discovery of new fields has remained remarkably constant at 15–20 Gb/yr.

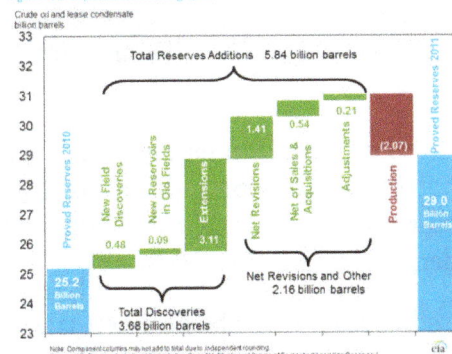

Although US proved oil reserves grew by 3.8 billion barrels in 2011, even after deducting 2.07 billion barrels of production, only 8 percent of the 5.84 billion barrels of the newly booked oil was because of new field discoveries (U.S. EIA)

But despite the fall-off in new field discoveries, and record-high production rates, the reported proved reserves of crude oil remaining in the ground in 2014, which totaled 1,490 billion barrels, not counting Canadian heavy oil sands, were more than quadruple the 1965 proved reserves of 354 billion barrels. A researcher for the U.S. Energy Information Administration has pointed out that after the first wave of discoveries in an area, most oil and natural gas reserve growth comes not from discoveries of new fields, but from extensions and additional gas found within existing fields.

A report by the UK Energy Research Centre noted that "discovery" is often used ambiguously, and explained the seeming contradiction between falling discovery rates since the 1960s and increasing reserves by the phenomenon of reserve growth. The report noted that increased reserves within a field may be discovered or developed by new technology years or decades after the original discovery. But because of the practice of "backdating," any new reserves within a field, even those to be discovered decades after the field discovery, are attributed to the year of initial field discovery, creating an illusion that discovery is not keeping pace with production.

Reserves

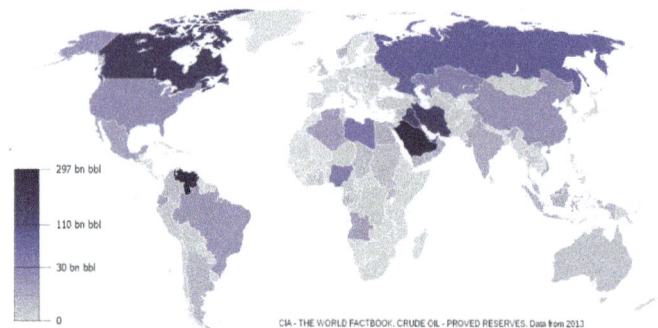

Proven oil reserves, 2013

Total possible conventional crude oil reserves include crude oil with 90% certainty of being technically able to be produced from reservoirs (through a wellbore using primary, secondary, improved, enhanced, or tertiary methods); all crude with a 50% probability of being produced in the future (probable); and discovered reserves that have a 10% possibility of being produced in the future (possible). Reserve estimates based on these are referred to as 1P, proven (at least 90% probability); 2P, proven and probable (at least 50% probability); and 3P, proven, probable and possible (at least 10% probability), respectively. This does not include liquids extracted from mined solids or gasses (oil sands, oil shale, gas-to-liquid processes, or coal-to-liquid processes).

Hubbert's 1956 peak projection for the United States depended on geological estimates of ultimate recoverable oil resources, but starting in his 1962 publication, he concluded that ultimate oil recovery was an output of his mathematical analysis, rather than an assumption. He regarded his peak oil calculation as independent of reserve estimates.

Many current 2P calculations predict reserves to be between 1150 and 1350 Gb, but some authors have written that because of misinformation, withheld information, and misleading reserve calculations, 2P reserves are likely nearer to 850–900 Gb. The Energy Watch Group wrote that actual reserves peaked in 1980, when production first surpassed new discoveries, that apparent increases in reserves since then are illusory, and concluded (in 2007): "Probably the world oil production has peaked already, but we cannot be sure yet."

Concerns Over Stated Reserves

[World] reserves are confused and in fact inflated. Many of the so-called reserves are in fact resources. They're not delineated, they're not accessible, they're not available for production.

— *Sadad I. Al-Husseini, former VP of Aramco, presentation to the Oil and Money conference,*
October 2007.

Al-Husseini estimated that 300 billion barrels (48×10^9 m³) of the world's 1,200 billion barrels (190×10^9 m³) of proven reserves should be recategorized as speculative resources.

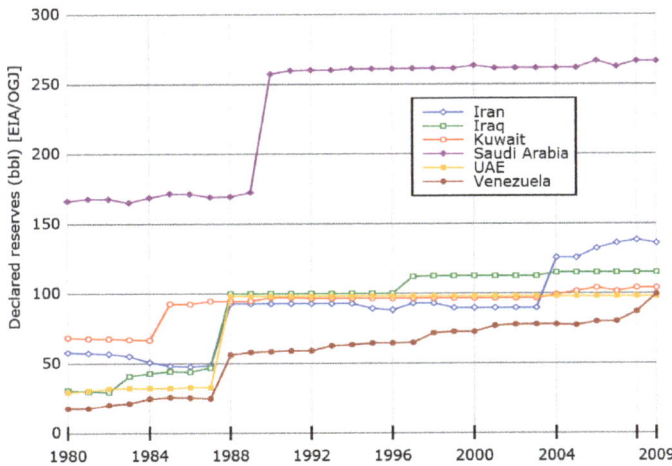

Graph of OPEC reported reserves showing jumps in stated reserves without associated discoveries, as well as the lack of depletion despite yearly production

One difficulty in forecasting the date of peak oil is the opacity surrounding the oil reserves classified as "proven". In many major producing countries, the majority of reserves claims have not been subject to outside audit or examination. Many worrying signs concerning the depletion of proven reserves have emerged in recent years. This was best exemplified by the 2004 scandal surrounding the "evaporation" of 20% of Shell's reserves.

For the most part, proven reserves are stated by the oil companies, the producer states and the consumer states. All three have reasons to overstate their proven reserves: oil companies may look to increase their potential worth; producer countries gain a stronger international stature; and governments of consumer countries may seek a means to foster sentiments of security and stability within their economies and among consumers.

Major discrepancies arise from accuracy issues with the self-reported numbers from the Organization of the Petroleum Exporting Countries (OPEC). Besides the possibility that these nations have overstated their reserves for political reasons (during periods of no substantial discoveries), over 70 nations also follow a practice of not reducing their reserves to account for yearly production. Analysts have suggested that OPEC member nations have economic incentives to exaggerate their reserves, as the OPEC quota system allows greater output for countries with greater reserves.

Kuwait, for example, was reported in the January 2006 issue of *Petroleum Intelligence Weekly* to have only 48 billion barrels (7.6×10^9 m³) in reserve, of which only 24 were fully proven. This report was based on the leak of a confidential document from Kuwait and has not been formally denied by the Kuwaiti authorities. This leaked document is from 2001, but excludes revisions or discoveries made since then. Additionally, the reported 1.5 billion barrels (240×10⁶ m³) of oil burned off by Iraqi soldiers in the First Persian Gulf War are conspicuously missing from Kuwait's figures.

On the other hand, investigative journalist Greg Palast argues that oil companies have an interest in making oil look more rare than it is, to justify higher prices. This view is contested by ecological journalist Richard Heinberg. Other analysts argue that oil producing countries understate the extent of their reserves to drive up the price.

The EUR reported by the 2000 USGS survey of 2,300 billion barrels (370×10⁹ m³) has been criticized for assuming a discovery trend over the next twenty years that would reverse the observed trend of the past 40 years. Their 95% confidence EUR of 2,300 billion barrels (370×10⁹ m³) assumed that discovery levels would stay steady, despite the fact that new-field discovery rates have declined since the 1960s. That trend of falling discoveries has continued in the ten years since the USGS made their assumption. The 2000 USGS is also criticized for other assumptions, as well as assuming 2030 production rates inconsistent with projected reserves.

Reserves of Unconventional Oil

Syncrude's Mildred Lake mine site and plant near Fort McMurray, Alberta

As conventional oil becomes less available, it can be replaced with production of liquids from unconventional sources such as tight oil, oil sands, ultra-heavy oils, gas-to-liquid technologies, coal-to-liquid technologies, biofuel technologies, and shale oil. In the 2007 and subsequent International Energy Outlook editions, the word "Oil" was replaced with "Liquids" in the chart of world energy consumption. In 2009 biofuels was included in "Liquids" instead of in "Renewables". The inclusion of natural gas liquids, a bi-product of natural gas extraction, in "Liquids" has been criticized as it is mostly a chemical feedstock which is generally not used as transport fuel.

Texas oil production declined since peaking in 1972 but has recently had a resurgence due to tight oil production

Reserve estimates are based on the oil price. Hence, unconventional sources such as heavy crude oil, oil sands, and oil shale may be included as new techniques reduce the cost of extraction. With rule changes by the SEC, oil companies can now book them as proven reserves after opening a strip mine or thermal facility for extraction. These unconventional sources are more labor and resource intensive to produce, however, requiring extra energy to refine, resulting in higher production costs and up to three times more greenhouse gas emissions per barrel (or barrel equivalent) on a "well to tank" basis or 10 to 45% more on a "well to wheels" basis, which includes the carbon emitted from combustion of the final product.

While the energy used, resources needed, and environmental effects of extracting unconventional sources have traditionally been prohibitively high, major unconventional oil sources being considered for large-scale production are the extra heavy oil in the Orinoco Belt of Venezuela, the Athabasca Oil Sands in the Western Canadian Sedimentary Basin, and the oil shale of the Green River Formation in Colorado, Utah, and Wyoming in the United States. Energy companies such as Syncrude and Suncor have been extracting bitumen for decades but production has increased greatly in recent years with the development of Steam Assisted Gravity Drainage and other extraction technologies.

Chuck Masters of the USGS estimates that, "Taken together, these resource occurrences, in the Western Hemisphere, are approximately equal to the Identified Reserves of conventional crude oil accredited to the Middle East." Authorities familiar with the resources believe that the world's ultimate reserves of unconventional oil are several times as large as those of conventional oil and will be highly profitable for companies as a result of higher prices in the 21st century. In October 2009, the USGS updated the Orinoco tar sands (Venezuela) recoverable "mean value" to 513 billion barrels (8.16×10^{10} m³), with a 90% chance of being within the range of 380-652 billion barrels (103.7×10^{9} m³), making this area "one of the world's largest recoverable oil accumulations".

Total World Oil Reserves

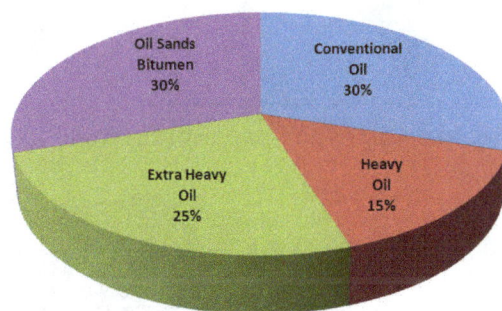

Unconventional resources are much larger than conventional ones

Despite the large quantities of oil available in non-conventional sources, Matthew Simmons argued in 2005 that limitations on production prevent them from becoming an effective substitute for conventional crude oil. Simmons stated "these are high energy intensity projects that can never reach high volumes" to offset significant losses from other sources. Another study claims that even under highly optimistic assumptions, "Canada's oil sands will not prevent peak oil," although production could reach 5,000,000 bbl/d (790,000 m³/d) by 2030 in a "crash program" development effort.

Moreover, oil extracted from these sources typically contains contaminants such as sulfur and heavy metals that are energy-intensive to extract and can leave tailings, ponds containing hydrocarbon sludge, in some cases. The same applies to much of the Middle East's undeveloped conventional oil reserves, much of which is heavy, viscous, and contaminated with sulfur and metals to the point of being unusable. However, high oil prices make these sources more financially appealing. A study by Wood Mackenzie suggests that by the early 2020s all the world's extra oil supply is likely to come from unconventional sources.

Production

The point in time when peak global oil production occurs defines peak oil. Some adherents of 'peak oil' believe that production capacity will remain the main limitation of supply, and that when production decreases, it will be the main bottleneck to the petroleum supply/demand equation. Others believe that the increasing industrial effort to extract oil will have a negative effect on global economic growth, leading to demand contraction and a price collapse, thereby causing production decline as some unconventional sources become uneconomical. Yet others believe that the peak may be to some extent led by declining demand as new technologies and improving efficiency shift energy usage away from oil.

Worldwide oil discoveries have been less than annual production since 1980. World population has grown faster than oil production. Because of this, oil production per capita peaked in 1979 (preceded by a plateau during the period of 1973–1979).

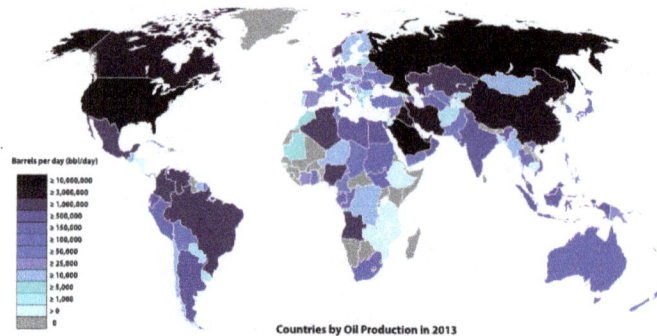

Countries producing oil 2013, bbl/day (CIA World Factbook)

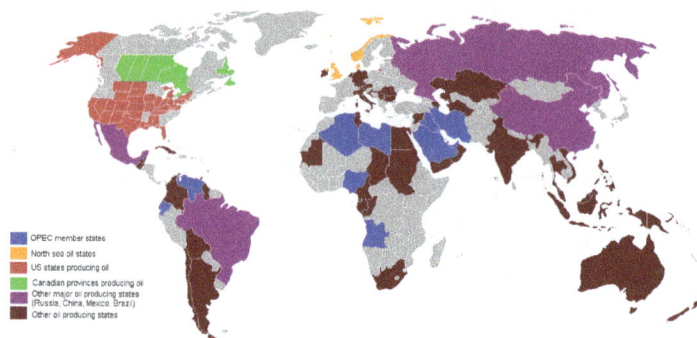

Oil producing countries

The increasing investment in harder-to-reach oil is a sign of oil companies' belief in the end of easy oil. Also, while it is widely believed that increased oil prices spur an increase in production, an increasing number of oil industry insiders were reportedly coming to believe that even with higher prices, oil production was unlikely to increase significantly. Among the reasons cited were both geological factors as well as "above ground" factors that are likely to see oil production plateau.

An important concept with regard to declining "easy oil" is energy returned on energy invested, also referred to as EROEI. A 2008 *Journal of Energy Security* analysis of the energy return on drilling effort in the United States concluded that there was extremely limited potential to increase production of both gas and (especially) oil. By looking at the historical response of production to variation in drilling effort, the analysis showed very little increase of production attributable to increased drilling. This was because of a tight quantitative relationship of diminishing returns with increasing drilling effort: as drilling effort increased, the energy obtained per active drill rig was reduced according to a severely diminishing power law. The study concluded that even an enormous increase of drilling effort was unlikely to significantly increase oil and gas production in a mature petroleum region such as the United States. However, contrary to the study's conclusion, since the analysis was published in 2008, US production of crude oil has increased 74%, and production of dry natural gas has increased 28% (2014 compared to 2008).

Anticipated Production by Major Agencies

Average yearly gains in global supply from 1987 to 2005 were 1.2 million barrels per day (190×10^3 m³/d) (1.7%). In 2005, the IEA predicted that 2030 production rates would reach 120,000,000 barrels per day (19,000,000 m³/d), but this number was gradually reduced to 105,000,000 barrels per day (16,700,000 m³/d). A 2008 analysis of IEA predictions questioned several underlying assumptions and claimed that a 2030 production level of 75,000,000 barrels per day (11,900,000 m³/d) (comprising 55,000,000 barrels (8,700,000 m³) of crude oil and 20,000,000 barrels (3,200,000 m³) of both non-conventional oil and natural gas liquids) was more realistic than the IEA numbers. More recently, the EIA's Annual Energy Outlook 2015 indicated no production peak out to 2040. However, this required a future Brent crude oil price of $US144/bbl (2013 dollars) "as growing demand leads to the development of more costly resources." Whether the world economy can grow and maintain demand for such a high oil price remains to be seen.

Oil Field Decline

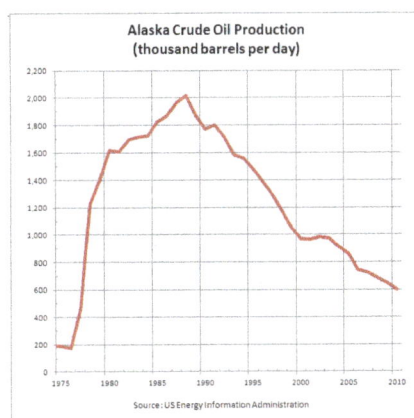

Alaska's oil production has declined 70% since peaking in 1988

In a 2013 study of 733 giant oil fields, only 32% of the ultimately recoverable oil, condensate and gas remained.Ghawar, which is the largest oil field in the world and responsible for approximately half of Saudi Arabia's oil production over the last 50 years, was in decline before 2009. The world's second largest oil field, the Burgan Field in Kuwait, entered decline in November 2005.

It is well established that once an oilfield reaches maximum production, it will decrease at a certain decline rate. For example, Mexico announced that production from its giant Cantarell Field began to decline in March 2006, reportedly at a rate of 13% per year. Also in 2006, Saudi Aramco Senior Vice President Abdullah Saif estimated that its existing fields were declining at a rate of 5% to 12% per year. According to a study of the largest 811 oilfields conducted in early 2008 by Cambridge Energy Research Associates, the average rate of field decline is 4.5% per year. The Association for the Study of Peak Oil and Gas agreed with their decline rates, but considered the rate of new fields coming online overly optimistic. The IEA stated in November 2008 that an analysis of 800 oilfields showed the decline in oil production to be 6.7% a year for fields past their peak, and that this would grow to 8.6% in 2030. A more rapid annual rate of decline of 5.1% in 800 of the world's largest oil fields weighted for production over their whole lives was reported by the International Energy Agency in their World Energy Outlook 2008. The 2013 study of 733 giant fields mentioned previously had an average decline rate 3.83% which was described as "conservative."

Control Over Supply

Entities such as governments or cartels can reduce supply to the world market by limiting access to the supply through nationalizing oil, cutting back on production, limiting drilling rights, imposing taxes, etc. International sanctions, corruption, and military conflicts can also reduce supply.

Nationalization of Oil Supplies

Another factor affecting global oil supply is the nationalization of oil reserves by producing nations. The nationalization of oil occurs as countries begin to deprivatize oil production and withhold exports. Kate Dourian, Platts' Middle East editor, points out that while estimates of oil reserves may vary, politics have now entered the equation of oil supply. "Some countries are becoming off limits. Major oil companies operating in Venezuela find themselves in a difficult position because of the growing nationalization of that resource. These countries are now reluctant to share their reserves."

According to consulting firm PFC Energy, only 7% of the world's estimated oil and gas reserves are in countries that allow companies like ExxonMobil free rein. Fully 65% are in the hands of state-owned companies such as Saudi Aramco, with the rest in countries such as Russia and Venezuela, where access by Western European and North American companies is difficult. The PFC study implies political factors are limiting capacity increases in Mexico, Venezuela, Iran, Iraq, Kuwait, and Russia. Saudi Arabia is also limiting capacity expansion, but because of a self-imposed cap, unlike the other countries. As a result of not having access to countries amenable to oil exploration, ExxonMobil is not making nearly the investment in finding new oil that it did in 1981.

OPEC Influence on Supply

OPEC is an alliance between 12 diverse oil producing countries (Algeria, Angola, Ecuador, Iran, Iraq, Kuwait, Libya, Nigeria, Qatar, Saudi Arabia, the United Arab Emirates, and Venezuela) to control the supply of oil. OPEC's power was consolidated as various countries nationalized their oil holdings, and wrested decision-making away from the "Seven Sisters," (Anglo-Iranian, Socony-Vacuum, Royal Dutch Shell, Gulf, Esso, Texaco, and Socal) and created their own oil companies to control the oil. OPEC tries to influence prices by restricting production. It does this by allocating each member country a quota for production. All 12 members agree to keep prices high by producing at lower levels than they otherwise would. There is no way to verify adherence to the quota, so every member faces the same incentive to "cheat" the cartel. United States policy of selling arms and providing security to Saudi Arabia is often seen as an attempt to influence the Saudis to increase oil production. According to sociology professor Michael Schwartz, the purpose for the second Iraq war was to break the back of OPEC and return control of the oil fields to western oil companies.

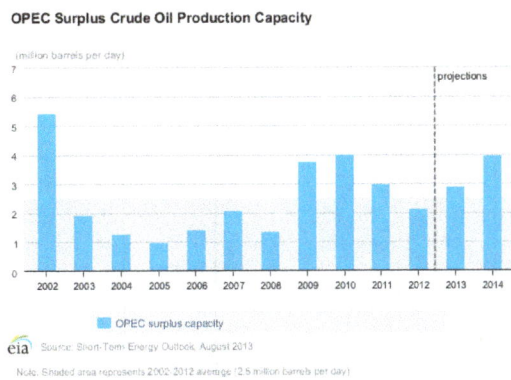

OPEC surplus crude oil production capacity (US EIA)

Alternatively, commodities trader Raymond Learsy, author of *Over a Barrel: Breaking the Middle East Oil Cartel*, contends that OPEC has trained consumers to believe that oil is a much more finite resource than it is. To back his argument, he points to past false alarms and apparent collaboration. He also believes that peak oil analysts are conspiring with OPEC and the oil companies to create a "fabricated drama of peak oil" to drive up oil prices and profits; oil had risen to a little over $30/barrel at that time. A counter-argument was given in the Huffington Post after he and Steve Andrews, co-founder of ASPO, debated on CNBC in June 2007.

Predictions of Peak Oil

Pub.	Made by	Peak year/range	Pub.	Made by	Peak year/range
1972	Esso	About 2000	1999	Parker	2040
1972	United Nations	By 2000	2000	A. A. Bartlett	2004 or 2019
1974	Hubbert	1991–2000	2000	Duncan	2006
1976	UK Dep. of Energy	About 2000	2000	EIA	2021–2067; 2037 most likely
1977	Hubbert	1996	2000	EIA (WEO)	Beyond 2020
1977	Ehrlich, et al.	2000	2001	Deffeyes	2003–2008
1979	Shell	Plateau by 2004	2001	Goodstein	2007
1981	World Bank	Plateau around 2000	2002	Smith	2010–2016
1985	J. Bookout	2020	2002	Campbell	2010

1989	Campbell	1989	2002	Cavallo	2025–2028
1994	L. F. Ivanhoe	OPEC plateau 2000–2050	2003	Greene, et al.	2020–2050
1995	Petroconsultants	2005	2003	Laherrère	2010–2020
1997	Ivanhoe	2010	2003	Lynch	No visible peak
1997	J. D. Edwards	2020	2003	Shell	After 2025
1998	IEA	2014	2003	Simmons	2007–2009
1998	Campbell &Laherrère	2004	2004	Bakhitari	2006–2007
1999	Campbell	2010	2004	CERA	After 2020
1999	Peter Odell	2060	2004	PFC Energy	2015–2020
A selection of estimates of the year of peak world oil production, compiled by the United States Energy Information Administration					

In 1962, Hubbert predicted that world oil production would peak at a rate of 12.5 billion barrels per year, around the year 2000. In 1974, Hubbert predicted that peak oil would occur in 1995 "if current trends continue." Those predictions proved incorrect. However, a number of industry leaders and analysts believe that world oil production will peak between 2015 and 2030, with a significant chance that the peak will occur before 2020. They consider dates after 2030 implausible. By comparison, a 2014 analysis of production and reserve data predicted a peak in oil production about 2035. Determining a more specific range is difficult due to the lack of certainty over the actual size of world oil reserves. Unconventional oil is not currently predicted to meet the expected shortfall even in a best-case scenario. For unconventional oil to fill the gap without "potentially serious impacts on the global economy", oil production would have to remain stable after its peak, until 2035 at the earliest.

Papers published since 2010 have been relatively pessimistic. A 2010 Kuwait University study predicted production would peak in 2014. A 2010 Oxford University study predicted that production will peak before 2015, but its projection of a change soon "… from a demand-led market to a supply constrained market …" was incorrect. A 2014 validation of a significant 2004 study in the journal *Energy* proposed that it is likely that conventional oil production peaked, according to various definitions, between 2005 and 2011. A set of models published in a 2014 Ph.D. thesis predicted that a 2012 peak would be followed by a drop in oil prices, which in some scenarios could turn into a rapid rise in prices thereafter. According to energy blogger Ron Patterson, the peak of world oil production was probably around 2010.

Major oil companies hit peak production in 2005. Several sources in 2006 and 2007 predicted that worldwide production was at or past its maximum.Fatih Birol, chief economist at the International Energy Agency, also stated that "crude oil production for the world has already peaked in 2006." However, in 2013 OPEC's figures showed that world crude oil production and remaining proven reserves were at record highs. According to Matthew Simmons, former Chairman of Simmons & Company International and author of *Twilight in the Desert: The Coming Saudi Oil Shock and the World Economy,* "peaking is one of these fuzzy events that you only know clearly when you see it through a rear view mirror, and by then an alternate resolution is generally too late."

Possible Consequences

The wide use of fossil fuels has been one of the most important stimuli of economic growth and prosperity since the industrial revolution, allowing humans to participate in takedown, or the consumption of energy at a greater rate than it is being replaced. Some believe that when oil production decreases, human culture, and modern technological society will be forced to change drastically. The impact of peak oil will depend heavily on the rate of decline and the development and adoption of effective alternatives.

In 2005, the United States Department of Energy published a report titled *Peaking of World Oil Production: Impacts, Mitigation, & Risk Management.* Known as the Hirsch report, it stated, "The peaking of world oil production presents the U.S. and the world with an unprecedented risk management problem. As peaking is approached, liquid fuel prices and price volatility will increase dramatically, and, without timely mitigation, the economic, social, and political costs will be unprecedented. Viable mitigation options exist on both the supply and demand sides, but to have substantial impact, they must be initiated more than a decade in advance of peaking." Some of the information was updated in 2007.

Oil Prices

Historical Oil Prices

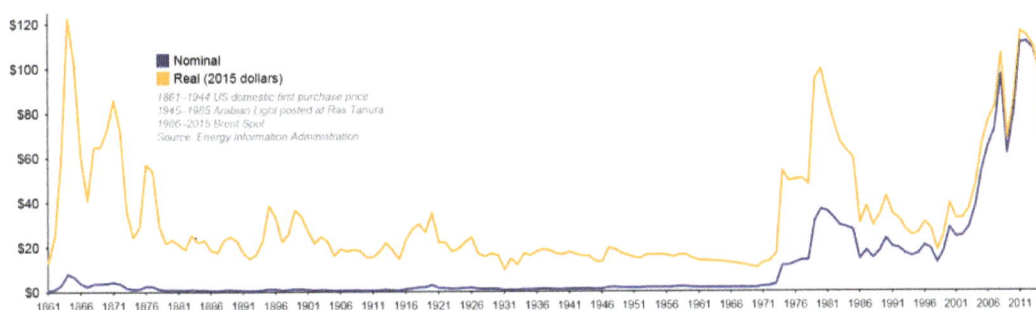

Long-term oil prices, 1861–2015 (top line adjusted for inflation)

The oil price historically was comparatively low until the 1973 oil crisis and the 1979 energy crisis when it increased more than tenfold during that six-year timeframe. Even though the oil price dropped significantly in the following years, it has never come back to the previous levels. Oil price began to increase again during the 2000s until it hit historical heights of $143 per barrel (2007 inflation adjusted dollars) on 30 June 2008. As these prices were well above those that caused the 1973 and 1979 energy crises, they contributed to fears of an economic recession similar to that of the early 1980s.

It is generally agreed that the main reason for the price spike in 2005–2008 was strong demand pressure. For example, global consumption of oil rose from 30 billion barrels (4.8×10^9 m³) in 2004 to 31 billion in 2005. The consumption rates were far above new discoveries in the period, which had fallen to only eight billion barrels of new oil reserves in new accumulations in 2004.

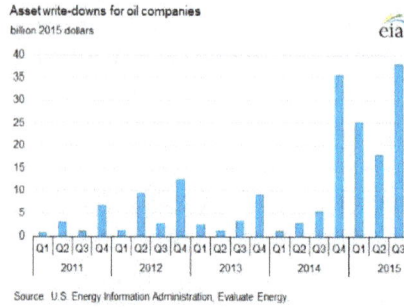

Asset write downs for oil companies 2015

Oil price increases were partially fueled by reports that petroleum production is at or near full capacity. In June 2005, OPEC stated that they would 'struggle' to pump enough oil to meet pricing pressures for the fourth quarter of that year. From 2007 to 2008, the decline in the U.S. dollar against other significant currencies was also considered as a significant reason for the oil price increases, as the dollar lost approximately 14% of its value against the Euro from May 2007 to May 2008.

Besides supply and demand pressures, at times security related factors may have contributed to increases in prices, including the War on Terror, missile launches in North Korea, the Crisis between Israel and Lebanon, nuclear brinkmanship between the U.S. and Iran, and reports from the U.S. Department of Energy and others showing a decline in petroleum reserves.

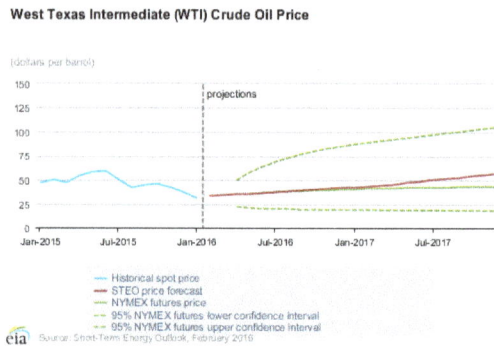

Depicts EIA projections for West Texas Intermediate crude oil price for 2016-2017

More recently, between 2011 and 2014 the price of crude oil was relatively stable, fluctuating around $US100 per barrel. It dropped sharply in late 2014 to below $US70 where it remained for most of 2015. In early 2016 it traded at a low of $US27. The price drop has been attributed to both oversupply and reduced demand as a result of the slowing global economy, OPEC reluctance to concede market share, and a stronger US dollar. These factors may be exacerbated by a combination of monetary policy and the increased debt of oil producers, who may increase production to maintain liquidity.

This price drop has placed many US tight oil producers under considerable financial pressure. As a result, there has been a reduction by oil companies in capital expenditure of over $US400 billion. It is anticipated that this will have effects on global production in the longer term, leading to statements of concern by the International Energy Agency that governments should not be complacent about energy security. Energy Information Agency projections anticipate market oversupply and prices below $US50 until late 2017.

Effects of Historical Oil Price Rises

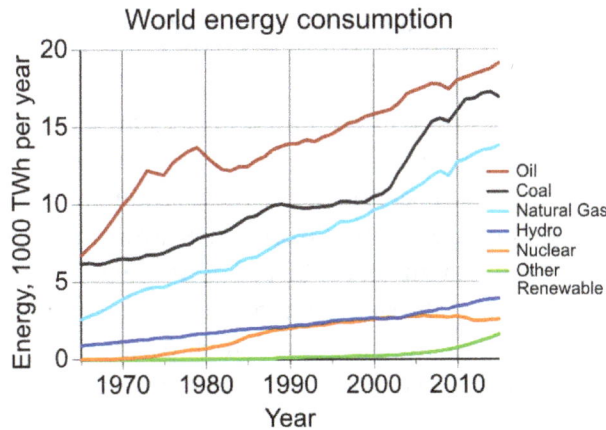

World consumption of primary energy by energy type

In the past, the price of oil has led to economic recessions, such as the 1973 and 1979 energy crises. The effect the price of oil has on an economy is known as a price shock. In many European countries, which have high taxes on fuels, such price shocks could potentially be mitigated somewhat by temporarily or permanently suspending the taxes as fuel costs rise. This method of softening price shocks is less useful in countries with much lower gas taxes, such as the United States. A baseline scenario for a recent IMF paper found oil production growing at 0.8% (as opposed to a historical average of 1.8%) would result in a small reduction in economic growth of 0.2–0.4%.

Researchers at the Stanford Energy Modeling Forum found that the economy can adjust to steady, gradual increases in the price of crude better than wild lurches.

Some economists predict that a substitution effect will spur demand for alternate energy sources, such as coal or liquefied natural gas. This substitution can be only temporary, as coal and natural gas are finite resources as well.

Prior to the run-up in fuel prices, many motorists opted for larger, less fuel-efficient sport utility vehicles and full-sized pickups in the United States, Canada, and other countries. This trend has been reversing because of sustained high prices of fuel. The September 2005 sales data for all vehicle vendors indicated SUV sales dropped while small cars sales increased. Hybrid and diesel vehicles are also gaining in popularity.

EIA published Household Vehicles Energy Use: Latest Data and Trends in Nov 2005 illustrating the steady increase in disposable income and $20–30 per barrel price of oil in 2004. The report notes "The average household spent $1,520 on fuel purchases for transport." According to CNBC that expense climbed to $4,155 in 2011.

In 2008, a report by Cambridge Energy Research Associates stated that 2007 had been the year of peak gasoline usage in the United States, and that record energy prices would cause an "enduring shift" in energy consumption practices. The total miles driven in the U.S. peaked in 2006.

The Export Land Model states that after peak oil petroleum exporting countries will be forced to reduce their exports more quickly than their production decreases because of internal demand growth. Countries that rely on imported petroleum will therefore be affected earlier and more dramatically than exporting countries. Mexico is already in this situation. Internal consumption grew by 5.9% in 2006 in the five biggest exporting countries, and their exports declined by over 3%. It was estimated that by 2010 internal demand would decrease worldwide exports by 2,500,000 barrels per day (400,000 m³/d).

Canadian economist Jeff Rubin has stated that high oil prices are likely to result in increased consumption in developed countries through partial manufacturing de-globalisation of trade. Manufacturing production would move closer to the end consumer to minimise transportation network costs, and therefore a demand decoupling from gross domestic product would occur. Higher oil prices would lead to increased freighting costs and consequently, the manufacturing industry would move back to the developed countries since freight costs would outweigh the current economic wage advantage of developing countries. Economic research carried out by the International Monetary Fund puts overall price elasticity of demand for oil at −0.025 short-term and −0.093 long term.

Agricultural Effects and Population Limits

Since supplies of oil and gas are essential to modern agriculture techniques, a fall in global oil supplies could cause spiking food prices and unprecedented famine in the coming decades.Geologist Dale Allen Pfeiffer contends that current population levels are unsustainable, and that to achieve a sustainable economy and avert disaster the United States population would have to be reduced by at least one-third, and world population by two-thirds.

The largest consumer of fossil fuels in modern agriculture is ammonia production (for fertilizer) via the Haber process, which is essential to high-yielding intensive agriculture. The specific fossil fuel input to fertilizer production is primarily natural gas, to provide hydrogen via steam reforming. Given sufficient supplies of renewable electricity, hydrogen can be generated without fossil fuels using methods such as electrolysis. For example, the Vemork hydroelectric plant in Norway used its surplus electricity output to generate renewable ammonia from 1911 to 1971.

Iceland currently generates ammonia using the electrical output from its hydroelectric and geothermal power plants, because Iceland has those resources in abundance while having no domestic hydrocarbon resources, and a high cost for importing natural gas.

Long-term Effects on Lifestyle

A majority of Americans live in suburbs, a type of low-density settlement designed around universal personal automobile use. Commentators such as James Howard Kunstler argue that because over 90% of transportation in the U.S. relies on oil, the suburbs' reliance on the automobile is an unsustainable living arrangement. Peak oil would leave many Americans unable to afford petroleum based fuel for their cars, and force them to use bicycles or electric vehicles. Additional options include telecommuting, moving to rural areas, or moving to higher density areas, where walking and public transportation are more viable options. In the latter two cases, suburbs may become the "slums of the future." The issue of petroleum supply and demand is also a concern for growing cities in developing countries (where urban areas are expected to absorb most of the world's projected 2.3 billion population increase by 2050). Stressing the energy component of future development plans is seen as an important goal.

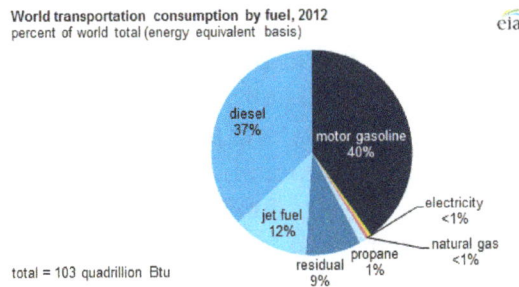

World transport energy use by fuel type 2012

Rising oil prices, if they occur, would also affect the cost of food, heating, and electricity. A high amount of stress would then be put on current middle to low income families as economies contract from the decline in excess funds, decreasing employment rates. The Hirsch/US DoE Report concludes that "without timely mitigation, world supply/demand balance will be achieved through massive demand destruction (shortages), accompanied by huge oil price increases, both of which would create a long period of significant economic hardship worldwide."

Methods that have been suggested for mitigating these urban and suburban issues include the use of non-petroleum vehicles such as electric cars, battery electric vehicles, transit-oriented development, carfree cities, bicycles, new trains, new pedestrianism, smart growth, shared space, urban consolidation, urban villages, and New Urbanism.

An extensive 2009 report on the effects of compact development by the United States National Research Council of the Academy of Sciences, commissioned by the United States Congress, stated six main findings. First, that compact development is likely to reduce "Vehicle Miles Traveled" (VMT) throughout the country. Second, that doubling residential density in a given area could reduce VMT by as much as 25% if coupled with measures such as increased employment density and improved public transportation. Third, that higher density, mixed-use developments would produce both direct reductions in CO_2 emissions (from less driving), and indirect reductions (such as from lower amounts of materials used per housing unit, higher efficiency climate control, longer vehicle lifespans, and higher efficiency delivery of goods and services). Fourth, that although short term reductions in energy use and CO_2 emissions would be modest, that these reductions would become more significant over time. Fifth, that a major obstacle to more compact development in the United States is political resistance from local zoning regulators, which would hamper efforts by state and regional governments to participate in land-use planning. Sixth, the committee agreed that changes in development that would alter driving patterns and building efficiency would have various secondary costs and benefits that are difficult to quantify. The report recommends that policies supporting compact development (and especially its ability to reduce driving, energy use, and CO_2 emissions) should be encouraged.

An economic theory that has been proposed as a remedy is the introduction of a steady state economy. Such a system could include a tax shifting from income to depleting natural resources (and pollution), as well as the limitation of advertising that stimulates demand and population growth. It could also include the institution of policies that move away from globalization and toward localization to conserve energy resources, provide local jobs, and maintain local decision-making authority. Zoning policies could be adjusted to promote resource conservation and eliminate sprawl.

Mitigation

To avoid the serious social and economic implications a global decline in oil production could entail, the Hirsch report emphasized the need to find alternatives, at least ten to twenty years before the peak, and to phase out the use of petroleum over that time. This was similar to a plan proposed for Sweden that same year. Such mitigation could include energy conservation, fuel substitution, and the use of unconventional oil. The timing of mitigation responses is critical. Premature initiation would be undesirable, but if initiated too late could be more costly and have more negative economic consequences.

Positive Aspects

Permaculture sees peak oil as holding tremendous potential for positive change, assuming countries act with foresight. The rebuilding of local food networks, energy production, and the general implementation of "energy descent culture" are argued to be ethical responses to the acknowledgment of finite fossil resources. Majorca is an island currently diversifying its energy supply from fossil fuels to alternative sources and looking back at traditional construction and permaculture methods.

The Transition Towns movement, started in Totnes, Devon and spread internationally by "The Transition Handbook" (Rob Hopkins) and Transition Network, sees the restructuring of society for more local resilience and ecological stewardship as a natural response to the combination of peak oil and climate change.

Criticisms

General Arguments

Opponents to the theory of peak oil often cite new oil reserves that have been found, which continue to forestall a peak oil event. In particular, some contend that oil production from these new oil reserves as well as from existing fields will continue to increase at a rate that outpaces demand, until alternate energy sources for our current fossil fuel dependence are found. As of 2015, analysts in both the petroleum and financial industries were concluding that the "age of oil" had already reached a new stage where the excess supply that appeared in late 2014 may continue to prevail in the future. A consensus appeared to be emerging that an international agreement will be reached to introduce measures to constrain the combustion of hydrocarbons in an effort to limit global temperature rise to the nominal 2 °C that is consensually predicted to limit environmental harm to tolerable levels.

Further criticism against peak oil is confidence in the various options and technologies for substituting oil. And indeed there are some promising approaches that seem to have the potential to reduce or even counterbalance the effects of a peak oil situation. For example, US federal funding has increased for algae fuels since the year 2000 due to rising fuel prices. Numerous more projects are being funded in Australia, New Zealand, Europe, the Middle East, and other parts of the world and private companies are entering the field.

Oil Industry Representatives

Oil industry representatives have criticised peak oil theory, at least as it has been presented by Matthew Simmons. The president of Royal Dutch Shell's U.S. operations John Hofmeister, while agreeing that conventional oil production would soon start to decline, criticized Simmons's analysis for being "overly focused on a single country: Saudi Arabia, the world's largest exporter and OPEC swing producer." He also pointed to the large reserves at the US outer continental shelf, which held an estimated 100 billion barrels (16×10^9 m^3) of oil and natural gas. However, only 15% of those reserves were currently exploitable, a good part of that off the coasts of Louisiana, Alabama, Mississippi, and Texas. Hofmeister also contended that Simmons erred in excluding unconventional sources of oil such as the oil sands of Canada, where Shell was active. The Canadian oil sands—a natural combination of sand, water, and oil found largely in Alberta and Saskatchewan—are believed to contain one trillion barrels of oil. Another trillion barrels are also said to be trapped in rocks in Colorado, Utah, and Wyoming, but are in the form of oil shale. These particular reserves present major environmental, social, and economic obstacles to recovery. Hofmeister also claimed that if oil companies were allowed to drill more in the United States enough to produce another 2 million barrels per day (320×10^3 m^3/d), oil and gas prices would not be as high as they were in the later part of the 2000 to 2010 decade. He thought in 2008 that high energy prices would cause social unrest similar to the 1992 Rodney King riots.

In 2009, Dr. Christoph Rühl, chief economist of BP, argued against the peak oil hypothesis:

Physical peak oil, which I have no reason to accept as a valid statement either on theoretical, scientific or ideological grounds, would be insensitive to prices. ... In fact the whole hypothesis of peak oil – which is that there is a certain amount of oil in the ground, consumed at a certain rate, and then it's finished – does not react to anything ... Therefore there will never be a moment when the world runs out of oil because there will always be a price at which the last drop of oil can clear the market. And you can turn anything into oil into if you are willing to pay the financial and environmental price ... (Global Warming) is likely to be more of a natural limit than all these peak oil theories combined. ... Peak oil has been predicted for 150 years. It has never happened, and it will stay this way.

According to Rühl, the main limitations for oil availability are "above ground" and are to be found in the availability of staff, expertise, technology, investment security, money and last but not least in global warming. The oil question is about price and not the basic availability. Rühl's views are shared by Daniel Yergin of CERA, who added that the recent high price phase might add to a future demise of the oil industry, not of complete exhaustion of resources or an apocalyptic shock but the timely and smooth setup of alternatives.

Clive Mather, CEO of Shell Canada, said the Earth's supply of bitumen hydrocarbons is "almost infinite", referring to hydrocarbons in oil sands.

Others

Economist Robert L. Bradley, Jr. wrote in a 2007 article in The Review of Austrian Economics that, "[a]n Austrian institutional theory is more robust for explaining changes in mineral-resource scarcity than neoclassical depletionism[.]" Using the writings of Erich Zimmermann and Julian Simon, Bradley also argued in 2012 that resources have subjective rather than objective existences in economics. He concluded that, "what resources come from the ground ultimately depend on the resources in the mind."

Attorney and mechanical engineer Peter W. Huber pointed out in 2006 that the world is just running out of "cheap oil." As oil prices rise, unconventional sources become economically viable. He predicted that, "[t]he tar sands of Alberta alone contain enough hydrocarbon to fuel the entire planet for over 100 years."

Industry blogger Steve Maley echoed some of the points of Yergin, Rühl, Mather and Hofmeister.

Environmental journalist George Monbiot responded to a 2012 report by Leonardo Maugeri by proclaiming that there is more than enough oil (from unconventional sources) for capitalism to "deep-fry" the world with climate change. Stephen Sorrell, senior lecturer Science and Technology Policy Research, Sussex Energy Group, and lead author of the UKERC Global Oil Depletion report, and Christophe McGlade, doctoral researcher at the UCL Energy Institute have criticized Maugeri's assumptions about decline rates.

Oil Well Control

Oil well control is the management of the dangerous effects caused by the unexpected release of formation fluid, such as natural gas and/or crude oil, upon surface equipment of oil or gas drilling rigs and escaping into the atmosphere. Technically, oil well control involves preventing the formation fluid, usually referred to as kick, from entering into the wellbore during drilling.

Formation fluid can enter the wellbore if the pressure exerted by the column of drilling fluid is not great enough to overcome the pressure exerted by the fluids in the formation being drilled. Oil well control also includes monitoring a well for signs of impending influx of formation fluid into the wellbore during drilling and procedures, to stop the well from flowing when it happens by taking proper remedial actions.

Failure to manage and control these pressure effects can cause serious equipment damage and injury, or loss of life. Improperly managed well control situations can cause blowouts, which are uncontrolled and explosive expulsions of formation fluid from the well, potentially resulting in a fire.

Importance of Oil Well Control

Oil well control is one of the most important aspects of drilling operations. Improper handling of kicks in oil well control can result in blowouts with very grave consequences, including the loss of valuable resources. Even though the cost of a blowout (as a result of improper/no oil well control) can easily reach several millions of US dollars, the monetary loss is not as serious as the other damages that can occur: irreparable damage to the environment, waste of valuable resources, ruined equipment, and most importantly, the safety and lives of personnel on the drilling rig.

In order to avert the consequences of blowout, the utmost attention must be given to oil well control. That is why oil well control procedures should be in place prior to the start of an abnormal situation noticed within the wellbore, and ideally when a new rig position is sited. In other words, this includes the time the new location is picked, all drilling, completion, workover, snubbing and any other drilling-related operations that should be executed with proper oil well control in mind. This type of preparation involves widespread training of personnel, the development of strict operational guidelines and the design of drilling programs — maximizing the probability of successfully regaining hydrostatic control of a well after a significant influx of formation fluid has taken place. dangers of oil drilling

Fundamental Concepts and Terminology

Pressure is a very important concept in the oil and gas industry. Pressure can be defined as: the force exerted per unit area. Its SI unit is newtons per square metre or pascals. Another unit, bar, is also widely used as a measure of pressure, with 1 bar equal to 100 kilopascals. Normally pressure is measured in the U.S. petroleum industry in units of pounds force per square inch of area, or psi. 1000 psi equals 6894.76 kilo-pascals.

Hydrostatic Pressure

Hydrostatic pressure (HSP), as stated, is defined as pressure due to a column of fluid that is not moving. That is, a column of fluid that is static, or at rest, exerts pressure due to local force of gravity on the column of the fluid.

The formula for calculating hydrostatic pressure in SI units (N/m^2) is:

$$\text{Hydrostatic pressure} = \text{Height (m)} \times \text{Density (kg/m}^3) \times \text{Gravity (m/s}^2).$$

All fluids in a wellbore exert hydrostatic pressure, which is a function of density and vertical height of the fluid column. In US oil field units, hydrostatic pressure can be expressed as:

$HSP = 0.052 \times MW \times TVD$', where MW (Mud Weight or density) is the drilling-fluid density in pounds per gallon (ppg), TVD is the true vertical depth in feet and HSP is the hydrostatic pressure in psi.

The 0.052 is needed as the conversion factor to psi unit of HSP.

To convert these units to SI units, one can use:

- 1 ppg = $\approx$ 119.8264273 kg/m^3

- 1 ft = 0.3048 metres

- 1 psi = 0.0689475729 bar

- 1 bar = 10^5 pascals

Pressure Gradient

The pressure gradient is described as the pressure per unit length. Often in oil well control, pressure exerted by fluid is expressed in terms of its pressure gradient. The SI unit is pascals/metre. The hydrostatic pressure gradient can be written as:

$$\text{Pressure gradient (psi/ft)} = HSP/TVD = 0.052 \times MW \text{ (ppg)}.$$

Formation Pressure

Formation pressure is the pressure exerted by the formation fluids, which are the liquids and gases contained in the geologic formations encountered while drilling for oil or gas. It can also be said to be the pressure contained within the pores of the formation or reservoir being drilled. Formation pressure is a result of the hydrostatic pressure of the formation fluids, above the depth of interest, together with pressure trapped in the formation. Under formation pressure, there are 3 levels: normally pressured formation, abnormal formation pressure, or subnormal formation pressure.

Normally pressured formation

Normally pressured formation has a formation pressure that is the same with the hydrostatic pressure of the fluids above it. As the fluids above the formation are usually some form of water, this pressure can be defined as the pressure exerted by a column of water from the formation's depth to sea level.

The normal hydrostatic pressure gradient for freshwater is 0.433 pounds per square inch per foot (psi/ft), or 9.792 kilopascals per meter (kPa/m), and 0.465 psi/ft for water with dissolved solids like in Gulf Coast waters, or 10.516 kPa/m. The density of formation water in saline or marine environments, such as along the Gulf Coast, is about 9.0 ppg or 1078.43 kg/m³. Since this is the highest for both Gulf Coast water and fresh water, a normally pressured formation can be controlled with a 9.0 ppg mud.

Sometimes the weight of the overburden, which refers to the rocks and fluids above the formation, will tend to compact the formation, resulting in pressure built-up within the formation if the fluids are trapped in place. The formation in this case will retain its normal pressure only if there is a communication with the surface. Otherwise, an *abnormal formation pressure* will result.

Abnormal formation pressure

As discussed above, once the fluids are trapped within the formation and not allow to escape there is a pressure build-up leading to abnormally high formation pressures. This will generally require a mud weight of greater than 9.0 ppg to control. Excess pressure, called "overpressure" or "geopressure", can cause a well to blow out or become uncontrollable during drilling.

Subnormal formation pressure

Subnormal formation pressure is a formation pressure that is less than the normal pressure for the given depth. It is common in formations that had undergone production of original hydrocarbon or formation fluid in them.

Overburden Pressure

Overburden pressure is the pressure exerted by the weight of the rocks and contained fluids above the zone of interest. Overburden pressure varies in different regions and formations. It is the force that tends to compact a formation vertically. The density of these usual ranges of rocks is about 18 to 22 ppg (2,157 to 2,636 kg/m³). This range of densities will generate an overburden pressure gradient of about 1 psi/ft (22.7 kPa/m). Usually, the 1 psi/ft is not applicable for shallow marine sediments or massive salt. In offshore however, there is a lighter column of sea water, and the column of underwater rock does not go all the way to the surface. Therefore, a lower overburden pressure is usually generated at an offshore depth, than would be found at the same depth on land.

Mathematically, overburden pressure can be derived as:

$$S = \rho_b \times D \times g$$

where

g = acceleration due to gravity

S = overburden pressure

ρ_b = average formation bulk density

D = vertical thickness of the overlying sediments

The bulk density of the sediment is a function of rock matrix density, porosity within the confines of the pore spaces, and porefluid density. This can be expressed as

$$\rho_b = \varphi\rho_f + (1 - \varphi)\rho_m$$

where

φ = rock porosity

ρ_f = formation fluid density

ρ_m = rock matrix density

Fracture Pressure

Fracture pressure can be defined as pressure required to cause a formation to fail or split. As the name implies, it is the pressure that causes the formation to fracture and the circulating fluid to be lost. Fracture pressure is usually expressed as a gradient, with the common units being psi/ft (kPa/m) or ppg (kg/m³).

To fracture a formation, three things are generally needed, which are:

1. Pump into the formation. This will require a pressure in the wellbore greater than formation pressure.

2. The pressure in the wellbore must also exceed the rock matrix strength.

3. And finally the wellbore pressure must be greater than one of the three principal stresses in the formation.

Pump Pressure (System Pressure Losses)

Pump pressure, which is also referred to as *system pressure loss*, is the sum total of all the pressure losses from the oil well surface equipment, the drill pipe, the drill collar, the drill bit, and annular friction losses around the drill collar and drill pipe. It measures the system pressure loss at the start of the circulating system and measures the total friction pressure.

Slow Pump Pressure (SPP)

Slow pump pressure is the circulating pressure (pressure used to pump fluid through the whole active fluid system, including the borehole and all the surface tanks that constitute the primary system during drilling) at a reduced rate. SPP is very important during a well kill operation in which circulation (a process in which drilling fluid is circulated out of the suction pit, down the drill pipe and drill collars, out the bit, up the annulus, and back to the pits while drilling proceeds) is done at a reduced rate to allow better control of circulating pressures and to enable the mud properties (density and viscosity) to be kept at desired values. The slow pump pressure can also be referred to as "kill rate pressure" or "slow circulating pressure" or "kill speed pressure" and so on.

Shut-in Drill Pipe Pressure

Shut-in drill pipe pressure (SIDPP), which is recorded when a well is shut in on a kick, is a measure of the difference between the pressure at the bottom of the hole and the hydrostatic pressure (HSP) in the drillpipe. During a well shut-in, the pressure of the wellbore stabilizes, and the formation pressure equals the pressure at the bottom of the hole. The drillpipe at this time should be full of known-density fluid. Therefore, the formation pressure can be easily calculated using the SIDPP. This means that the SIDPP gives a direct of formation pressure during a kick.

Shut-in Casing Pressure (SICP)

The *shut-in casing pressure* (SICP) is a measure of the difference between the formation pressure and the HSP in the annulus when a kick occurs.

The pressures encountered in the annulus can be estimated using the following mathematical equation:

$$FP = HSP_{mud} + HSP_{influx} + SICP$$

where

> FP = formation pressure (psi)
>
> HSP_{mud} = Hydrostatic pressure of the mud in the annulus (psi)
>
> HSP_{influx} = Hydrostatic pressure of the influx (psi)
>
> SICP = shut-in casing pressure (psi)

Bottom-hole Pressure (BHP)

Bottom-hole pressure (BHP) is the pressure at the bottom of a well. The pressure is usually measured at the bottom of the hole. This pressure may be calculated in a static, fluid-filled wellbore with the equation:

$$BHP = D \times \rho \times C,$$

where

BHP = bottom-hole pressure

D = the vertical depth of the well

ρ = density

C = units conversion factor

(or, in the English system, $BHP = D \times MWD \times 0.052$).

In Canada the formula is depth in meters x density in kgs x the constant gravity factor (0.00981), which will give the hydrostatic pressure of the well bore or (hp) hp=bhp with pumps off. The bottom-hole pressure is dependent on the following:

- Hydrostatic pressure (HSP)

- Shut-in surface pressure (SIP)

- Friction pressure

- Surge pressure (occurs when transient pressure increases the bottom-hole pressure)

- Swab pressure (occurs when transient pressure reduces the bottom-hole pressure)

Therefore, BHP can be said to be the sum of all pressures at the bottom of the wellhole, which equals:

$$BHP = HSP + SIP + friction + Surge - swab$$

Basic Calculations in Oil Well Control

There are some basic calculations that need to be carried during oil well control. A few of these essential calculations will be discussed below. Most of the units here are in US oil field units, but these units can be converted to their SI units equivalent by using this Conversion of units link.

Capacity

The capacity of drill string is an essential issue in oil well control. The capacity of drillpipe, drill collars or hole is the volume of fluid that can be contained within them.

The capacity formula is as shown below:

$$Capacity = ID^2/1029.4$$

where

Capacity = Volume in barrels per foot(bbl/ft)

ID = Inside diameter in inches

1029.4 = Units conversion factor

Also the total pipe or hole volume is given by :

Volume in barrels (bbls) = *Capacity* (bbl/ft) × length (ft)

Feet of pipe occupied by a given volume is given by:

Feet of pipe (ft) = *Volume of mud* (bbls) / *Capacity* (bbls/ft)

Capacity calculation is important in oil well control due to the following:

- Volume of the drillpipe and the drill collars must be pumped to get kill weight mud to the bit during kill operation.

- It is used to spot pills and plugs at various depths in the wellbore.

Annular Capacity

This is the volume contained between the inside diameter of the hole and the outside diameter of the pipe. Annular capacity is given by :

Annular capacity (bbl/ft) = $(ID_{hole}^2 - OD_{pipe}^2)$ / 1029.4

where

ID_{hole}^2 = Inside diameter of the casing or open hole in inches

OD_{pipe}^2 = Outside diameter of the pipe in inches

Similarly

Annular volume (bbls) = *Annular capacity* (bbl/ft) × length (ft)

and

Feet occupied by volume of mud in annulus = *Volume of mud* (bbls) / *Annular Capacity* (bbls/ft).

Fluid Level Drop

Fluid level drop is the distance the mud level will drop when a dry string(a bit that is not plugged) is being pulled from the wellbore and it is given by:

Fluid level drop = Bbl disp / (CSG cap - Pipe disp)

or

Fluid level drop = Bbl disp / (Ann cap + Pipe cap)

and the resulting loss of HSP is given by:

Lost HSP = 0.052 × MW × *Fluid drop*

where

> *Fluid drop* = distance the fluid falls (ft)
>
> *Bbl disp* = displacement of the pulled pipe (bbl)
>
> *CSG cap* = casing capacity (bbl/ft)
>
> *Pipe disp* = pipe displacement (bbl/ft)
>
> *Ann cap* = Annular capacity between casing and pipe (bbl/ft)
>
> *Pipe cap* = pipe capacity
>
> *Lost HSP* = Lost hydrostatic pressure (psi)
>
> *MW* = mud weight (ppg)

When pulling a wet string (the bit is plugged) and the fluid from the drillpipe is not returned to the hole. The fluid drop is then changed to the following:

> *Fluid level drop = Bbl disp / Ann cap*

Kill Weight Fluid

Kill weight fluid which can also be called *Kill weight Mud* is the density of the mud required to balance formation pressure during kill operation. The Kill Weight Mud can be calculated by:

> *KWM = SIDPP/(0.052 × TVD) + OWM*

where

> *KWM* = kill weight mud (ppg)
>
> *SIDPP* = shut-in drillpipe pressure (psi)
>
> *TVD* = true vertical depth (ft)
>
> *OWM* = original weight mud (ppg)

But when the formation pressure can be determined from data sources such as bottom hole pressure, then *KWM* can be calculated as follows:

> *KWM = FP / 0.052 × TVD*

where *FP* = Formation pressure.

Kicks

Kick is the entry of formation fluid into the wellbore during drilling operations. It occurs because the pressure exerted by the column of drilling fluid is not great enough to overcome the pressure exerted by the fluids in the formation drilled. The whole essence of oil well control is to prevent kick from occurring and if it happens to prevent it from developing into blowout. An uncontrolled kick usually results from not deploying the proper equipment, using poor practices, or a lack of training of the rig crews. Loss of oil well control may lead into blowout, which represents one of the most severe threats associated with the exploration of petroleum resources involving the risk of lives and environmental and economic consequences.

Ixtoc I oil well blowout

Causes of Kicks

A kick will occur when the bottom hole pressure(BHP) of a well falls below the formation pressure and the formation fluid flows into the wellbore.There are usually causes for kicks some of which are:

- Failure to keep the hole full during a trip

- Swabbing while tripping

- Lost circulation

- Insufficient density of fluid

- Abnormal pressure

- Drilling into an adjacent well

- Lost control during drill stem test

Failure to keep the Hole Full During a Trip

Tripping is the complete operation of removing the drillstring from the wellbore and running it back in the hole. This operation is typically undertaken when the bit (which is the tool used to crush or cut rock during drilling) becomes dull or broken, and no longer drills the rock efficiently. A typical drilling operation of deep oil or gas wells may require up to 8 or more trips of the drill string to replace a dull rotary bit for one well.

Tripping out of the hole means that the entire volume of steel (of drillstring) is being removed, or has been removed, from the well. This displacement of the drill string (the steel) will leave out a volume of space that must be replaced with an equal volume of mud. If the replacement is not done, the fluid level in the wellbore will drop, resulting in a loss of hydrostatic pressure (HSP) and bottom hole pressure (BHP). If this bottom hole pressure reduction goes below the formation pressure, a kick will definitely occur.

Swabbing while Tripping

Swabbing occurs when bottom hole pressure is reduced due to the effects of pulling the drill string upward in the bored hole. During the tripping out of the hole, the space formed by the drillpipe, drill collar, or tubing (which are being removed) must be replaced by something, usually mud. If the rate of tripping out is greater than the rate the mud is being pumped into the void space (created by the removal of the drill string), then swab will occur. If the reduction in bottom hole pressure caused by swabbing is below formation pressure, then a kick will occur.

Lost Circulation

Lost circulation usually occurs when the hydrostatic pressure fractures an open formation. When this occurs, there is loss in circulation, and the height of the fluid column decreases, leading to lower HSP in the wellbore. A kick can occur if steps are not taken to keep the hole full. Lost circulation can be caused by:

- excessive mud weights
- excessive annular friction loss
- excessive surge pressure during trips, or "spudding" the bit
- excessive shut-in pressures.

Insufficient Density of Fluid

If the density of the drilling fluid or mud in the well bore is not sufficient to keep the formation pressure in check, then a kick can occur. Insufficient density of the drilling fluid can be as a result of the following :

- attempting to drill by using an underbalanced weight solution
- excessive dilution of the mud
- heavy rains in the pits
- barite settling in the pits
- spotting low density pills in the well.

Abnormal Pressure

Another cause of kicks is drilling accidentally into abnormally-pressured permeable zones. The increased formation pressure may be greater than the bottom hole pressure, resulting in a kick.

Drilling into an Adjacent Well

Drilling into an adjacent well is a potential problem, particularly in offshore drilling where a large number of directional wells are drilled from the same platform. If the drilling well penetrates the production string of a previously completed well, the formation fluid from the completed well will enter the wellbore of the drilling well, causing a kick. If this occurs at a shallow depth, it is an extremely dangerous situation and could easily result in an uncontrolled blowout with little to no warning of the event.

Lost Control During Drill Stem Test

A drill-stem test is performed by setting a packer above the formation to be tested, and allowing the formation to flow. During the course of the test, the bore hole or casing below the packer, and at least a portion of the drill pipe or tubing, is filled with formation fluid. At the conclusion of the test, this fluid must be removed by proper well control techniques to return the well to a safe condition. Failure to follow the correct procedures to kill the well could lead to a blowout.

Kick Warning Signs

Deepwater Horizon drilling rig blowout, 21 April 2010

In oil well control, a kick should be able to be detected promptly, and if a kick is detected, proper kick prevention operations must be taken immediately to avoid a blowout. There are various telltale signs that signal an alert crew that a kick is about to start. Knowing these signs will keep a kicking oil well under control, and avoid a blowout:

Sudden Increase in Drilling Rate

A sudden increase in penetration rate (drilling break) is usually caused by a change in the type of formation being drilled. However, it may also signal an increase in formation pore pressure, which may indicate a possible kick.

Increase in Annulus Flow Rate

If the rate at which the pumps are running is held constant, then the flow from the annulus should be constant. If the annulus flow increases without a corresponding change in pumping rate, the additional flow is caused by formation fluid(s) feeding into the well bore or gas expansion. This will indicate an impending kick.

Gain in Pit Volume

If there is an unexplained increase in the volume of surface mud in the pit (a large tank that holds drilling fluid on the rig), it could signify an impending kick. This is because as the formation fluid feeds into the wellbore, it causes more drilling fluid to flow from the annulus than is pumped down the drill string, thus the volume of fluid in the pit(s) increases.

Change in Pump Speed/Pressure

A decrease in pump pressure or increase in pump speed can happen as a result of a decrease in hydrostatic pressure of the annulus as the formation fluids enters the wellbore. As the lighter formation fluid flows into the wellbore, the hydrostatic pressure exerted by the annular column of fluid decreases, and the drilling fluid in the drill pipe tends to U-tube into the annulus. When this occurs, the pump pressure will drop, and the pump speed will increase. The lower pump pressure and increase in pump speed symptoms can also be indicative of a hole in the drill string, commonly referred to as a washout. Until a confirmation can be made whether a washout or a well kick has occurred, a kick should be assumed.

Improper Fill on Trips

Improper fill on trip occurs when the volume of drilling fluid to keep the hole full on a Trip (complete operation of removing the drillstring from the wellbore and running it back in the hole) is less than that calculated or less than Trip Book Record. This condition is usually caused by formation fluid entering the wellbore due to the swabbing action of the drill string, and, if action is not taken soon, the well will enter a kick state.

Categories of Oil Well Control

There are basically three types of oil well control which are: primary oil well control, secondary oil well control, and tertiary oil well control. Those types are explained below.

Primary Oil Well Control

Primary oil well control is the process which maintains a hydrostatic pressure in the wellbore greater than the pressure of the fluids in the formation being drilled, but less than formation fracture pressure. It uses the mud weight to provide sufficient pressure to prevent an influx of formation fluid into the wellbore. If hydrostatic pressure is less than formation pressure, then formation fluids will enter the wellbore. If the hydrostatic pressure of the fluid in the wellbore exceeds the fracture pressure of the formation, then the fluid in the well could be lost. In an extreme case of lost circulation, the formation pressure may exceed hydrostatic pressure, allowing formation fluids to enter into the well.

Secondary Oil Well Control

Secondary oil well control is done after the Primary oil well control has failed to prevent formation fluids entering the wellbore. This process is stopped using a "blow out preventer", a BOP, to prevent the escape of wellbore fluids from the well. As the rams and choke of the BOP remain closed, a pressure built up test is carried out and a kill mud weight calculated and pumped inside the well to kill the kick and circulate it out.

Tertiary (or Shearing) Oil Well Control

Tertiary oil well control describes the third line of defense, where the formation cannot be controlled by primary or secondary well control (hydrostatic and equipment). This happens in underground blowout situations. The following are examples of tertiary well control:

- Drill a relief well to hit an adjacent well that is flowing and kill the well with heavy mud

- Rapid pumping of heavy mud to control the well with equivalent circulating density

- Pump barite or heavy weighting agents to plug the wellbore in order to stop flowing

- Pump cement to plug the wellbore

Shut-in Procedures

Using shut-in procedures is one of the oil-well-control measures to curtail kicks and prevent a blowout from occurring. Shut-in procedures are specific procedures for closing a well in case of a kick. When any positive indication of a kick is observed, such as a sudden increase in flow, or an increase in pit level, then the well should be shut-in immediately. If a well shut-in is not done promptly, a blowout is likely to happen.

Shut-in procedures are usually developed and practiced for every rig activity, such as drilling, tripping, logging, running tubular, performing a drill stem test, and so on. The primary purpose of a specific shut-in procedure is to minimize kick volume entering into a wellbore when a kick occurs, regardless of what phase of rig activity is occurring. However, a shut-in procedure is a company-specific procedure, and the policy of a company will dictate how a well should be shut-in.

They are generally two type of Shut-in procedures which are: soft shut-in, or hard shut-in.

Of these two methods, the hard shut-in is the fastest method to shut in the well; therefore, it will minimize the volume of kick allowed into the wellbore.

Well Kill Procedures

A well kill procedure is an oil well control method. Once the well has been shut-in on a kick, proper kill procedures must be done immediately. The general idea in well kill procedure is to circulate out any formation fluid already in the wellbore during kick, and then circulate a satisfactory weight of kill mud called Kill Weight Mud (KWM) into the well without allowing further fluid into the hole. If this can be done, then once the kill mud has been fully circulated around the well, it is possible to open up the well and restart normal operations. Generally, a kill mud (KWM) mix, which provides just hydrostatic balance for formation pressure, is circulated. This allows approximately constant bottom hole pressure, which is slightly greater than formation pressure to be maintained, as the kill circulation proceeds because of the additional small circulating friction pressure loss. After circulation, the well is opened up again.

The major well kill procedures used in oil well control are listed below:

- Wait and Weight

- Driller method

- Circulate and Weight

- Concurrent Method

- Reverse Circulation

- Dynamic Kill procedure

- Bullheading

- Volumetric Method

- Lubricate and Bleed

Oil Well Control Incidents - Root Causes

There will always be potential oil well control problems, as long as there are drilling operations anywhere in the world. Most of these well control problems are as a result of some errors and can be eliminated, even though some are actually unavoidable. Since we know the consequences of failed well control are severe, efforts should be made to prevent some human errors which are the root causes of these incidents. These causes include:

- Lack of knowledge and skills of rig personnel

- Improper work practices

- Lack of understanding of oil well control training

- Lack of application of policies, procedures, and standards

- Inadequate risk management

Organizations for Building Well-control Culture

Good oil-well-control culture requires personnel involved in oil well control to develop a core value for it by doing the proper thing at the proper time. A good well-control culture will definitely minimize well control incidents. Building well-control culture would involve developing competent personnel that are able to recognize well-control problems and know what to do to mitigate against them. This is usually done through quality-assurance programs and training. These programs are done by organizations such as the International Association of Drilling Contractors (IADC) or International Well Control Forum (IWCF).

IADC operates the Well-Control accreditation Program (WellCAP), which is a training program aimed at providing the necessary knowledge and practical skills critical to successful well control and to develop competent rig personnel. This training starts with floor-hand level and continues to the most-experienced drilling personnel.

IWCF is an NGO whose main aim is to develop and administer well-control certification programs for personnel employed in oil-well drilling, workover and well-intervention operations.

Blowout (Well Drilling)

The Lucas Gusher at Spindletop, Texas (1901)

A blowout is the uncontrolled release of crude oil and/or natural gas from an oil well or gas well after pressure control systems have failed. Modern wells have blowout preventers intended to prevent such an occurrence.

Prior to the advent of pressure control equipment in the 1920s, the uncontrolled release of oil and gas from a well while drilling was common and was known as an oil gusher, gusher or wild well. An accidental spark during a blowout can lead to a catastrophic oil or gas fire.

History

Gushers were an icon of oil exploration during the late 19th and early 20th centuries. During that era, the simple drilling techniques such as cable-tool drilling and the lack of blowout preventers meant that drillers could not control high-pressure reservoirs. When these high pressure zones were breached, the oil or natural gas would travel up the well at a high rate, forcing out the drill string and creating a gusher. A well which began as a gusher was said to have "blown in": for instance, the Lakeview Gusher*blew in* in 1910. These uncapped wells could produce large amounts of oil, often shooting 200 feet (60 m) or higher into the air. A blowout primarily composed of natural gas was known as a *gas gusher*.

Despite being symbols of new-found wealth, gushers were dangerous and wasteful. They killed workmen involved in drilling, destroyed equipment, and coated the landscape with thousands of barrels of oil; additionally, the explosive concussion released by the well when it pierces an oil/gas reservoir has been responsible for a number of oilmen losing their hearing entirely; standing too near to the drilling rig at the moment it drills into the oil reservoir is extremely hazardous. The impact on wildlife is very hard to quantify, but can only be estimated to be mild in the most optimistic models—realistically, the ecological impact is estimated by scientists across the ideological spectrum to be severe, profound, and lasting.

To complicate matters further, the free flowing oil was—and is—in danger of igniting. One dramatic account of a blowout and fire reads,

With a roar like a hundred express trains racing across the countryside, the well blew out, spewing oil in all directions. The derrick simply evaporated. Casings wilted like lettuce out of water, as heavy machinery writhed and twisted into grotesque shapes in the blazing inferno.

The development of rotary drilling techniques where the density of the drilling fluid is sufficient to overcome the downhole pressure of a newly penetrated zone meant that gushers became avoidable. If however the fluid density was not adequate or fluids were lost to the formation, then there was still a significant risk of a well blowout.

In 1924 the first successful blowout preventer was brought to market. The BOP valve affixed to the wellhead could be closed in the event of drilling into a high pressure zone, and the well fluids contained. Well control techniques could be used to regain control of the well. As the technology developed, blowout preventers became standard equipment, and gushers became a thing of the past.

In the modern petroleum industry, uncontrollable wells became known as blowouts and are comparatively rare. There has been significant improvement in technology, well control techniques, and personnel training which has helped to prevent their occurring. From 1976 to 1981, 21 blowout reports are available.

Notable Gushers

1. The earliest known oil gusher, in 1815, actually resulted from an attempt to drill for salt, not for oil. Joseph Eichar and his team were digging west of the town of Wooster, Ohio, along Killbuck Creek, when they struck oil. In a written retelling by Eichar's daughter, Eleanor, the strike produced "a spontaneous outburst, which shot up high as the tops of the highest trees!"

2. The Shaw Gusher in Oil Springs, Ontario, was North America's (and possibly the world's) first oil gusher when actually drilling for oil. On January 16, 1862, it shot oil from over 60 metres (200 ft) below ground to above the treetops at a rate of 3,000 barrels (480 m³) per day, triggering the oil boom in Lambton County.

3. Lucas Gusher at Spindletop in Beaumont, Texas in 1901 flowed at 100,000 barrels (16,000 m³) per day at its peak, but soon slowed and was capped within nine days. The well tripled U.S. oil production overnight and marked the start of the Texas oil industry.

4. Masjed Soleiman, Iran in 1908 marked the first major oil strike recorded in the Middle East.

5. Dos Bocas in the State of Veracruz, Mexico, was a famous 1908 Mexican blowout that formed a large crater. It leaked oil from the main reservoir for many years, continuing even after 1938 (when Pemex nationalized the Mexican oil industry).

6. Lakeview Gusher on the Midway-Sunset Oil Field in Kern County, California of 1910 is believed to be the largest-ever U.S. gusher. At its peak, more than 100,000 barrels (16,000 m³) of oil per day flowed out, reaching as high as 200 feet (60 m) in the air. It remained uncapped for 18 months, spilling over 9 million barrels (1,400,000 m³) of oil, less than half of which was recovered.

7. A short-lived gusher at Alamitos #1 in Signal Hill, California in 1921 marked the discovery of the Long Beach Oil Field, one of the most productive oil fields in the world.

8. The Barroso 2 well in Cabimas, Venezuela in December 1922 flowed at around 100,000 barrels (16,000 m³) per day for nine days, plus a large amount of natural gas.

9. Baba Gurgur near Kirkuk, Iraq, an oilfield known since antiquity, erupted at a rate of 95,000 barrels (15,100 m³) a day in 1927.

10. The Wild Mary Sudik gusher in Oklahoma City, Oklahoma in 1930 flowed at a rate of 72,000 barrels (11,400 m³) per day.

11. The Daisy Bradford gusher in 1930 marked the discovery of the East Texas Oil Field, the largest oilfield in the contiguous United States.

12. The largest known 'wildcat' oil gusher blew near Qom, Iran on August 26, 1956. The uncontrolled oil gushed to a height of 52 m (170 ft), at a rate of 120,000 barrels (19,000 m³) per day. The gusher was closed after 90 days' work by Bagher Mostofi and Myron Kinley (USA).

13. One of the most troublesome gushers happened on June 23, 1985 at well #37 at the Tengiz field in Atyrau, Kazakh SSR, Soviet Union, where the 4209-metre deep well blew out and the 200-metre high gusher self-ignited two days later. Oil pressure up to 800 atm and high hydrogen sulfide content had led to the gusher being capped only on 27 July 1986. The total volume of erupted material measured at 4.3 million metric tons of oil, 1.7 billion m³ of natural gas, and the burning gusher resulted in 890 tons of various mercaptans and more than 900,000 tons of soot released into the atmosphere.

14. The largest underwater blowout in U.S. history occurred on April 20, 2010, in the Gulf of Mexico at the Macondo Prospect oil field. The blowout caused the explosion of the Deepwater Horizon, a mobile offshore drilling platform owned by Transocean and under lease to BP at the time of the blowout. While the exact volume of oil spilled is unknown, as of June 3, 2010, the United States Geological Survey (USGS) Flow Rate Technical Group has placed the estimate at between 35,000 to 60,000 barrels (5,600 to 9,500 m³) of crude oil per day.

Cause of Blowouts

Reservoir Pressure

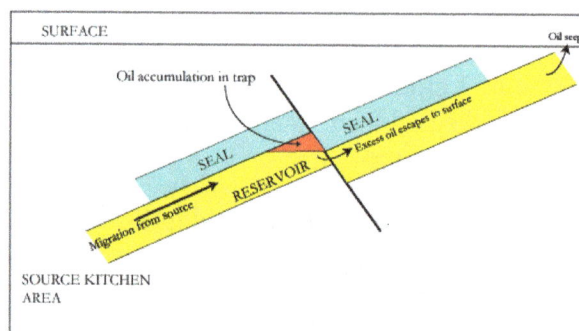

A petroleum trap. An irregularity (the *trap*) in a layer of impermeable rocks (the *seal*) retains upward-flowing petroleum, forming a reservoir.

Petroleum or crude oil is a naturally occurring, flammable liquid consisting of a complex mixture of hydrocarbons of various molecular weights, and other organic compounds, that are found in geologic formations beneath the Earth's surface. Because most hydrocarbons are lighter than rock or water, they often migrate upward and occasionally laterally through adjacent rock layers until either reaching the surface or becoming trapped within porous rocks (known as reservoirs) by impermeable rocks above. When hydrocarbons are concentrated in a trap, an oil field forms, from which the liquid can be extracted by drilling and pumping. The down hole pressure in the rock structures changes depending upon the depth and the characteristic of the source rock.Natural gas (mostly methane) may be present also, usually above the oil within the reservoir, but sometimes dissolved in the oil at reservoir pressure and temperature. This dissolved gas often evolves as free gas as the pressure is reduced either under controlled production operations or in a kick or in an uncontrolled blowout. The hydrocarbon in some reservoirs may be essentially all natural gas.

Formation Kick

The downhole fluid pressures are controlled in modern wells through the balancing of the hydrostatic pressure provided by the mud column. Should the balance of the drilling mud pressure be incorrect (i.e., the mud pressure gradient is less than the formation pore pressure gradient), then formation fluids (oil, natural gas and/or water) can begin to flow into the wellbore and up the annulus (the space between the outside of the drill string and the wall of the open hole or the inside of the casing), and/or inside the drill pipe. This is commonly called a *kick*. Ideally, mechanical barriers such as blowout preventers (BOPs) can be closed to isolate the well while the hydrostatic balance is regained through circulation of fluids in the well. But if the well is not shut in (common term for the closing of the blow-out preventer), a kick can quickly escalate into a blowout when the formation fluids reach the surface, especially when the influx contains gas that expands rapidly with the reduced pressure as it flows up the wellbore, further decreasing the effective weight of the fluid.

Early warning signs of an impending well kick while drilling are:

- Sudden change in drilling rate;

- Reduction in drillpipe weight;

- Change in pump pressure;

- Change in drilling fluid return rate.

Other warning signs during the drilling operation are:

- Returning mud "cut" by (i.e., contaminated by) gas, oil or water;

- Connection gases, high background gas units, and high bottoms-up gas units detected in the mudlogging unit.

The primary means of detecting a kick while drilling is a relative change in the circulation rate back up to the surface into the mud pits. The drilling crew or mud engineer keeps track of the level in the mud pits and/or closely monitors the rate of mud returns versus the rate that is being pumped down the drill pipe. Upon encountering a zone of higher pressure than is being exerted by the hydrostatic head of the drilling mud (including the small additional frictional head while circulating) at the bit, an increase in mud return rate would be noticed as the formation fluid influx blends in with the circulating drilling mud. Conversely, if the rate of returns is slower than expected, it means that a certain amount of the mud is being lost to a thief zone somewhere below the last casing shoe. This does not necessarily result in a kick (and may never become one); however, a drop in the mud level might allow influx of formation fluids from other zones if the hydrostatic head at is reduced to less than that of a full column of mud.

Well Control

The first response to detecting a kick would be to isolate the wellbore from the surface by activating the blow-out preventers and closing in the well. Then the drilling crew would attempt to circulate in a heavier *kill fluid* to increase the hydrostatic pressure (sometimes with the assistance of a well control company). In the process, the influx fluids will be slowly circulated out in a controlled manner, taking care not to allow any gas to accelerate up the wellbore too quickly by controlling casing pressure with chokes on a predetermined schedule.

This effect will be minor if the influx fluid is mainly salt water. And with an oil-based drilling fluid it can be masked in the early stages of controlling a kick because gas influx may dissolve into the oil under pressure at depth, only to come out of solution and expand rather rapidly as the influx nears the surface. Once all the contaminant has been circulated out, the casing pressure should have reached zero.

Capping stacks are used for controlling blowouts. The cap is an open valve that is closed after bolted on.

Types of Blowouts

Well blowouts can occur during the drilling phase, during well testing, during well completion, during production, or during workover activities.

Surface Blowouts

Blowouts can eject the drill string out of the well, and the force of the escaping fluid can be strong enough to damage the drilling rig. In addition to oil, the output of a well blowout might include natural gas, water, drilling fluid, mud, sand, rocks, and other substances.

Blowouts will often be ignited from sparks from rocks being ejected, or simply from heat generated by friction. A well control company then will need to extinguish the well fire or cap the well, and replace the casing head and other surface equipment. If the flowing gas contains poisonous hydrogen sulfide, the oil operator might decide to ignite the stream to convert this to less hazardous substances.

Sometimes blowouts can be so forceful that they cannot be directly brought under control from the surface, particularly if there is so much energy in the flowing zone that it does not deplete significantly over time. In such cases, other wells (called relief wells) may be drilled to intersect the well or pocket, in order to allow kill-weight fluids to be introduced at depth. When first drilled in the 1930s relief wells were drilled to inject water into the main drill well hole. Contrary to what might be inferred from the term, such wells generally are not used to help relieve pressure using multiple outlets from the blowout zone.

Subsea Blowouts

The two main causes of a subsea blowout are equipment failures and imbalances with encountered subsurface reservoir pressure.Subsea wells have pressure control equipment located on the seabed or between the riser pipe and drilling platform. Blowout preventers (BOPs) are the primary safety devices designed to maintain control of geologically driven well pressures. They contain hydraulic-powered cut-off mechanisms to stop the flow of hydrocarbons in the event of a loss of well control.

Even with blowout prevention equipment and processes in place, operators must be prepared to respond to a blowout should one occur. Before drilling a well, a detailed well construction design plan, an Oil Spill Response Plan as well as a Well Containment Plan must be submitted, reviewed and approved by BSEE and is contingent upon access to adequate well containment resources in accordance to NTL 2010-N10.

The Deepwater Horizon well blowout in the Gulf of Mexico in April 2010 occurred at a 5,000 feet (1,500 m) water depth. Current blowout response capabilities in the U.S. Gulf of Mexico meet capture and process rates of 130,000 barrels of fluid per day and a gas handling capacity of 220 million cubic feet per day at depths through 10,000 feet.

Underground Blowouts

An underground blowout is a special situation where fluids from high pressure zones flow uncontrolled to lower pressure zones within the wellbore. Usually this is from deeper higher pressure zones to shallower lower pressure formations. There may be no escaping fluid flow at the wellhead. However, the formation(s) receiving the influx can become overpressured, a possibility that future drilling plans in the vicinity must consider.

Blowout Control Companies

Myron M. Kinley was a pioneer in fighting oil well fires and blowouts. He developed many patents and designs for the tools and techniques of oil firefighting. His father, Karl T. Kinley, attempted to extinguish an oil well fire with the help of a massive explosion — a method that remains a common technique for fighting oil fires. The first oil well put out with explosives by Myron Kinley and his father, was in 1913. Kinley would later form the M.M. Kinley Company in 1923. Asger "Boots" Hansen and Edward Owen "Coots" Matthews also begin their careers under Kinley.

Paul N. "Red" Adair joined the M.M. Kinley Company in 1946, and worked 14 years with Myron Kinley before starting his own company, Red Adair Co., Inc., in 1959.

Red Adair Co. has helped in controlling offshore blowouts, including:

- CATCO fire in the Gulf of Mexico in 1959
- "The Devil's Cigarette Lighter" in 1962 in Gassi Touil, Algeria, in the Sahara Desert
- The Ixtoc I oil spill in Mexico's Bay of Campeche in 1979
- The Piper Alpha disaster in the North Sea in 1988
- The Kuwaiti oil fires following the Gulf War in 1991.

The 1968 American film, Hellfighters, which starred John Wayne, is about a group of oil well fire-fighters, based loosely on the life of Adair, who served as a technical advisor on the film, along with his associates, "Boots" Hansen, and "Coots" Matthews.

In 1994, Adair retired and sold his company to Global Industries. Management of Adair's company left and created International Well Control (IWC). In 1997, they would buy the company Boots & Coots International Well Control, Inc., which was founded by Hansen and Matthews in 1978.

Methods of Quenching Blowouts

Subsea Well Containment

After the Macondo-1 blowout on the Deepwater Horizon, the offshore industry collaborated with government regulators to develop a framework to respond to future subsea incidents. As a result, all energy companies operating in the deep-water U.S. Gulf of Mexico must submit an OPA 90 required Oil Spill Response Plan with the addition of a Regional Containment Demonstration Plan prior to any drilling activity. In the event of a subsea blowout, these plans are immediately activated, drawing on some of the equipment and processes effectively used to contain the Deepwater Horizon well as well as others that have been developed in its aftermath.

In order to regain control of a subsea well, the Responsible Party would first secure the safety of all personnel on board the rig and then begin a detailed evaluation of the incident site. Remotely operated underwater vehicles (ROVs) would be dispatched to inspect the condition of the wellhead, Blowout Preventer (BOP) and other subsea well equipment. The debris removal process would begin immediately to provide clear access for a capping stack.

Once lowered and latched on the wellhead, a capping stack uses stored hydraulic pressure to close a hydraulic ram and stop the flow of hydrocarbons. If shutting in the well could introduce unstable geological conditions in the wellbore, a cap and flow procedure would be used to contain hydrocarbons and safely transport them to a surface vessel.

The Responsible Party works in collaboration with BSEE and the United States Coast Guard to oversee response efforts, including source control, recovering discharged oil and mitigating environmental impact.

Several not-for-profit organizations provide a solution to effectively contain a subsea blowout. HWCG LLC and Marine Well Containment Company operate within the U.S. Gulf of Mexico waters, while cooperatives like Oil Spill Response Limited offer support for international operations.

Use of Nuclear Explosions

On Sep. 30, 1966 the Soviet Union in Urta-Bulak, an area about 80 kilometers from Bukhara, Uzbekistan, experienced blowouts on five natural gas wells. It was claimed in Komsomoloskaya Pravda that after years of burning uncontrollably they were able to stop them entirely. The Soviets lowered a specially made 30 kiloton nuclear bomb into a 6 kilometres (20,000 ft) borehole drilled 25 to 50 metres (82 to 164 ft) away from the original (rapidly leaking) well. A nuclear explosive was deemed necessary because conventional explosive both lacked the necessary power and would also require a great deal more space underground. When the bomb was set off, it proceeded to crush the original pipe that was carrying the gas from the deep reservoir to the surface, as well as to glassify all the surrounding rock. This caused the leak and fire at the surface to cease within approximately one minute of the explosion, and proved over the years to have been a permanent solution. A second attempt on a similar well was not as successful and other tests were for such experiments as oil extraction enhancement (Stavropol, 1969) and the creation of gas storage reservoirs (Orenburg, 1970).

Notable Offshore Well Blowouts

Data from industry information.

Year	Rig Name	Rig Owner	Type	Damage / details
1955	S-44	Chevron Corporation	Sub Recessed pontoons	Blowout and fire. Returned to service.
1959	C. T. Thornton	Reading & Bates	Jackup	Blowout and fire damage.
1964	C. P. Baker	Reading & Bates	Drill barge	Blowout in Gulf of Mexico, vessel capsized, 22 killed.
1965	Trion	Royal Dutch Shell	Jackup	Destroyed by blowout.
1965	Paguro	SNAM	Jackup	Destroyed by blowout and fire.
1968	Little Bob	Coral	Jackup	Blowout and fire, killed 7.
1969	Wodeco III	Floor drilling	Drilling barge	Blowout
1969	Sedco 135G	Sedco Inc	Semi-submersible	Blowout damage
1969	Rimrick Tidelands	ODECO	Submersible	Blowout in Gulf of Mexico
1970	Stormdrill III	Storm Drilling	Jackup	Blowout and fire damage.
1970	Discoverer III	Offshore Co.	Drillship	Blowout (S. China Seas)
1971	Big John	Atwood Oceanics	Drill barge	Blowout and fire.
1971	Wodeco II	Floor Drilling	Drill barge	Blowout and fire off Peru, 7 killed.
1972	J. Storm II	Marine Drilling Co.	Jackup	Blowout in Gulf of Mexico
1972	M. G. Hulme	Reading & Bates	Jackup	Blowout and capsize in Java Sea.
1972	Rig 20	Transworld Drilling	Jackup	Blowout in Gulf of Martaban.
1973	Mariner I	Sante Fe Drilling	Semi-sub	Blowout off Trinidad, 3 killed.
1975	Mariner II	Sante Fe Drilling	Semi-submersible	Lost BOP during blowout.
1975	J. Storm II	Marine Drilling Co.	Jackup	Blowout in Gulf of Mexico.
1976	Petrobras III	Petrobras	Jackup	No info.
1976	W. D. Kent	Reading & Bates	Jackup	Damage while drilling relief well.
1977	Maersk Explorer	Maersk Drilling	Jackup	Blowout and fire in North Sea
1977	Ekofisk Bravo	Phillips Petroleum	Platform	Blowout during well workover.
1978	Scan Bay	Scan Drilling	Jackup	Blowout and fire in the Persion Gulf.
1979	Salenergy II	Salen Offshore	Jackup	Blowout in Gulf of Mexico
1979	Sedco 135F	Sedco Drilling	Semi-submersible	Blowout and fire in Bay of Campeche Ixtoc I well.
1980	Sedco 135G	Sedco Drilling	Semi-submersible	Blowout and fire of Nigeria.

1980	Discoverer 534	Offshore Co.	Drillship	Gas escape caught fire.
1980	Ron Tappmeyer	Reading & Bates	Jackup	Blowout in Persian Gulf, 5 killed.
1980	Nanhai II	Peoples Republic of China	Jackup	Blowout of Hainan Island.
1980	Maersk Endurer	Maersk Drilling	Jackup	Blowout in Red Sea, 2 killed.
1980	Ocean King	ODECO	Jackup	Blowout and fire in Gulf of Mexico, 5 killed.
1980	Marlin 14	Marlin Drilling	Jackup	Blowout in Gulf of Mexico
1981	Penrod 50	Penrod Drilling	Submersible	Blowout and fire in Gulf of Mexico.
1985	West Vanguard	Smedvig	Semi-submersible	Shallow gas blowout and fire in Norwegian sea, 1 fatality.
1981	Petromar V	Petromar	Drillship	Gas blowout and capsize in S. China seas.
1983	Bull Run	Atwood Oceanics	Tender	Oil and gas blowout Dubai, 3 fatalities.
1988	Ocean Odyssey	Diamond Offshore Drilling	Semi-submersible	Gas blowout at BOP and fire in the UK North Sea, 1 killed.
1988	PCE-1	Petrobras	Jackup	Blowout at Petrobras PCE-1 (Brazil) in April 24. Fire burned for 31 days. No fatalities.
1989	Al Baz	Sante Fe	Jackup	Shallow gas blowout and fire in Nigeria, 5 killed.
1993	M. Naqib Khalid	Naqib Co.	Naqib Drilling	fire and explosion. Returned to service.
1993	Actinia	Transocean	Semi-submersible	Sub-sea blowout in Vietnam. .
2001	Ensco 51	Ensco	Jackup	Gas blowout and fire, Gulf of Mexico, no casualties
2002	Arabdrill 19	Arabian Drilling Co.	Jackup	Structural collapse, blowout, fire and sinking.
2004	Adriatic IV	Global Sante Fe	Jackup	Blowout and fire at Temsah platform, Mediterranean Sea
2007	Usumacinta	PEMEX	Jackup	Storm forced rig to move, causing well blowout on Kab 101 platform, 22 killed.
2009	West Atlas / Montara	Seadrill	Jackup / Platform	Blowout and fire on rig and platform in Australia.
2010	Deepwater Horizon	Transocean	Semi-submersible	Blowout and fire on the rig, subsea well blowout, killed 11 in explosion.
2010	Vermilion Block 380	Mariner Energy	Platform	Blowout and fire, 13 survivors, 1 injured.
2012	KS Endeavour	KS Energy Services	Jack-Up	Blowout and fire on the rig, collapsed, killed 2 in explosion.

References

- Forbes, R.J. (1970). A Short History of the Art of Distillation from the Beginnings Up to the Death of Cellier Blumenthal. Brill Publishers. pp. 41–42. ISBN 978-90-04-00617-1. Retrieved 2009-06-02.

- Cane, R.F. (1976). "The origin and formation of oil shale". In Teh Fu Yen; Chilingar, George V. Oil Shale. Amsterdam: Elsevier. p. 56. ISBN 978-0-444-41408-3. Retrieved 2009-06-05.

- Dyni, John R. (2010). "Oil Shale". In Clarke, Alan W.; Trinnaman, Judy A. Survey of energy resources (PDF) (22 ed.). World Energy Council. pp. 93–123. ISBN 978-0-946121-02-1.

- Prien, Charles H. (1976). "Survey of oil-shale research in last three decades". In Teh Fu Yen; Chilingar, George V. Oil Shale. Amsterdam: Elsevier. pp. 237–243. ISBN 978-0-444-41408-3. Retrieved 2009-06-05.

- Speight, James G. (2008). Synthetic Fuels Handbook: Properties, Process, and Performance. McGraw-Hill. pp.

13; 182; 186. ISBN 978-0-07-149023-8. Retrieved 2009-03-14.

- Lee, Sunggyu; Speight, James G.; Loyalka, Sudarshan K. (2007). Handbook of Alternative Fuel Technologies. CRC Press. p. 290. ISBN 978-0-8247-4069-6. Retrieved 2009-03-14.

- Plunkett, Jack W. (2008). Plunkett's Energy Industry Almanac 2009: The Only Comprehensive Guide to the Energy & Utilities Industry. Plunkett Research, Ltd. p. 71. ISBN 978-1-59392-128-6. Retrieved 2009-03-14.

- Environmental consequences of, and control processes for, energy technologies. Argonne National Laboratory. William Andrew Inc. 1990. p. 104. ISBN 978-0-8155-1231-8. Retrieved 2008-08-19.

- Mills, Robin M. (2008). The myth of the oil crisis: overcoming the challenges of depletion, geopolitics, and global warming. Greenwood Publishing Group. pp. 158–159. ISBN 978-0-313-36498-3.

- Deffeyes, Kenneth S (2002). Hubbert's Peak: The Impending World Oil Shortage. Princeton University Press. ISBN 0-691-09086-6.

- Madureira, Nuno Luis (2014). Key Concepts in Energy. London: Springer International Publishing. pp. 125–6. doi:10.1007/978-3-319-04978-6_6. ISBN 978-3-319-04977-9.

- P. Crabbè, North Atlantic Treaty Organization. Scientific Affairs Division (2000). "Implementing ecological integrity: restoring regional and global environmental and human health". Springer. p.411. ISBN 0-7923-6351-5

- Kunstler, James Howard (1994). Geography of Nowhere: The Rise And Decline of America's Man-Made Landscape. New York: Simon & Schuster. ISBN 0-671-88825-0

- Micheal Nelkon & Philip Parker, Advanced Level Physics, 7th Edition, New Delhi, India, CBS Publishers, 1995, pp. 103–105, ISBN 81-239-0400-2

- "Short - Term Energy Outlook March 2016 1 March 2016 Short - Term Energy Outlook (STEO)" (PDF). U.S. Energy Information Administration. Retrieved 12 March 2016.

- Baumeister and Kilian. "Understanding the Decline in the Price of Oil since June 2014". Social Science Research Network. CFS Working Paper No. 501. Retrieved 12 March 2016.

- Tokic, Damir (October 2015). "The 2014 oil bust: Causes and consequences". Energy Policy. 85: 162–169. doi:10.1016/j.enpol.2015.06.005. Retrieved 12 March 2016.

- Caruana, Jaime (February 5, 2016). "Credit, commodities and currencies" (PDF). Bank of International Settlements. Retrieved 12 March 2016.

- Tverberg, Gail. "What's Ahead? Lower Oil Prices, Despite Higher Extraction Costs". Our Finite World. Retrieved 10 January 2016.

- Etherington, John; et al. "Comparison of Selected Reserves and Resource Classifications and Associated Definitions" (PDF). Society of Petroleum Engineers. Retrieved 26 September 2016.

- "the Guinness Book of Records 1990 - Norris Mcwhirter, Donald McFarlan - Google Books". Books.google.com. 1994-04-01. Retrieved 2016-01-30.

Hydraulic Fracturing: An Overview

Hydraulic fracturing is a technique of fracturing a rock by using pressuring liquid. The topics explained in this chapter are waterless fracturing, hydraulic fracturing proppants, regulation of hydraulic fracturing, the uses of radioactivity in oil and gas wells etc. The section on hydraulic fracturing is an overview of the subject matter incorporating all the major aspects of hydraulic fracturing.

Hydraulic Fracturing

Hydraulic fracturing (also fraccing, fracking, hydrofracturing or hydrofracking) is a well stimulation technique in which rock is fractured by a pressurized liquid. The process involves the high-pressure injection of 'fracking fluid' (primarily water, containing sand or other proppants suspended with the aid of thickening agents) into a wellbore to create cracks in the deep-rock formations through which natural gas, petroleum, and brine will flow more freely. When the hydraulic pressure is removed from the well, small grains of hydraulic fracturing proppants (either sand or aluminium oxide) hold the fractures open.

Hydraulic fracturing began as an experiment in 1947, and the first commercially successful application followed in 1950. As of 2012, 2.5 million "frac jobs" had been performed worldwide on oil and gas wells; over one million of those within the U.S. Such treatment is generally necessary to achieve adequate flow rates in shale gas, tight gas, tight oil, and coal seam gas wells. Some hydraulic fractures can form naturally in certain veins or dikes.

Hydraulic fracturing is highly controversial in many countries. Its proponents advocate the economic benefits of more extensively accessible hydrocarbons. However, opponents argue that these are outweighed by the potential environmental impacts, which include risks of ground and surface water contamination, air and noise pollution, and the triggering of earthquakes, along with the consequential hazards to public health and the environment.

Increases in seismic activity following hydraulic fracturing along dormant or previously unknown faults are sometimes caused by the deep-injection disposal of hydraulic fracturing flowback (a byproduct of hydraulically fractured wells), and produced formation brine (a byproduct of both fractured and nonfractured oil and gas wells). For these reasons, hydraulic fracturing is under international scrutiny, restricted in some countries, and banned altogether in others. Some countries have banned the practice or put moratoria in place, while others have adopted an approach involving tight regulation. The European Union is drafting regulations that would permit the controlled application of hydraulic fracturing.

Geology

Halliburton fracturing operation in the Bakken Formation, North Dakota, United States

A fracturing operation in progress

Mechanics

Fracturing rocks at great depth frequently becomes suppressed by pressure due to the weight of the overlying rock strata and the cementation of the formation. This suppression process is particularly significant in "tensile" (Mode 1) fractures which require the walls of the fracture to move against this pressure. Fracturing occurs when effective stress is overcome by the pressure of fluids within the rock. The minimum principal stress becomes tensile and exceeds the tensile strength of the material. Fractures formed in this way are generally oriented in a plane perpendicular to the minimum principal stress, and for this reason, hydraulic fractures in well bores can be used to determine the orientation of stresses. In natural examples, such as dikes or vein-filled fractures, the orientations can be used to infer past states of stress.

Veins

Most mineral vein systems are a result of repeated natural fracturing during periods of relatively high pore fluid pressure. The impact of high pore fluid pressure on the formation process of mineral vein systems is particularly evident in "crack-seal" veins, where the vein material is part of a series of discrete fracturing events, and extra vein material is deposited on each occasion. One example of long-term repeated natural fracturing is in the effects of seismic activity. Stress levels rise and fall episodically, and earthquakes can cause large volumes of connate water to be expelled from fluid-filled fractures. This process is referred to as "seismic pumping".

Dikes

Minor intrusions in the upper part of the crust, such as dikes, propagate in the form of fluid-filled cracks. In such cases, the fluid is magma. In sedimentary rocks with a significant water content, fluid at fracture tip will be steam.

History

Precursors

Fracturing as a method to stimulate shallow, hard rock oil wells dates back to the 1860s. Dynamite or nitroglycerin detonations were used to increase oil and natural gas production from petroleum bearing formations. On 25 April 1865, US Civil War veteran Col. Edward A. L. Roberts received a patent for an "exploding torpedo". It was employed in Pennsylvania, New York, Kentucky, and West Virginia using liquid and also, later, solidified nitroglycerin. Later still the same method was applied to water and gas wells. Stimulation of wells with acid, instead of explosive fluids, was introduced in the 1930s. Due to acid etching, fractures would not close completely resulting in further productivity increase.

Oil and Gas Wells

The relationship between well performance and treatment pressures was studied by Floyd Farris of Stanolind Oil and Gas Corporation. This study was the basis of the first hydraulic fracturing experiment, conducted in 1947 at the Hugoton gas field in Grant County of southwestern Kansas by Stanolind. For the well treatment, 1,000 US gallons (3,800 l; 830 imp gal) of gelled gasoline (essentially napalm) and sand from the Arkansas River was injected into the gas-producing limestone formation at 2,400 feet (730 m). The experiment was not very successful as deliverability of the well did not change appreciably. The process was further described by J.B. Clark of Stanolind in his paper published in 1948. A patent on this process was issued in 1949 and exclusive license was granted to the Halliburton Oil Well Cementing Company. On 17 March 1949, Halliburton performed the first two commercial hydraulic fracturing treatments in Stephens County, Oklahoma, and Archer County, Texas. Since then, hydraulic fracturing has been used to stimulate approximately one million oil and gas wells in various geologic regimes with good success.

In contrast with large-scale hydraulic fracturing used in low-permeability formations, small hydraulic fracturing treatments are commonly used in high-permeability formations to remedy "skin damage", a low-permeability zone that sometimes forms at the rock-borehole interface. In such cases the fracturing may extend only a few feet from the borehole.

In the Soviet Union, the first hydraulic proppant fracturing was carried out in 1952. Other countries in Europe and Northern Africa subsequently employed hydraulic fracturing techniques including Norway, Poland, Czechoslovakia, Yugoslavia, Hungary, Austria, France, Italy, Bulgaria, Romania, Turkey, Tunisia, and Algeria.

Massive Fracturing

Well head where fluids are injected into the ground

Massive hydraulic fracturing (also known as high-volume hydraulic fracturing) is a technique first applied by Pan American Petroleum in Stephens County, Oklahoma, USA in 1968. The definition of massive hydraulic fracturing varies, but generally refers to treatments injecting over 150 short tons, or approximately 300,000 pounds (136 metric tonnes), of proppant.

Well head after all the hydraulic fracturing equipment has been taken off location

American geologists gradually became aware that there were huge volumes of gas-saturated sandstones with permeability too low (generally less than 0.1 millidarcy) to recover the gas economically. Starting in 1973, massive hydraulic fracturing was used in thousands of gas wells in the San Juan Basin, Denver Basin, the Piceance Basin, and the Green River Basin, and in other hard rock formations of the western US. Other tight sandstone wells in the US made economically viable by massive hydraulic fracturing were in the Clinton-Medina Sandstone (Ohio, Pennsylvania, and New York), and Cotton Valley Sandstone (Texas and Louisiana).

Massive hydraulic fracturing quickly spread in the late 1970s to western Canada, Rotliegend and Carboniferous gas-bearing sandstones in Germany, Netherlands (onshore and offshore gas fields), and the United Kingdom in the North Sea.

Horizontal oil or gas wells were unusual until the late 1980s. Then, operators in Texas began completing thousands of oil wells by drilling horizontally in the Austin Chalk, and giving massive *slickwater* hydraulic fracturing treatments to the wellbores. Horizontal wells proved much more effective than vertical wells in producing oil from tight chalk; sedimentary beds are usually nearly horizontal, so horizontal wells have much larger contact areas with the target formation.

Shales

Hydraulic fracturing of shales goes back at least to 1965, when some operators in the Big Sandy gas field of eastern Kentucky and southern West Virginia started hydraulically fracturing the Ohio Shale and Cleveland Shale, using relatively small fracs. The frac jobs generally increased production, especially from lower-yielding wells.

In 1976, the United States government started the Eastern Gas Shales Project, which included numerous public-private hydraulic fracturing demonstration projects. During the same period, the Gas Research Institute, a gas industry research consortium, received approval for research and funding from the Federal Energy Regulatory Commission.

In 1997, Nick Steinsberger, an engineer of Mitchell Energy (now part of Devon Energy), applied the slickwater fracturing technique, using more water and higher pump pressure than previous fracturing techniques, which was used in East Texas by Union Pacific Resources (now part of Anadarko Petroleum Corporation), in the Barnett Shale of north Texas. In 1998, the new technique proved to be successful when the first 90 days gas production from the well called S.H. Griffin No. 3 exceeded production of any of the company's previous wells. This new completion technique made gas extraction widely economical in the Barnett Shale, and was later applied to other shales. George P. Mitchell has been called the "father of fracking" because of his role in applying it in shales. The first horizontal well in the Barnett Shale was drilled in 1991, but was not widely done in the Barnett until it was demonstrated that gas could be economically extracted from vertical wells in the Barnett.

As of 2013, massive hydraulic fracturing is being applied on a commercial scale to shales in the United States, Canada, and China. Several additional countries are planning to use hydraulic fracturing.

Process

According to the United States Environmental Protection Agency (EPA), hydraulic fracturing is a process to stimulate a natural gas, oil, or geothermal well to maximize extraction. The EPA defines the broader process to include acquisition of source water, well construction, well stimulation, and waste disposal.

Method

A hydraulic fracture is formed by pumping fracturing fluid into a wellbore at a rate sufficient to increase pressure at the target depth (determined by the location of the well casing perforations), to exceed that of the fracture *gradient* (pressure gradient) of the rock. The fracture gradient is defined as pressure increase per unit of depth relative to density, and is usually measured in pounds per square inch, per square foot, or bars. The rock cracks, and the fracture fluid permeates the rock extending the crack further, and further, and so on. Fractures are localized as pressure drops off with the rate of frictional loss, which is relevant to the distance from the well. Operators typically try to maintain "fracture width", or slow its decline following treatment, by introducing a proppant into the injected fluid – a material such as grains of sand, ceramic, or other particulate, thus preventing the fractures from closing when injection is stopped and pressure removed. Consideration of proppant strength and prevention of proppant failure becomes more important at greater depths where pressure and stresses on fractures are higher. The propped fracture is permeable enough to allow the flow of gas, oil, salt water and hydraulic fracturing fluids to the well.

During the process, fracturing fluid leakoff (loss of fracturing fluid from the fracture channel into the surrounding permeable rock) occurs. If not controlled, it can exceed 70% of the injected volume. This may result in formation matrix damage, adverse formation fluid interaction, and altered fracture geometry, thereby decreasing efficiency.

The location of one or more fractures along the length of the borehole is strictly controlled by various methods that create or seal holes in the side of the wellbore. Hydraulic fracturing is performed in cased wellbores, and the zones to be fractured are accessed by perforating the casing at those locations.

Hydraulic-fracturing equipment used in oil and natural gas fields usually consists of a slurry blender, one or more high-pressure, high-volume fracturing pumps (typically powerful triplex or quintuplex pumps) and a monitoring unit. Associated equipment includes fracturing tanks, one or more units for storage and handling of proppant, high-pressure treating iron, a chemical additive unit (used to accurately monitor chemical addition), low-pressure flexible hoses, and many gauges and meters for flow rate, fluid density, and treating pressure. Chemical additives are typically 0.5% percent of the total fluid volume. Fracturing equipment operates over a range of pressures and injection rates, and can reach up to 100 megapascals (15,000 psi) and 265 litres per second (9.4 cu ft/s) (100 barrels per minute).

Well Types

A distinction can be made between conventional, low-volume hydraulic fracturing, used to stimulate high-permeability reservoirs for a single well, and unconventional, high-volume hydraulic fracturing, used in the completion of tight gas and shale gas wells. High-volume hydraulic fracturing usually requires higher pressures than low-volume fracturing; the higher pressures are needed to push out larger volumes of fluid and proppant that extend farther from the borehole.

Horizontal drilling involves wellbores with a terminal drillhole completed as a "lateral" that extends parallel with the rock layer containing the substance to be extracted. For example, laterals extend 1,500 to 5,000 feet (460 to 1,520 m) in the Barnett Shale basin in Texas, and up to 10,000 feet (3,000 m) in the Bakken formation in North Dakota. In contrast, a vertical well only accesses the thickness of the rock layer, typically 50–300 feet (15–91 m). Horizontal drilling reduces surface disruptions as fewer wells are required to access the same volume of rock.

Drilling often plugs up the pore spaces at the wellbore wall, reducing permeability at and near the wellbore. This reduces flow into the borehole from the surrounding rock formation, and partially seals off the borehole from the surrounding rock. Low-volume hydraulic fracturing can be used to restore permeability.

Fracturing Fluids

Water tanks preparing for hydraulic fracturing

The main purposes of fracturing fluid are to extend fractures, add lubrication, change gel strength, and to carry proppant into the formation. There are two methods of transporting proppant in the fluid – high-rate and high-viscosity. High-viscosity fracturing tends to cause large dominant fractures, while high-rate (slickwater) fracturing causes small spread-out micro-fractures.

Water-soluble gelling agents (such as guar gum) increase viscosity and efficiently deliver proppant into the formation.

Process of mixing water with hydraulic fracturing fluids to be injected into the ground

Fluid is typically a slurry of water, proppant, and chemical additives. Additionally, gels, foams, and compressed gases, including nitrogen, carbon dioxide and air can be injected. Typically, 90% of the fluid is water and 9.5% is sand with chemical additives accounting to about 0.5%. However, fracturing fluids have been developed using liquefied petroleum gas (LPG) and propane in which water is unnecessary.

The proppant is a granular material that prevents the created fractures from closing after the fracturing treatment. Types of proppant include silica sand, resin-coated sand, bauxite, and man-made ceramics. The choice of proppant depends on the type of permeability or grain strength needed. In some formations, where the pressure is great enough to crush grains of natural silica sand, higher-strength proppants such as bauxite or ceramics may be used. The most commonly used proppant is silica sand, though proppants of uniform size and shape, such as a ceramic proppant, are believed to be more effective.

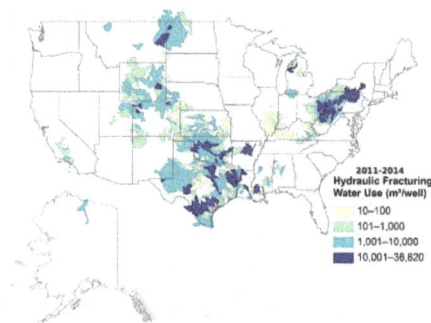

USGS map of water use from hydraulic fracturing between 2011 and 2014. One cubic meter of water is 264.172 gallons.

The fracturing fluid varies depending on fracturing type desired, and the conditions of specific wells being fractured, and water characteristics. The fluid can be gel, foam, or slickwater-based. Fluid choices are tradeoffs: more viscous fluids, such as gels, are better at keeping proppant in suspension; while less-viscous and lower-friction fluids, such as slickwater, allow fluid to be pumped at higher rates, to create fractures farther out from the wellbore. Important material properties of the fluid include viscosity, pH, various rheological factors, and others.

Water is mixed with sand and chemicals to create fracking fluid. Approximately 40,000 gallons of chemicals are used per fracturing. A typical fracture treatment uses between 3 and 12 additive chemicals. Although there may be unconventional fracturing fluids, typical chemical additives can include one or more of the following:

- Acids—hydrochloric acid or acetic acid is used in the pre-fracturing stage for cleaning the perforations and initiating fissure in the near-wellbore rock.

- Sodium chloride (salt)—delays breakdown of gel polymer chains.

- Polyacrylamide and other friction reducers decrease turbulence in fluid flow and pipe friction, thus allowing the pumps to pump at a higher rate without having greater pressure on the surface.

- Ethylene glycol—prevents formation of scale deposits in the pipe.

- Borate salts—used for maintaining fluid viscosity during the temperature increase.

- Sodium and potassium carbonates—used for maintaining effectiveness of crosslinkers.

- Anaerobic, Biocide, BIO—Glutaraldehyde used as disinfectant of the water (bacteria elimination).

- Guar gum and other water-soluble gelling agents—increases viscosity of the fracturing fluid to deliver proppant into the formation more efficiently.

- Citric acid—used for corrosion prevention.

- Isopropanol—used to winterize the chemicals to ensure it doesn't freeze.

The most common chemical used for hydraulic fracturing in the United States in 2005–2009 was methanol, while some other most widely used chemicals were isopropyl alcohol, 2-butoxyethanol, and ethylene glycol.

Typical fluid types are:

- Conventional linear gels. These gels are cellulose derivative (carboxymethyl cellulose, hydroxyethyl cellulose, carboxymethyl hydroxyethyl cellulose, hydroxypropyl cellulose, hydroxyethyl methyl cellulose), guar or its derivatives (hydroxypropyl guar, carboxymethyl hydroxypropyl guar), mixed with other chemicals.

- Borate-crosslinked fluids. These are guar-based fluids cross-linked with boron ions (from aqueous borax/boric acid solution). These gels have higher viscosity at pH 9 onwards and are used to carry proppant. After the fracturing job, the pH is reduced to 3–4 so that the cross-links are broken, and the gel is less viscous and can be pumped out.

- Organometallic-crosslinked fluids - zirconium, chromium, antimony, titanium salts - are known to crosslink guar-based gels. The crosslinking mechanism is not reversible, so once the proppant is pumped down along with cross-linked gel, the fracturing part is done. The gels are broken down with appropriate breakers.

- Aluminium phosphate-ester oil gels. Aluminium phosphate and ester oils are slurried to form cross-linked gel. These are one of the first known gelling systems.

For slickwater fluids the use of sweeps is common. Sweeps are temporary reductions in the proppant concentration, which help ensure that the well is not overwhelmed with proppant. As the fracturing process proceeds, viscosity-reducing agents such as oxidizers and enzyme breakers are sometimes added to the fracturing fluid to deactivate the gelling agents and encourage flowback. Such oxidizer react with and break down the gel, reducing the fluid's viscosity and ensuring that no proppant is pulled from the formation. An enzyme acts as a catalyst for breaking down the gel. Sometimes pH modifiers are used to break down the crosslink at the end of a hydraulic fracturing job, since many require a pH buffer system to stay viscous. At the end of the job, the well is commonly flushed with water under pressure (sometimes blended with a friction reducing chemical.) Some (but not all) injected fluid is recovered. This fluid is managed by several methods, including underground injection control, treatment, discharge, recycling, and temporary storage in pits or containers. New technology is continually developing to better handle waste water and improve re-usability.

Fracture Monitoring

Measurements of the pressure and rate during the growth of a hydraulic fracture, with knowledge of fluid properties and proppant being injected into the well, provides the most common and simplest method of monitoring a hydraulic fracture treatment. This data along with knowledge of the underground geology can be used to model information such as length, width and conductivity of a propped fracture.

Injection of radioactive tracers along with the fracturing fluid is sometimes used to determine the injection profile and location of created fractures.Radiotracers are selected to have the readily detectable radiation, appropriate chemical properties, and a half life and toxicity level that will minimize initial and residual contamination. Radioactive isotopes chemically bonded to glass (sand) and/or resin beads may also be injected to track fractures. For example, plastic pellets coated with 10 GBq of Ag-110mm may be added to the proppant, or sand may be labelled with Ir-192, so that the proppant's progress can be monitored. Radiotracers such as Tc-99m and I-131 are also used to measure flow rates. The Nuclear Regulatory Commission publishes guidelines which list a wide range of radioactive materials in solid, liquid and gaseous forms that may be used as tracers and limit the amount that may be used per injection and per well of each radionuclide.

A new technique in well-monitoring involves fiber-optic cables outside the casing. Using the fiber optics, temperatures can be measured every foot along the well - even while the wells are being fracked and pumped. By monitoring the temperature of the well, engineers can determine how much fracking fluid different parts of the well use as well as how much natural gas they collect.

Microseismic Monitoring

For more advanced applications, microseismic monitoring is sometimes used to estimate the size and orientation of induced fractures. Microseismic activity is measured by placing an array of geophones in a nearby wellbore. By mapping the location of any small seismic events associated with the growing fracture, the approximate geometry of the fracture is inferred. Tiltmeter arrays deployed on the surface or down a well provide another technology for monitoring strain

Microseismic mapping is very similar geophysically to seismology. In earthquake seismology, seismometers scattered on or near the surface of the earth record S-waves and P-waves that are released during an earthquake event. This allows for motion along the fault plane to be estimated and its location in the earth's subsurface mapped. Hydraulic fracturing, an increase in formation stress proportional to the net fracturing pressure, as well as an increase in pore pressure due to leakoff. Tensile stresses are generated ahead of the fracture's tip, generating large amounts of shear stress. The increases in pore water pressure and in formation stress combine and affect weaknesses near the hydraulic fracture, like natural fractures, joints, and bedding planes.

Different methods have different location errors and advantages. Accuracy of microseismic event mapping is dependent on the signal-to-noise ratio and the distribution of sensors. Accuracy of events located by seismic inversion is improved by sensors placed in multiple azimuths from the monitored borehole. In a downhole array location, accuracy of events is improved by being close to the monitored borehole (high signal-to-noise ratio).

Monitoring of microseismic events induced by reservoir stimulation has become a key aspect in evaluation of hydraulic fractures, and their optimization. The main goal of hydraulic fracture monitoring is to completely characterize the induced fracture structure, and distribution of conductivity within a formation. Geomechanical analysis, such as understanding a formations material properties, in-situ conditions, and geometries, helps monitoring by providing a better definition of the environment in which the fracture network propagates. The next task is to know the location of proppant within the fracture and the distribution of fracture conductivity. This can be monitored using multiple types of techniques to finally develop a reservoir model than accurately predicts well performance.

Horizontal Completions

Since the early 2000s, advances in drilling and completion technology have made horizontal wellbores much more economical. Horizontal wellbores allow far greater exposure to a formation than conventional vertical wellbores. This is particularly useful in shale formations which do not have sufficient permeability to produce economically with a vertical well. Such wells, when drilled onshore, are now usually hydraulically fractured in a number of stages, especially in North America. The type of wellbore completion is used to determine how many times a formation is fractured, and at what locations along the horizontal section.

In North America, shale reservoirs such as the Bakken, Barnett, Montney, Haynesville, Marcellus, and most recently the Eagle Ford, Niobrara and Utica shales are drilled horizontally through the producing interval(s), completed and fractured. The method by which the fractures are placed along the wellbore is most commonly achieved by one of two methods, known as "plug and perf" and "sliding sleeve".

The wellbore for a plug-and-perf job is generally composed of standard steel casing, cemented or uncemented, set in the drilled hole. Once the drilling rig has been removed, a wireline truck is used to perforate near the bottom of the well, and then fracturing fluid is pumped. Then the wireline truck sets a plug in the well to temporarily seal off that section so the next section of the wellbore can be treated. Another stage is pumped, and the process is repeated along the horizontal length of the wellbore.

The wellbore for the sliding sleeve technique is different in that the sliding sleeves are included at set spacings in the steel casing at the time it is set in place. The sliding sleeves are usually all closed at this time. When the well is due to be fractured, the bottom sliding sleeve is opened using one of several activation techniques and the first stage gets pumped. Once finished, the next sleeve is opened, concurrently isolating the previous stage, and the process repeats. For the sliding sleeve method, wireline is usually not required.

Sleeves

These completion techniques may allow for more than 30 stages to be pumped into the horizontal section of a single well if required, which is far more than would typically be pumped into a vertical well that had far fewer feet of producing zone exposed.

Uses

Hydraulic fracturing is used to increase the rate at which fluids, such as petroleum, water, or natural gas can be recovered from subterranean natural reservoirs. Reservoirs are typically porous sandstones, limestones or dolomite rocks, but also include "unconventional reservoirs" such as shale rock or coal beds. Hydraulic fracturing enables the extraction of natural gas and oil from rock formations deep below the earth's surface (generally 2,000–6,000 m (5,000–20,000 ft)), which is greatly below typical groundwater reservoir levels. At such depth, there may be insufficient permeability or reservoir pressure to allow natural gas and oil to flow from the rock into the wellbore at high economic return. Thus, creating conductive fractures in the rock is instrumental in extraction from naturally impermeable shale reservoirs. Permeability is measured in the microdarcy to nanodarcy range. Fractures are a conductive path connecting a larger volume of reservoir to the well. So-called "super fracking," creates cracks deeper in the rock formation to release more oil and gas, and increases efficiency. The yield for typical shale bores generally falls off after the first year or two, but the peak producing life of a well can be extended to several decades.

While the main industrial use of hydraulic fracturing is in stimulating production from oil and gas wells, hydraulic fracturing is also applied:

- To stimulate groundwater wells

- To precondition or induce rock cave-ins mining

- As a means of enhancing waste remediation, usually hydrocarbon waste or spills

- To dispose waste by injection deep into rock

- To measure stress in the Earth

- For electricity generation in enhanced geothermal systems

- To increase injection rates for geologic sequestration of CO_2

Since the late 1970s, hydraulic fracturing has been used, in some cases, to increase the yield of drinking water from wells in a number of countries, including the United States, Australia, and South Africa.

Economic Effects

Hydraulic fracturing has been seen as one of the key methods of extracting unconventional oil and unconventional gas resources. According to the International Energy Agency, the remaining technically recoverable resources of shale gas are estimated to amount to 208 trillion cubic metres (7,300 trillion cubic feet), tight gas to 76 trillion cubic metres (2,700 trillion cubic feet), and coal-bed methane to 47 trillion cubic metres (1,700 trillion cubic feet). As a rule, formations of these resources have lower permeability than conventional gas formations. Therefore, depending on the geological characteristics of the formation, specific technologies such as hydraulic fracturing are required. Although there are also other methods to extract these resources, such as conventional drilling or horizontal drilling, hydraulic fracturing is one of the key methods making their extraction economically viable. The multi-stage fracturing technique has facilitated the development of shale gas and light tight oil production in the United States and is believed to do so in the other countries with unconventional hydrocarbon resources.

Some studies call into question the claim that hydraulic fracturing of shale gas wells has a significant macro-economic impact. A study released in the beginning of 2014 by the Institute for Sustainable Development and International Relations states that, on the long-term as well as on the short-run, the "shale gas revolution" due to hydraulic fracturing in the United States has had very little impact on economic growth and competitiveness. The same report concludes that in Europe, using hydraulic fracturing would have very little advantage in terms of competitiveness and energy security. Indeed, for the period 2030-2035, shale gas is estimated to cover 3 to 10% of EU projected energy demand, which is not enough to have a significant impact on energetic independence and competitiveness.

Research suggests that hydraulic fracturing wells have an adverse impact on agricultural productivity in the vicinity of the wells. One paper found "that productivity of an irrigated crop decreases by 5% when a well is drilled during the agriculturally active months within 11-20 km radius of a producing township. This effect becomes smaller and weaker as the distance between township and wells increases." The findings imply that the introduction of fracking wells to Alberta cost the province $3.3 million in 2014 due to the decline in the crop productivity,

Public Debate

Poster against hydraulic fracturing in Vitoria-Gasteiz, Spain, October 2012

Politics and Public Policy

An anti-fracking movement has emerged both internationally with involvement of international environmental organizations and nations such as France and locally in affected areas such as Balcombe in Sussex where the Balcombe drilling protest was in progress during mid-2013. The considerable opposition against hydraulic fracturing activities in local townships in the United States has led companies to adopt a variety of public relations measures to reassure the public, including the employment of former military personnel with training in psychological warfare operations. According to Matt Pitzarella, the communications director at Range Resources, employees trained in the Middle East have been valuable to Range Resources in Pennsylvania, when dealing with emotionally charged township meetings and advising townships on zoning and local ordinances dealing with hydraulic fracturing.

There have been many protests directed at hydraulic fracturing. For example, ten people were arrested in 2013 during an anti-fracking protest near New Matamoras, Ohio, after they illegally entered a development zone and latched themselves to drilling equipment. Though usually non-violent, some protestors use violence and intimidation. In northwest Pennsylvania, there was a drive-by shooting at a well site, in which someone shot two rounds of a small-caliber rifle in the direction of a drilling rig, before shouting profanities at the site and fleeing the scene. In Washington County, Pennsylvania, a contractor working on a gas pipeline found a pipe bomb that had been placed where a pipeline was to be constructed, which local authorities said would have caused a "catastrophe" had they not discovered and detonated it.

In 2014 a number of European officials suggested that several major European protests against fracking (with mixed success in Lithuania and Ukraine) may be partially sponsored by Gazprom, Russia's state-controlled gas company. The New York Times suggested that Russia saw its natural gas exports to Europe as a key element of its geopolitical influence, and that this market would diminish if fracking is adopted in Eastern Europe, as it opens up significant shale gas reserves in the region. Russian officials have on numerous occasions made public statements to the effect that fracking "poses a huge environmental problem".

Documentary Films

Josh Fox's 2010 Academy Award nominated film Gasland became a center of opposition to hydraulic fracturing of shale. The movie presented problems with groundwater contamination near well sites in Pennsylvania, Wyoming, and Colorado.*Energy in Depth*, an oil and gas industry lobbying group, called the film's facts into question. In response, a rebuttal of *Energy in Depth's* claims of inaccuracy was posted on *Gasland's* website.

The Director of the Colorado Oil and Gas Conservation Commission (COGCC) offered to be interviewed as part of the film if he could review what was included from the interview in the final film but Fox declined the offer.Exxon Mobil, Chevron Corporation and ConocoPhillips aired advertisements during 2011 and 2012 that claimed to describe the economic and environmental benefits of natural gas and argue that hydraulic fracturing was safe.

The film Promised Land, starring Matt Damon, takes on hydraulic fracturing. The gas industry is making plans to try to counter the film's criticisms of hydraulic fracturing with informational flyers, and Twitter and Facebook posts.

In January 2013 Northern Irish journalist and filmmaker Phelim McAleer released a crowdfunded documentary called FrackNation as a response to the statements made by Fox in *Gasland*. *FrackNation* premiered on Mark Cuban's AXS TV. The premiere corresponded with the release of *Promised Land*.

In April 2013, Josh Fox released *Gasland 2*, a documentary that states that the gas industry's portrayal of natural gas as a clean and safe alternative to oil is a myth, and that hydraulically fractured wells inevitably leak over time, contaminating water and air, hurting families, and endangering the earth's climate with the potent greenhouse gas methane.

In 2014, Vido Innovations released the documentary *The Ethics of Fracking*. The film covers the politics, spiritual, scientific, medical and professional points of view on hydraulic fracturing. It also digs into the way the gas industry portrays fracking in their advertising.

In 2015, the Canadian documentary film Fractured Land had its world premiere at the Hot Docs Canadian International Documentary Festival.

Research Issues

Typically the funding source of the research studies is a focal point of controversy. Concerns have been raised about research funded by foundations and corporations, or by environmental groups, which can at times lead to at least the appearance of unreliable studies. Several organizations, researchers, and media outlets have reported difficulty in conducting and reporting the results of studies on hydraulic fracturing due to industry and governmental pressure, and expressed concern over possible censoring of environmental reports. Some have argued there is a need for more research into the environmental and health effects of the technique.

However, it is important to note that many of the most-cited studies over the last decade are either government estimates, environmentalist group reports, or peer-reviewed papers from academic scientists, including a 2013 EPA report that significantly lowered estimates of methane leakage compared to previous estimates, and a study commissioned by the environmentalist group Environmental Defense Fund and published in the Proceedings of the National Academy of Sciences, which similarly showed that the environmental effects of natural gas production were overestimated. Similarly, a study from the environmentalist Natural Resources Defense Council was cited to show that a previous highly cited study "significantly overestimate[s] the fugitive emissions associated with unconventional gas extraction."

Health Risks

There is concern over the possible adverse public health implications of hydraulic fracturing activity. A 2013 review on shale gas production in the United States stated, "with increasing numbers of drilling sites, more people are at risk from accidents and exposure to harmful substances used at fractured wells." A 2011 hazard assessment recommended full disclosure of chemicals used for hydraulic fracturing and drilling as many have immediate health effects, and many may have long-term health effects.

In June 2014 Public Health England published a review of the potential public health impacts of exposures to chemical and radioactive pollutants as a result of shale gas extraction in the UK, based on the examination of literature and data from countries where hydraulic fracturing already occurs. The executive summary of the report stated: "An assessment of the currently available evidence indicates that the potential risks to public health from exposure to the emissions associated with shale gas extraction will be low if the operations are properly run and regulated. Most evidence suggests that contamination of groundwater, if it occurs, is most likely to be caused by leakage through the vertical borehole. Contamination of groundwater from the underground hydraulic fracturing process itself (ie the fracturing of the shale) is unlikely. However, surface spills of hydraulic fracturing fluids or wastewater may affect groundwater, and emissions to air also have the potential to impact on health. Where potential risks have been identified in the literature, the reported problems are typically a result of operational failure and a poor regulatory environment."

A 2012 report prepared for the European Union Directorate-General for the Environment identified potential risks to humans from air pollution and ground water contamination posed by hydraulic fracturing. This led to a series of recommendations in 2014 to mitigate these concerns. A 2012 guidance for pediatric nurses in the US said that hydraulic fracturing had a potential negative impact on public health and that pediatric nurses should be prepared to gather information on such topics so as to advocate for improved community health.

Environmental Impacts

The potential environmental impacts of hydraulic fracturing include air emissions and climate change, high water consumption, water contamination, land use, risk of earthquakes, noise pollution, and health effects on humans. Air emissions are primarily methane that escapes from wells, along with industrial emissions from equipment used in the extraction process. Modern UK and EU regulation requires zero emissions of methane, a potent greenhouse gas. Escape of methane is a bigger problem in older wells than in ones built under more recent EU legislation.

Hydraulic fracturing uses between 1.2 and 3.5 million US gallons (4,500 and 13,200 m³) of water per well, with large projects using up to 5 million US gallons (19,000 m³). Additional water is used when wells are refractured. An average well requires 3 to 8 million US gallons (11,000 to 30,000 m³) of water over its lifetime. According to the Oxford Institute for Energy Studies, greater volumes of fracturing fluids are required in Europe, where the shale depths average 1.5 times greater than in the U.S.Surface water may be contaminated through spillage and improperly built and maintained waste pits, and ground water can be contaminated if the fluid is able to escape the formation being fractured (through, for example, abandoned wells) or by produced water (the returning fluids, which also contain dissolved constituents such as minerals and brine waters). Produced water is managed by underground injection, municipal and commercialwastewater treatment and discharge, self-contained systems at well sites or fields, and recycling to fracture future wells. Typically less than half of the produced water used to fracture the formation is recovered.

About 3.6 hectares (8.9 acres) of land is needed per each drill pad for surface installations. Well pad and supporting structure construction significantly fragments landscapes which likely has negative effects on wildlife. These sites need to be remediated after wells are exhausted. Each well pad (in average 10 wells per pad) needs during preparatory and hydraulic fracturing process about 800 to 2,500 days of noisy activity, which affect both residents and local wildlife. In addition, noise is created by continuous truck traffic (sand, etc.) needed in hydraulic fracturing. Research is underway to determine if human health has been affected by air and water pollution, and rigorous following of safety procedures and regulation is required to avoid harm and to manage the risk of accidents that could cause harm.

In July 2013, the US Federal Railroad Administration listed oil contamination by hydraulic fracturing chemicals as "a possible cause" of corrosion in oil tank cars.

Hydraulic fracturing sometimes causes induced seismicity or earthquakes. The magnitude of these events is usually too small to be detected at the surface, although tremors attributed to fluid injection into disposal wells have been large enough to have often been felt by people, and to have caused property damage and possibly injuries.

Microseismic events are often used to map the horizontal and vertical extent of the fracturing. A better understanding of the geology of the area being fracked and used for injection wells can be helpful in mitigating the potential for significant seismic events.

Regulations

Countries using or considering use of hydraulic fracturing have implemented different regulations, including developing federal and regional legislation, and local zoning limitations. In 2011, after public pressure France became the first nation to ban hydraulic fracturing, based on the precautionary principle as well as the principle of preventive and corrective action of environmental hazards. The ban was upheld by an October 2013 ruling of the Constitutional Council. Some other countries such as Scotland have placed a temporary moratorium on the practice due to public health concerns and strong public opposition. Countries like the United Kingdom and South Africa have lifted their bans, choosing to focus on regulation instead of outright prohibition. Germany has announced draft regulations that would allow using hydraulic fracturing for the exploitation of shale gas deposits with the exception of wetland areas. In China, regulation on shale gas still faces hurdles, as it has complex interrelations with other regulatory regimes, especially trade.

The European Union has adopted a recommendation for minimum principles for using high-volume hydraulic fracturing. Its regulatory regime requires full disclosure of all additives. In the United States, the Ground Water Protection Council launched FracFocus.org, an online voluntary disclosure database for hydraulic fracturing fluids funded by oil and gas trade groups and the

U.S. Department of Energy. Hydraulic fracturing is excluded from the Safe Drinking Water Act's underground injection control's regulation, except when diesel fuel is used. The EPA assures surveillance of the issuance of drilling permits when diesel fuel is employed.

In 2012, Vermont became the first state in the United States to ban hydraulic fracturing. On 17 December 2014, New York became the second state to issue a complete ban on any hydraulic fracturing due to potential risks to human health and the environment.

Waterless Fracturing

Waterless fracturing is an alternative to hydraulic fracturing in which liquefied petroleum gas (LPG) which uses propane liquefied into gel. Propane is pumped into shale rock formation instead of water. Using propane can be considered more beneficial than water, because it does not block all pathways and therefore more natural gas is released. In addition, it does not carry poisonous chemicals and underground radioactivity back to the surface, and when it does comes back to the surface it can have another usage or be used again for LPG. The only drawback is the cost as propane is more expensive than water. Another drawback is that the possible leak of propane may lead to flash fire, so fracturing with propane requires more safety instructions and monitoring equipments in order to reduce possible risks.

According to Nathan Janiczek, the new era of waterless fracturing is growing and it is called Liquefied Petroleum Gas, or LPG fracturing. LPG is always pumped in a well and because of that pumping, rocks are destroyed and gas is released. Besides that, LPG is fully converted into gas when it is pumped up to the surface and that guarantees a one hundred percent retrieval rate. If we compare conventional fracking and fracking using propane gel, LPG fracking does not produce waste, nearly 100% of propane gas is pumped back while 50% of hydraulic frack fluid remains underground. It has lower viscosity, less surface tension and lower specific gravity, which makes the fluid more effective in transport and less dependent on pressure. LPG has many advantages; however, it is a new technique, which needs further research. Liquefied petroleum gas fracturing was developed by a GasFrac energy company, based in Calgary, Alberta. The Chief Technology Officer of the GasFrac company said that it is being used since 2008 in gas wells of Canada in Alberta, British Columbia, New Brunswick, Texas, Pennsylvania, Colorado, Oklahoma and New Mexico.

Hydraulic Fracturing Proppants

A proppant is a solid material, typically sand, treated sand or man-made ceramic materials, designed to keep an induced hydraulic fracture open, during or following a fracturing treatment. It is added to a *fracking fluid* which may vary in composition depending on the type of fracturing used, and can be gel, foam or slickwater–based. In addition, there may be unconventional fracking fluids. Fluids make tradeoffs in such material properties as viscosity, where more viscous fluids can carry more concentrated proppant; the energy or pressure demands to maintain a certain flux pump rate (flow velocity) that will conduct the proppant appropriately; pH, various rheological factors, among others. In addition, fluids may be used in low-volume well stimulation of high-permeability sandstone wells (20k to 80k gallons per well) to the high-volume operations such as shale gas and tight gas that use millions of gallons of water per well.

Conventional wisdom has often vacillated about the relative superiority of gel, foam and slickwater fluids with respect to each other, which is in turn related to proppant choice. For example, Zuber, Kuskraa and Sawyer (1988) found that gel-based fluids seemed to achieve the best results for coalbed methane operations, but as of 2012, slickwater treatments are more popular.

Other than proppant, slickwater fracturing fluids are mostly water, generally 99% or more by volume, but gel-based fluids can see polymers and surfactants comprising as much as 7 vol% , ignoring other additives. Other common additives include hydrochloric acid (low pH can etch certain rocks, dissolving limestone for instance), friction reducers, guar gum, biocides, emulsion breakers, emulsifiers, 2-butoxyethanol, and radioactive tracer isotopes.

Proppant Permeability and Mesh Size

Sand used for fracturing, USGS, 2012

Proppants used should be permeable or permittive to gas under high pressures; the interstitial space between particles should be sufficiently large, yet have the mechanical strength to withstand closure stresses to hold fractures open after the fracturing pressure is withdrawn. Large mesh proppants have greater permeability than small mesh proppants at low closure stresses, but will mechanically fail (i.e. get crushed) and produce very fine particulates ("fines") at high closure stresses such that smaller-mesh proppants overtake large-mesh proppants in permeability after a certain threshold stress.

Though sand is a common proppant, untreated sand is prone to significant fines generation; fines generation is often measured in wt% of initial feed. A commercial newsletter from Momentive cites untreated sand fines production to be 23.9% compared with 8.2% for lightweight ceramic and 0.5% for their product. One way to maintain an ideal mesh size (i.e. permeability) while having sufficient strength is to choose proppants of sufficient strength; sand might be coated with resin,- to form CRCS (Curable Resin Coated Sand) or PRCS (Pre-Cured Resin Coated Sands). In certain situations a different proppant material might be chosen altogether—popular alternatives include ceramics and sintered bauxite.

Proppant Weight and Strength

Increased strength often comes at a cost of increased density, which in turn demands higher flow rates, viscosities or pressures during fracturing, which translates to increased fracturing costs, both environmentally and economically. Lightweight proppants conversely are designed to be lighter than sand (~2.5 g/cm^3) and thus allow pumping at lower pressures or fluid velocities. Light proppants are less likely to settle. Porous materials can break the strength-density trend, or even afford greater gas permeability. Proppant geometry is also important; certain shapes or forms amplify stress on proppant particles making them especially vulnerable to crushing (a sharp discontinuity can classically allow infinite stresses in linear elastic materials).

Proppant Deposition and Post-treatment Behaviours

Proppant mesh size also affects fracture length: proppants can be "bridged out" if the fracture width decreases to less than twice the size of the diameter of the proppant. As proppants are deposited in a fracture, proppants can resist further fluid flow or the flow of other proppants, inhibiting further growth of the fracture. In addition, closure stresses (once external fluid pressure is released) may cause proppants to reorganise or "squeeze out" proppants, even if no fines are generated, resulting in smaller effective width of the fracture and decreased permeability. Some companies try to cause weak bonding at rest between proppant particles in order to prevent such reorganisation. The modelling of fluid dynamics and rheology of fracturing fluid and its carried proppants is a subject of active research by the industry.

Proppant Costs

Though good proppant choice positively impacts output rate and overall ultimate recovery of a well, commercial proppants are also constrained by cost. Transport costs from supplier to site form a significant component of the cost of proppants.

Other Components of Fracturing Fluids

Other than proppant, slickwater fracturing fluids are mostly water, generally 99% or more by volume, but gel-based fluids can see polymers and surfactants comprising as much as 7 vol% , ignoring other additives. Other common additives include hydrochloric acid (low pH can etch certain rocks, dissolving limestone for instance), friction reducers, guar gum,biocides, emulsion breakers, emulsifiers, and 2-Butoxyethanol.

Radioactive tracer isotopes are sometimes included in the hydrofracturing fluid to determine the injection profile and location of fractures created by hydraulic fracturing. Patents describe in detail how several tracers are typically used in the same well. Wells are hydraulically fractured in different stages. Tracers with different half-lives are used for each stage. Their half-lives range from 40.2 hours (lanthanum-140) to 5.27 years (cobalt-60). Amounts per injection of radionuclide are listed in The US Nuclear Regulatory Commission (NRC) guidelines. The NRC guidelines also list a wide range or radioactive materials in solid, liquid and gaseous forms that are used as field flood or enhanced oil and gas recovery study applications tracers used in single and multiple wells.

In the US, except for diesel-based additive fracturing fluids, noted by the American Environmental Protection Agency to have a higher proportion of volatile organic compounds and carcinogenic BTEX, use of fracturing fluids in hydraulic fracturing operations was explicitly excluded from regulation under the American Clean Water Act in 2005, a legislative move that has since attracted controversy for being the product of special interests lobbying.

Regulation of Hydraulic Fracturing

Countries using or considering to use hydraulic fracturing have implemented different regulations, including developing federal and regional legislation, and local zoning limitations. In 2011, after public pressure France became the first nation to ban hydraulic fracturing, based on the precautionary principle as well as the principal of preventive and corrective action of environmental hazards. The ban was upheld by an October 2013 ruling of the Constitutional Council. Some other countries have placed a temporary moratorium on the practice. Countries like the United Kingdom and South Africa, have lifted their bans, choosing to focus on regulation instead of outright prohibition. Germany has announced draft regulations that would allow using hydraulic fracturing for the exploitation of shale gas deposits with the exception of wetland areas.

The European Union has adopted a recommendation for minimum principles for using high-volume hydraulic fracturing. Its regulatory regime requires full disclosure of all additives. In the United States, the Ground Water Protection Council launched FracFocus.org, an online voluntary disclosure database for hydraulic fracturing fluids funded by oil and gas trade groups and the U.S. Department of Energy. Hydraulic fracturing is excluded from the Safe Drinking Water Act's underground injection control's regulation, except when diesel fuel is used. The EPA assures surveillance of the issuance of drilling permits when diesel fuel is employed.

On 17 December 2014, New York state issued a statewide ban on hydraulic fracturing, becoming the second state in the United States to issue such a ban after Vermont.

Approaches

Risk-based Approach

The main tool used by this approach is risk assessment. A risk assessment method, based on experimenting and assessing risk ex-post, once the technology is in place. In the context of hydraulic fracturing, it means that drilling permits are issued and exploitation conducted before the potential risks on the environment and human health are known. The risk-based approach mainly relies on a discourse that sacralizes technological innovations as an intrinsic good, and the analysis of such innovations, such as hydraulic fracturing, is made on a sole cost-benefit framework, which does not allow prevention or ex-ante debates on the use of the technology. This is also referred to as "learning-by-doing". A risk assessment method has for instance led to regulations that exist in the hydraulic fracturing in the United States (EPA will release its study on the effect of hydraulic fracturing on groundwater in 2014, though hydraulic fracturing has been used for more than 60 years. Commissions that have been implemented in the US to regulate the use of hydraulic fracturing have been created after hydraulic fracturing had started in their area of regulation. This is for instance the case in the Marcellus shale area where three regulatory committees were implemented ex-post.

Academic scholars who have studied the perception of hydraulic fracturing in the North of England have raised two main critiques of this approach. Firstly, it takes scientific issues out of the public debate since there is no debate on the use of a technology but on its effects. Secondly, it does not prevent environmental harm from happening since risks are taken then assessed instead of evaluated then taken as it would be the case with a precautionary approach to scientific debates. The relevance and reliability of risk assessments in hydraulic fracturing communities has also been debated amongst environmental groups, health scientists, and industry leaders. A study has epitomized this point: the participants to regulatory committees of the Marcellus shale have, for a majority, raised concerns about public health although nobody in these regulatory committees had expertise in public health. That highlights a possible underestimation of public health risks due to hydraulic fracturing. Moreover, more than a quarter of the participants raised concerns about the neutrality of the regulatory committees given the important weigh of the hydraulic fracturing industry. The risks, to some like the participants of the Marcellus Shale regulatory committees, are overplayed and the current research is insufficient in showing the link between hydraulic fracturing and adverse health effects, while to others like local environmental groups the risks are obvious and risk assessment is underfunded.

Precaution-based Approach

The second approach relies on the precautionary principle and the principal of preventive and corrective action of environmental hazards, using the best available techniques with an acceptable economic cost to insure the protection, the valuation, the restoration, management of spaces, resources and natural environments, of animal and vegetal species, of ecological diversity and equilibriums. The precautionary approach has led to regulations as implemented in France and Vermont, banning hydraulic fracturing.

Such an approach is called upon by social sciences and the public as studies have shown in the North of England and Australia. Indeed, in Australia, the anthropologist who studied the use of hydraulic fracturing concluded that the risk-based approach was closing down the debate on the ethics of such a practice, therefore avoiding questions on broader concerns that merely the risks implied by hydraulic fracturing. In the North of England, levels of concerns registered in the deliberative focus groups studied were higher regarding the framing of the debate, meaning the fact that people did not have a voice in the energetic choices that were made, including the use of hydraulic fracturing. Concerns relative to risks of seismicity and health issues were also important to the public, but less than this. A reason for that is that being withdrawn the right to participate in the decision-making triggered opposition of both supporters and opponents of hydraulic fracturing.

The points made to defend such an approach often relate to climate change and the impact on the direct environment; related to public concerns on the rural landscape for instance in the UK. Energetic choices indeed affect climate change since greenhouse gas emissions from fossil fuels extraction such as shale gas and oil contribute to climate change. Therefore, people have in the UK raised concerns about the exploitation of these resources, not just hydraulic fracturing as a method. They would hence prefer a precaution-based approach to decide whether or not, regarding the issue of climate change, they want to exploit shale gas and oil.

Framing of the Debate

There are two main areas of interest regarding how debates on hydraulic fracturing for the exploitation of unconventional oil and gas have been conducted.

"Learning-by-doing" and the Displacement of Ethics

A risk-based approach is often referred to as "learning-by-doing" by social sciences. Social sciences have raised two main critiques of this approach. Firstly, it takes scientific issues out of the public debate since there is no debate on the use of a technology but on its impacts. Secondly, it does not prevent environmental harm from happening since risks are taken then assessed instead of evaluated then taken. Public concerns are shown to be really linked to these issues of scientific approach. Indeed, the public in the North of England for instance fears "the denial of the deliberation of the values embedded in the development and application of that technology, as well as the future it is working towards" more than risks themselves. The legitimacy of the method is only questioned after its implementation, not before. This vision separates risks and effects from the values entitled by a technology. For instance, hydraulic fracturing entitles a transitional fuel for its supporters whereas for its opponents it represents a fossil fuel exacerbating the greenhouse effect and global warming. Not asking these questions leads to seeing only the mere economic cost-benefit analysis.

This is linked to a pattern of preventing non-experts from taking part in scientific-technological debates, including their ethical issues. An answer to that problem is seen to be increased public participation so as to have the public deciding which issues to address and what political and ethical norms to adopt as a society. Another public concern with the "learning-by-doing" approach is that the speed of innovation may exceed the speed of regulation and since innovation is seen as serving private interests, potentially at the expense of social good, it is a matter of public concern. Science and Technology Studies have theorized "slowing-down" and the precautionary principle as answers. The claim is that the possibility of an issue is legitimate and should be taken into account before any action is taken.

Variations in Risk-assessment of Environmental Effects of Hydraulic Fracturing

Issues also exist regarding the way risk assessment is conducted and whether it reflects some interests more than others. Firstly, an issue exists about whether risk assessment authorities are able to judge the impact of hydraulic fracturing in public health. A study conducted on the advisory committees of the Marcellus Shale gas area has shown that not a single member of these committees had public health expertise and that some concern existed about whether the commissions were not biased in their composition. Indeed, among 51 members of the committees, there is no evidence that a single one has any expertise in environmental public health, even after enlarging the category of experts to "include medical and health professionals who could be presumed to have some health background related to environmental health, however minimal". This cannot be explained by the purpose of the committee since all three executive orders of the different committees mentioned environmental public health related issues. Another finding of the authors is that a quarter of the opposed comments mentioned the possibility of bias in favor of gas industries in the composition of committees. The authors conclude saying that political leaders may not want to raise public health concerns not to handicap further economic development due to hydraulic fracturing.

Secondly, the conditions to allow hydraulic fracturing are being increasingly strengthened due to the move from governmental agencies' authority over the issue to elected officials' authority over it. The Shale Gas Drilling Safety Review Act of 2014 issued in Maryland forbids the issuance of drilling permits until a high standard "risk assessment of public health and environmental hazards relating to hydraulic fracturing activities" is conducted for at least 18 months based on the Governor's executive order.

Institutional Discourse and the Public

A qualitative study using deliberative focus groups has been conducted in the North of England, where there is a big shale gas reservoir exploited by hydraulic fracturing. These group discussions reflect many concerns on the issue of the use of unconventional oil and gas. There is a concern about trust linked with a doubt on the ability or will of public authorities to work for the greater social good since private interests and profits of industrial companies are seen as corruptive powers. Alienation is also a concern since the feeling of a game rigged against the public rises due to "decision making being made on your behalf without being given the possibility to voice an opinion". Exploitation also arises since economic rationality that is seen as favoring short-termism is accused of seducing policy-makers and industry. Risk is accentuated by what is hydraulic fracturing as well as what is at stake, and "blind spots" of current knowledge as well as risk assessment analysis are accused of increasing the potentiality of negative outcomes. Uncertainty and ignorance are seen as too important in the issue of hydraulic fracturing and decisions are therefore perceived as rushed, which is why participants favored some form of precautionary approach. There is a major fear on the possible disconnection between the public's and the authorities' visions of what is a good choice for the good reasons.

It also appears that media coverage and institutional responses are widely inaccurate to answer public concerns. Indeed, institutional responses to public concerns are mostly inadequate since they focus on risk assessment and giving information to the public that is considered anxious because ignorant. But public concerns are much wider and it appears that public knowledge on hydraulic fracturing is rather good.

The hydraulic fracturing industry has lobbied for permissive regulation in Europe, the US federal government, and US states. On March 20, 2015 the rules for disclosing the chemicals used were tightened by the Obama administration. The new rules give companies involved 30 days from the beginning of an operation on federal land to disclose those chemicals.

Hydro-slotted Perforation

Hydro-slotting perforation technology is the process of opening the productive formation through the casing and cement sheath to produce the oil or gas product flow (intensification, stimulation). The process has been used for industrial drilling since 1980, and involves the use of an underground hydraulic slotting engine (tool, equipment). The technology helps to minimize compressive stress following drilling in the well-bore zone (which reduces the permeability in the zone).

Overview

Since ancient times, when there were the first coal mines, it was observed, that increasing the depth of the development the coal tunnel, under the action of overburden pressure, surrounding rocks become harder and little-permeable. To solve this problem they developed a cavern of a certain form in the rock. More modern mining geo-mechanics explain the reason for the occurrence of this effect in relation to drilling wells. During any drilling process in the well there is formed the annular compressive stress conditions around the wellbore zone. The deeper the well, the more overburden pressure, which means the greater the annular compressive stress conditions. On the rocks lying at depths of 3–5 km the compressive stresses may reach up-to 75–125 MPa. In the near-well zone, as a result of concentration these stresses increase and sometimes become equal to double 150–250 MPa. If the tectonic stresses is several times higher than stresses from the weight of rocks, the stresses in the near-well zone may be even greater.

Under the action of stress conditions and high overburden pressure occurs a significant reduction in permeability in the near wellbore zone, in some cases close to zero. Oil or gas flow can not penetrate to the well. Traditional methods of opening the productive layer formation (cumulative, jet perforation, sand jet perforation, abrasive jetting perforation and other similar methods) did not consider this complicated situation in the near-well zone and therefore was not the effective. Porous and fractured formations are subjected to compression, that deforms the rock mass and reduces its permeability. The greater the depth, the stronger the effect can be.

Hydro-slotting perforation is quite different from jet (hydro-jetting or sand-blast) perforation. The energy of working fluid, consisting from water (layer water) and sand (abrasive quartz sand) pressure in the hydraulic engine, is divided into two components: five percent of energy goes to the creation of smooth uniform rectilinear motion of the working rod with the perforator and nozzles (between two and six nozzles) without participation in the process the multimeter tubing or coil-tubing. Ninety-five percent of energy goes to the cutting of continued and geometrically correct deep slots (up to five feet deep and between three and five slots at the same time). Slot length is equal to the length of the working engine shaft, usually 1.64 feet (0.00050 km).

The hydro-slotting perforation process does not deform the casing, does not create cracks in the cement, and does not clog-up the borders in the formation.

The geometry and depth of the slots creates the conditions for occurrence of the effect of unloading the circular stress conditions in the near wellbore zone (from 50 to 100 percent) and accordingly the increase of permeability (up to 30 to 50 percent) in this zone. In addition to this it forms a large area of the penetration (31.5 square feet (2.93 m²) area for one cut with two nozzles only), that provides a very good hydrodynamic connection of the productive layer with the well.

The cutting speed may be corrected with the temperature in the borehole, temperature of the working fluid, concentration, flow and pressure. (these components are enough to completely control the depth and length of the cut and thus forming the slots), to instantly cut through the steel casing, through the cement to delve into the productive formation and keep the jets in this state while moving along the borehole, keeping the same depth of cut. At the end of the cutting continuous slot process the engine is set up to the initial position and ready for the next cutting interval. The process of hydro-slotting perforation and the depth of cut is controlled by the working fluid supply, pressure and concentration. The equipment can be operated without lifting on the surface for 11–15 hours.

Hydro-slotting perforation is the ecologically safe, environmentally friendly and effective affordable method for intensifying the operation in oil, gas, injection and hydro-geological wells. Now this method is widely used in Azerbaijan, Brazil, China, Eastern and Western Siberia, Jordan, Kazakhstan, Komi Republic, North Caucasus, Russia, Udmurtia, Ukraine, Urals, Uzbekistan and Yemen. The first mention regarding the hydro-slotting perforation in America, was in 1987 at the oil and gas conference in Texas. The first use of hydro-slotting perforation in the United States dates back to 1996, when together with Shell E & P Technology Company, discovered two wells (Abrasive Hydro jet Technology in Albert Load, Michigan). After that the hydro-slotting perforation was highly appreciated by the Department of Geophysics at Stanford University and by Division of Shell Exploration and Production by Shell E&P Technology Company. Hydro-slotting perforation was used in California, Kansas, Michigan, Montana, Nebraska, New York, Pennsylvania, Texas and Wyoming states. In Canada it has been successfully applied in Saskatchewan.

General Concepts

For opening of any productive layer it is necessary to open the casing, cement sheath and productive layer formation. Geophysics and mining geo-mechanics dictated the next requirements:

- Zone of cement sheath should be opened completely and not have cracks (to prevent possible overflows of water);

- Productive layer formation should be opened to maximum and on the maximum depth. At the same time productive formation should not have clogging, plugging, grouting, occlusive and cinder borders to produce excellent hydrodynamic connections of the productive layer with the well. Encompassing unloading the circular stress conditions around the wellbore, formed as a result of drilling, and increasing the permeability (50–100 percent) in the near wellbore zone (as a consequence of the first)

In the early 1970s, the Ministry of Geology of the USSR placed the Government order to scientific research institutions of the Country for the solution of the annular stress conditions and increase the permeability problem in the drilling wells. It was necessary to create the technology of opening the productive layer formation taking into account of uploading the annular stress conditions and increase the permeability in the near wellbore zone. The work to study this problem were assigned to the Institute of Oceanology and VNIMI (St. Petersburg, Russia). During the study there was done hundreds of experiments and mathematical models. It was determined, that if creating a geometrically correct, extended slot, directed along the wellbore and perpendicular to it on the distance from around 0.7 inches (18 mm) to 3.5 feet (1.1 m), in the zone of around the wellbore, there occurs the unloading of the annular compressive stress conditions from 50 to 100 percent, that are redirected to the far plane of the surface of formed slot, parallel to the wellbore surface. At the same time the permeability in this zone increased 30 to 50 percent. The holes after cumulative, jet perforation, sand jet perforation, abrasive jetting perforation and other similar methods, do not give the effect. The spot perforation did not create a slot in the casing and did not reach the required (unloading effect) depth, because the reverse jet interfered with direct jet and the maximum depth of the hole could not exceed 0.65 feet. When perforation with the movement occurs, the direct jet does not intersect with the reverse jet and depth of cutting can be much more (up to five feet) which is known as the excavation effect. Later it was proven mathematically.

It was necessary to create a device, that could make the continuation, along the borehole, slots in the casing, cement and go further into the productive formation. The tests with the movement of the multimeter tubing were not been successful, showing it was impossible to create geometrically correct extended slots with moving tubing. It was necessary to create an apparatus, that created a movement of cutting jets by itself, independently from tubing and located on the end of the tubing, directly in the leveled area. independent movement of the cutting jets could only be done mechanically, electrically or hydraulically. After another six months of research and testing it was decided to use mechanics and hydraulics as the base. The first prototype of hydro-slotting perforation device was created in 1972. The technology of hydro-slotting perforation was never sold to anyone. The hydro-slotting perforation technology was transformed into the category of performance techniques (as the technique of conducting the drilling, cumulative perforation, hydraulic fracturing, logging, pumping and so on).

The finalization of the device (prototype) in the end of 1972 was tasked to the special laboratory of the Research Institute of Oceanology of PSU "Sevmorgeo". From the beginning of the work for the revision the existing device was carried out in two directions: hydro-slotting perforation and hydro-mechanical slotting perforation. The second variant differs from the first in that at the beginning the opening of the casing is produced with a circular saw, and then the rock eroded by working fluid (water and sand) jets. The works were done over three years. The work for improvements of the hydro-mechanical slotting perforation were terminated in the result of their further inexpedient. Firstly, it was not necessary to divide the process into two operations: cutting the casing with the circular saw and a further jet-slotting perforation, because the cutting of casing with jet-slotting perforation takes place in a matter of seconds. Second, the mechanism of the circular saw takes up a lot of space in the housing unit, it was impossible to use the energy of working fluid to full power for getting deep slots, the slots get small and not deep (not enough for occurs the unloading of the annular compressive stress conditions and increase the permeability in the near wellbore zone). The further project was focused for finishing the hydro-slotting perforation device only.

In 1975, the scientific research laboratory of the Research Institute of Oceanology of PSU Sevmorgeo completed the project to improve the prototype of hydro-slotting perforation tool and this tool has been able to operate independently of the tubing movement. The equipment was 16 feet (4.9 m) long, 4.02 inches (10.2 cm) OD, weight 300 pounds (140 kg) and stroke length of 0.5 feet (0.15 m) only, and it worked on the following principle: the energy of working fluid pressure was divided into two components. Part of energy was used for the motion creation for the working rod with the perforator and nozzles; the other part of the energy was use for the cutting process (creating the continued slots along the wellbore through the casing and cement into the productive formation). The form and depth of the slots allowed the device to perform its main task, unloading the annular stress conditions and increase the permeability. The first practical tests in the wells were successfully made at the end of 1975 on "Archeda" field (Volgograd, Russia).

Benefits

Ability to increase area of development

- Very deep penetration from three to six feet
- Vertical permeability
- Porosity increases four to five times

- Permeability increases 15 times

- Drainage volume increases six times

Ability to access reserves which are otherwise inaccessible

- In reservoirs located in close proximity to water, gas, and oil contacts

- In weakly permeable, tightly-cemented reservoirs

- In missed layers, or in layers covered by two or more columns

Gentle approach with the ability to repair well-bore damage

- In carbonates, dislodges clay particles and fines

- In sandstones, reduces sand mobility problems

- In deep gas sands, relieves overpressure damage from mud weight systems

- Does not crack casing or cement

- Maintains hydraulic integrity with no detonation impacts

- Redistributes stresses away from the near-well-bore zone

Development

During the period from the date of the first prototype of hydro-slotting perforation tool to present day, the type and technological characteristics of the equipment was significantly improved. The modern underground hydro-slotting equipment represents the devices, capable to instantly cut through the steel casing, through the cement to delve into the productive formation and keep the jets in this state while moving along the borehole, keeping the same depth of cut. Hydro-slotting equipment made of special high-strength materials, 12 feet (3.7 m) long, 3.5 feet (110 cm) OD, weight 180 pounds (82 kg), cutting speed from the point of perforation to 0.7 inches (1.8 cm) per minute, working stroke length 1.65 feet (0.50 m) (4.92 feet (1.50 m) x 1.64 feet (0.50 m) x 1.97 inches (5.0 cm) each slot), depth of slots five feet, continued and geometrically correct slots, opening area 63 square feet (5.9 m²) per cut with four nozzles, can apply streamlined perforators between two and six nozzles, unloading the annular stress conditions in the near wellbore zone 50 to 100 percent, and increase the permeability 30 to 50 percent. The continuous time without lifting to the surface is 11–15 hours (nozzles lifespan ~ 15 hours, perforator ~ seven wells, hydraulic engine ~ 40 wells).

Without lifting to the surface with the hydro-slotting tool can also:

- cut on the previous perforation (cumulative perforation)

- colmatation treatment

- cut the thin-interbedded layers

- mini hydraulic fracture stimulation

- create the continuous slot

- cut the shale

- accurate cut near the water reservoir or opposite in the injection wells

- bypass the water layers

- bypass the casing collars

- cut a few casings

- chemical treatment

- sealing, direct and reverse flushing

- tubing pressure testing

- cut the casing at abandonment

The hydro-slotting perforation process does not deform the casing, does not create cracks in the cement and does not clog up the borders in the formation. The process of hydro-slotting perforation is controlled. The cutting speed and depth of cutting may be corrected with the temperature in the borehole, temperature of the working fluid, concentration, flow and pressure. At the end of the cutting process of a continuous slot the engine is set up to the initial position and ready for the next cutting interval. Hydro-slotting perforation sets the perfect geometry for the subsequent fracturing. Hydro-slotting perforation can be applied in any formation: shale, carbonates, sandstone and so on.

Further improvement of the equipment for hydro-slotting perforation must follow the scientific and technical progress in this technology, not on the way of mindless increase of the holes in the hydro jets pipe. It is necessary to make the underground hydraulic engine for horizontal wells, which must be sealed to prevent the ingress of sand and mud inside and maintain the centerline position relative to the wellbore. It is necessary to make a self-orientation perforator (a particularly important issue of orientation in horizontal wells). For the orientation of the tool it is necessary there is communication with the tool (preferably two-sided) and surface of the well. Taking into account the specific conditions of hydro-slotting perforation process, signaling from the tool and back possibly using ultrasound only. Then the cutting process can be fully controlled from the surface, and it will be possible to change the speed and depth of cutting the slots regardless of the temperature inside the well.

Patents

Over the years this method has not undergone much change, but there are many patents on the method of hydro-slotting perforation. With the development of technological progress there has been continuously improved and refined equipment, but patents, regarding the hydro-slotting equipment in full is not so much, there are a few patents on parts.

- United States patents for complete hydro-slotting perforation equipment: US 8240369 B1, US 31,084

- Similar United States patents: US3130786, US4227582, US5337825, US6651741, US7073587, US7140429, US7568525, US20070187086, US20090101414, USRE21085, 166/55.2, 166/298, and E21B43/114

- United States patent for method of hydro-slotting perforation: US 20130105163 A1

- Similar United States patents: US3130786, US4047569, US4134453, US5445220, US6564868, US7568525, and US20050269100

Uses of Radioactivity in Oil and Gas Wells

Radioactive sources are used for logging formation parameters. Radioactive tracers, along with the other substances in hydraulic-fracturing fluid, are sometimes used to determine the injection profile and location of fractures created by hydraulic fracturing.

Use of Radioactive Sources for Logging

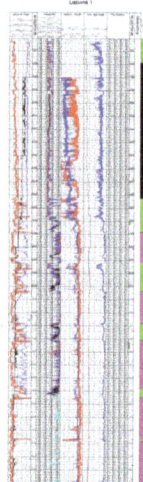

Composite wireline log for the Lisburne 1 well, Alaska - the neutron and density logs used radioactive sources

Sealed radioactive sources are routinely used in formation evaluation of both hydraulically fractured and non-fracked wells. The sources are lowered into the borehole as part of the well logging tools, and are removed from the borehole before any hydraulic fracturing takes place. Measurement of formation density is made using a sealed caesium-137 source. This bombards the formation with high energy gamma rays. The attenuation of these gamma rays gives an accurate measure of formation density; this has been a standard oilfield tool since 1965. Another source is americium berylium (Am-Be) neutron source used in evaluation of the porosity of the formation, and this has been used since 1950. In a drilling context, these sources are used by trained personnel, and radiation exposure of those personnel is monitored. Usage is covered by licenses from International Atomic Energy Agency (IAEA) guidelines, SU or European Union protocols, and the Environment Agency in the UK. Licenses are required for access, transport, and use of radioactive sources. These sources are very large, and the potential for their use in a 'dirty bomb' means security issues are considered as important. There is no risk to the public, or to water supplies under normal usage.

They are transported to a well site in shielded containers, which means exposure to the public is very low, much lower than the background radiation dose in one day.

Radiotracers and Markers

The oil and gas industry in general uses unsealed radioactive solids (powder and granular forms), liquids and gases to investigate or trace the movement of materials. The most common use of these radiotracers is at the well head for the measurement of flow rate for various purposes. A 1995 study found that radioactive tracers were used in over 15% of stimulated oil and gas wells.

Use of these radioactive tracers is strictly controlled. It is recommended that the radiotracer is chosen to have readily detectable radiation, appropriate chemical properties, and a half life and toxicity level that will minimize initial and residual contamination. Operators are to ensure that licensed material will be used, transported, stored, and disposed of in such a way that members of the public will not receive more than 1 mSv (100 mrem) in one year, and the dose in any unrestricted area will not exceed 0.02 mSv (2 mrem) in any one hour. They are required to secure stored licensed material from access, removal, or use by unauthorized personnel and control and maintain constant surveillance of licensed material when in use and not in storage. Federal and state nuclear regulatory agencies keep records of the radionuclides used.

As of 2003 the isotopes Antimony-124, argon-41, cobalt-60, iodine-131, iridium-192, lanthanum-140, manganese-56, scandium-46, sodium-24, silver-110m, technetium-99m, and xenon-133 were most commonly used by the oil and gas industry because they are easily identified and measured.Bromine-82, Carbon-14, hydrogen-3, iodine-125 are also used.

Examples of amounts used are:

Nuclide	Form	Activity
Iodine-131	Gas	100 millicuries (3.7 GBq) total, not to exceed 20 mCi (0.74 GBq) per injection
Iodine-131	Liquid	50 millicuries (1.9 GBq) total, not to exceed 10 mCi (0.37 GBq) per injection
Iridium-192	"Labeled" frac sand	200 millicuries (7.4 GBq) total, not to exceed 15 mCi (0.56 GBq) per injection
Silver-110m	Liquid	200 millicuries (7.4 GBq) total, not to exceed 10 mCi (0.37 GBq) per injection

In hydraulic fracturing, plastic pellets coated with Silver-110m or sand labelled with Iridium-192with may be added to a proppant when it is required to evaluate whether a fracturing process has penetrated rocks in the pay zone. Some radioactivity may by brought to the surface at the well head during testing to determine the injection profile and location of fractures. Typically this uses very small (50 kBq) Cobalt-60 sources and dilution factors are such that the activity concentrations will be very low in the topside plant and equipment.

Regulation in the US

The NRC and approved state agencies regulate the use of injected radionuclides in hydraulic fracturing in the United States.

The US EPA sets radioactivity standards for drinking water. Federal and state regulators do not require sewage treatment plants that accept gas well wastewater to test for radioactivity. In Pennsylvania, where the hydraulic fracturing drilling boom began in 2008, most drinking-water intake plants downstream from those sewage treatment plants have not tested for radioactivity since before 2006. The EPA has asked the Pennsylvania Department of Environmental Protection to require community water systems in certain locations, and centralized wastewater treatment facilities to conduct testing for radionuclides.

References

- Charlez, Philippe A. (1997). Rock Mechanics: Petroleum Applications. Paris: Editions Technip. p. 239. ISBN 9782710805861. Retrieved 2012-05-14.

- Fjaer, E. (2008). "Mechanics of hydraulic fracturing". Petroleum related rock mechanics. Developments in petroleum science (2nd ed.). Elsevier. p. 369. ISBN 978-0-444-50260-5. Retrieved 2012-05-14.

- Price, N. J.; Cosgrove, J. W. (1990). Analysis of geological structures. Cambridge University Press. pp. 30–33. ISBN 978-0-521-31958-4. Retrieved 5 November 2011.

- Zoback, M.D. (2007). Reservoir geomechanics. Cambridge University Press. p. 18. ISBN 9780521146197. Retrieved 6 March 2012.

- Gill, R. (2010). Igneous rocks and processes: a practical guide. John Wiley and Sons. p. 102. ISBN 978-1-4443-3065-6. Retrieved 5 November 2011.

- Mader, Detlef (1989). Hydraulic Proppant Fracturing and Gravel Packing. Elsevier. pp. 173–174; 202. ISBN 9780444873521.

- Gold, Russell (2014). The Boom: How Fracking Ignited the American Energy Revolution and Changed the World. New York: Simon & Schuster. pp. 115–121. ISBN 978-1-4516-9228-0.

- Chilingar, George V.; Robertson, John O.; Kumar, Sanjay (1989). Surface Operations in Petroleum Production. 2. Elsevier. pp. 143–152. ISBN 9780444426772.

- Wan Renpu (2011). Advanced Well Completion Engineering. Gulf Professional Publishing. p. 424. ISBN 9780123858689.

- Gidley, John L. (1989). Recent Advances in Hydraulic Fracturing. SPE Monograph. 12. SPE. p. ?. ISBN 9781555630201.

- Ching H. Yew (1997). Mechanics of Hydraulic Fracturing. Gulf Professional Publishing. p. ?. ISBN 9780884154747.

- Brown, Edwin Thomas (2007) [2003]. Block Caving Geomechanics (2nd ed.). Indooroopilly, Queensland: Julius Kruttschnitt Mineral Research Centre, UQ. ISBN 978-0-9803622-0-6. Retrieved 2012-05-14.

- Bell, Frederic Gladstone (2004). Engineering Geology and Construction. Taylor & Francis. p. 670. ISBN 9780415259392.

- Miller, Bruce G. (2005). Coal Energy Systems. Sustainable World Series. Academic Press. p. 380. ISBN 9780124974517.

- Zukerman, Gregory (2013-11-06). "Breakthrough: The Accidental Discovery That Revolutionized American Energy". The Atlantis. Retrieved 2016-09-18.

- "New Waterless Fracking Method Avoids Pollution Problems, But Drillers Slow to Embrace It". insideclimatenews.org. Retrieved 2016-05-05.

Petroleum Refinery: A Comprehensive Study

Petroleum refinery is an industrial process plant; it is used in processing crude oil. Some of the useful products produced by this process are diesel fuel, asphalt base, heating oil and kerosene. This chapter helps the readers in understanding the basic concepts and processes of petroleum refinery.

Oil Refinery

An oil refinery or petroleum refinery is an industrial processplant where crude oil is processed and refined into more useful products such as petroleum naphtha, gasoline, diesel fuel, asphalt base, heating oil, kerosene, and liquefied petroleum gas. Oil refineries are typically large, sprawling industrial complexes with extensive piping running throughout, carrying streams of fluids between large chemical processing units. In many ways, oil refineries use much of the technology of, and can be thought of, as types of chemical plants. The crude oil feed stock has typically been processed by an oil production plant. There is usually an oil depot (tank farm) at or near an oil refinery for the storage of incoming crude oil feedstock as well as bulk liquid products.

Anacortes Refinery (Tesoro), on the north end of March Point southeast of Anacortes, Washington

An oil refinery is considered an essential part of the downstream side of the petroleum industry.

Operation

Raw or unprocessed crude oil is not generally useful in industrial applications, although "light, sweet" (low viscosity, low sulfur) crude oil has been used directly as a burner fuel to produce steam for the propulsion of seagoing vessels. The lighter elements, however, form explosive vapors in the fuel tanks and are therefore hazardous, especially in warships. Instead, the hundreds of different hydrocarbon molecules in crude oil are separated in a refinery into components which can be used as fuels, lubricants, and as feedstocks in petrochemical processes that manufacture such products as plastics, detergents, solvents, elastomers and fibers such as nylon and polyesters.

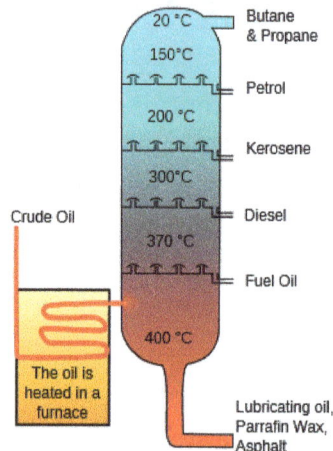

Crude oil is separated into fractions by fractional distillation. The fractions at the top of the fractionating column have lower boiling points than the fractions at the bottom. The heavy bottom fractions are often cracked into lighter, more useful products. All of the fractions are processed further in other refining units.

Petroleumfossil fuels are burned in internal combustion engines to provide power for ships, automobiles, aircraft engines, lawn mowers, dirt bikes, and other machines. Different boiling points allow the hydrocarbons to be separated by distillation. Since the lighter liquid products are in great demand for use in internal combustion engines, a modern refinery will convert heavy hydrocarbons and lighter gaseous elements into these higher value products.

The oil refinery in Haifa, Israel is capable of processing about 9 million tons (66 million barrels) of crude oil a year. Its two cooling towers are landmarks of the city's skyline.

Oil can be used in a variety of ways because it contains hydrocarbons of varying molecular masses, forms and lengths such as paraffins, aromatics, naphthenes (or cycloalkanes), alkenes, dienes, and alkynes. While the molecules in crude oil include different atoms such as sulfur and nitrogen, the hydrocarbons are the most common form of molecules, which are molecules of varying lengths and complexity made of hydrogen and carbonatoms, and a small number of oxygen atoms. The differences in the structure of these molecules account for their varying physical and chemical properties, and it is this variety that makes crude oil useful in a broad range of several applications.

Once separated and purified of any contaminants and impurities, the fuel or lubricant can be sold without further processing. Smaller molecules such as isobutane and propylene or butylenes can be recombined to meet specific octane requirements by processes such as alkylation, or more commonly, dimerization. The octane grade of gasoline can also be improved by catalytic reforming, which involves removing hydrogen from hydrocarbons producing compounds with higher octane ratings such as aromatics. Intermediate products such as gasoils can even be reprocessed to break a heavy, long-chained oil into a lighter short-chained one, by various forms of cracking such as fluid catalytic cracking, thermal cracking, and hydrocracking. The final step in gasoline production is the blending of fuels with different octane ratings, vapor pressures, and other properties to meet product specifications. Another method for reprocessing and upgrading these intermediate products (residual oils) uses a devolatilization process to separate usable oil from the waste asphaltene material.

Oil refineries are large scale plants, processing about a hundred thousand to several hundred thousand barrels of crude oil a day. Because of the high capacity, many of the units operate continuously, as opposed to processing in batches, at steady state or nearly steady state for months to years. The high capacity also makes process optimization and advanced process control very desirable.

Major Products

Petroleum products are usually grouped into four categories: light distillates (LPG, gasoline, naphtha), middle distillates (kerosene, jet fuel, diesel), heavy distillates and residuum (heavy fuel oil, lubricating oils, wax, asphalt). This classification is based on the way crude oil is distilled and separated into fractions (called distillates and residuum) as in the above drawing.

- Liquified petroleum gas (LPG)

- Gasoline (also known as petrol)

- Naphtha

- Kerosene and related jet aircraft fuels

- Diesel fuel

- Fuel oils

- Lubricating oils

- Paraffin wax

- Asphalt and tar

- Petroleum coke

- Sulfur

- Olefines

- Heat and electrical energy

Oil refineries also produce various intermediate products such as hydrogen, light hydrocarbons, reformate and pyrolysis gasoline. These are not usually transported but instead are blended or processed further on-site. Chemical plants are thus often adjacent to oil refineries or a number of further chemical processes are integrated into it. For example, light hydrocarbons are steam-cracked in an ethylene plant, and the produced ethylene is polymerized to produce polyethene.

Because technical reasons and environment protection demand a very low sulfur content in all but the most heavy products, it is transformed to hydrogen sulfide via catalytic Hydrodesulfurization and removed from the product stream via Amine gas treating. Using the so-called Claus process, hydrogen sulfide is afterwards transformed to elementary sulfur to be sold to the chemical industry. The rather large heat energy freed by this process is directly used in the other parts of the refinery. Often an electrical power plant is combined into the whole refinery process to take up the excess heat.

Common Process Units Found in a Refinery

Storage tanks and towers at Shell Puget Sound Refinery (Shell Oil Company), Anacortes, Washington

- Desalter unit washes out salt from the crude oil before it enters the atmospheric distillation unit.

- Atmospheric distillation unit distills crude oil into fractions.

- Vacuum distillation unit further distills residual bottoms after atmospheric distillation.

- Naphtha hydrotreater unit uses hydrogen to desulfurize naphtha from atmospheric distillation. Must hydrotreat the naphtha before sending to a catalytic reformer unit.

- Catalytic reformer unit is used to convert the naphtha-boiling range molecules into higher octane reformate (reformer product). The reformate has higher content of aromatics and cyclic hydrocarbons). An important byproduct of a reformer is hydrogen released during the catalyst reaction. The hydrogen is used either in the hydrotreaters or the hydrocracker.

- Distillate hydrotreater desulfurizes distillates (such as diesel) after atmospheric distillation.

- Fluid Catalytic Cracker (FCC) unit upgrades heavier fractions into lighter, more valuable products.

- Hydrocracker unit uses hydrogen to upgrade heavier fractions into lighter, more valuable products.

- Visbreaking unit upgrades heavy residual oils by thermally cracking them into lighter, more valuable reduced viscosity products.

- Merox unit treats LPG, kerosene or jet fuel by oxidizing mercaptans to organic disulfides.

- Alternative processes for removing mercaptans are known, e.g. doctor sweetening process and caustic washing.

- Coking units (delayed coking, fluid coker, and flexicoker) process very heavy residual oils into gasoline and diesel fuel, leaving petroleum coke as a residual product.

- Alkylation unit uses sulfuric acid or hydrofluoric acid to produce high-octane components for gasoline blending.

- Dimerization unit converts olefins into higher-octane gasoline blending components. For example, butenes can be dimerized into isooctene which may subsequently be hydrogenated to form isooctane. There are also other uses for dimerization. Gasoline produced through dimerization is highly unsaturated and very reactive. It tends spontaneously to form gums. For this reason the effluent from the dimerization need to be blended into the finished gasoline pool immediately or hydrogenated.

- Isomerization unit converts linear molecules to higher-octane branched molecules for blending into gasoline or feed to alkylation units.

- Steam reforming unit produces hydrogen for the hydrotreaters or hydrocracker.

- Liquified gas storage vessels store propane and similar gaseous fuels at pressure sufficient to maintain them in liquid form. These are usually spherical vessels or "bullets" (i.e., horizontal vessels with rounded ends).

- Storage tanks store crude oil and finished products, usually cylindrical, with some sort of vapor emission control and surrounded by an earthen berm to contain spills.

- Amine gas treater, Claus unit, and tail gas treatment convert hydrogen sulfide from hydrodesulfurization into elemental sulfur.

- Utility units such as cooling towers circulate cooling water, boiler plants generates steam, and instrument air systems include pneumatically operated control valves and an electrical substation.

- Wastewater collection and treating systems consist of API separators, dissolved air flotation (DAF) units and further treatment units such as an activated sludge biotreater to make water suitable for reuse or for disposal.

- Solvent refining units use solvent such as cresol or furfural to remove unwanted, mainly aromatics from lubricating oil stock or diesel stock.

- Solvent dewaxing units remove the heavy waxy constituents petrolatum from vacuum distillation products.

Flow Diagram of Typical Refinery

The image below is a schematic flow diagram of a typical oil refinery that depicts the various unit processes and the flow of intermediate product streams that occurs between the inlet crude oil feedstock and the final end products. The diagram depicts only one of the literally hundreds of different oil refinery configurations. The diagram also does not include any of the usual refinery facilities providing utilities such as steam, cooling water, and electric power as well as storage tanks for crude oil feedstock and for intermediate products and end products.

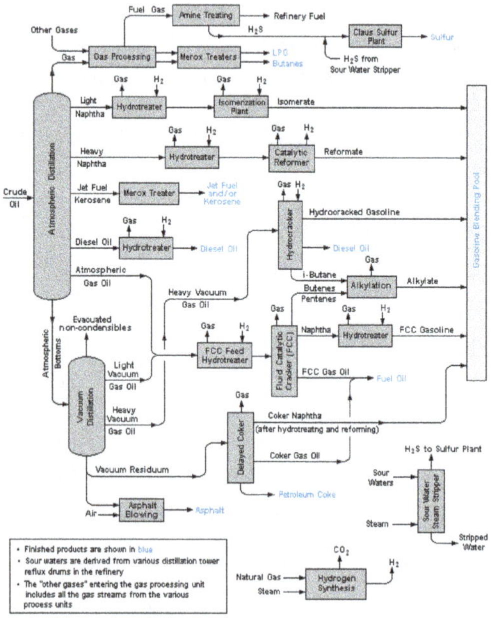

Schematic flow diagram of a typical oil refinery

There are many process configurations other than that depicted above. For example, the vacuum distillation unit may also produce fractions that can be refined into endproducts such as: spindle oil used in the textile industry, light machinery oil, motor oil, and various waxes.

The Crude Oil Distillation Unit

The crude oil distillation unit (CDU) is the first processing unit in virtually all petroleum refineries. The CDU distills the incoming crude oil into various fractions of different boiling ranges, each of which are then processed further in the other refinery processing units. The CDU is often referred to as the *atmospheric distillation unit* because it operates at slightly above atmospheric pressure.

Below is a schematic flow diagram of a typical crude oil distillation unit. The incoming crude oil is preheated by exchanging heat with some of the hot, distilled fractions and other streams. It is then desalted to remove inorganic salts (primarily sodium chloride).

Following the desalter, the crude oil is further heated by exchanging heat with some of the hot, distilled fractions and other streams. It is then heated in a fuel-fired furnace (fired heater) to a temperature of about 398 °C and routed into the bottom of the distillation unit.

The cooling and condensing of the distillation tower overhead is provided partially by exchanging heat with the incoming crude oil and partially by either an air-cooled or water-cooled condenser. Additional heat is removed from the distillation column by a pumparound system as shown in the diagram below.

As shown in the flow diagram, the overhead distillate fraction from the distillation column is naphtha. The fractions removed from the side of the distillation column at various points between the column top and bottom are called *sidecuts*. Each of the sidecuts (i.e., the kerosene, light gas oil and heavy gas oil) is cooled by exchanging heat with the incoming crude oil. All of the fractions (i.e., the overhead naphtha, the sidecuts and the bottom residue) are sent to intermediate storage tanks before being processed further.

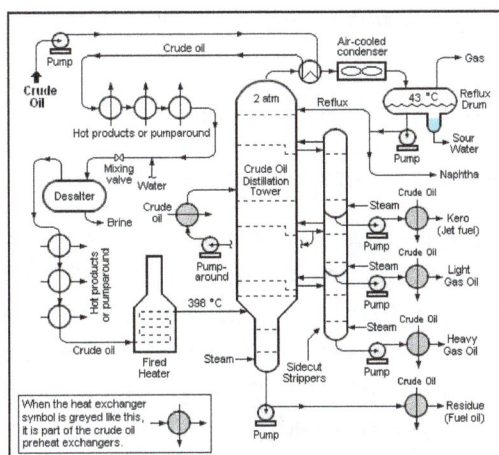

Schematic flow diagram of a typical crude oil distillation unit as used in petroleum crude oil refineries.

Specialty and Products

These require blending various feedstocks, mixing appropriate additives, providing short term storage, and preparation for bulk loading to trucks, barges, product ships, and railcars:

- Gaseous fuels such as propane, stored and shipped in liquid form under pressure in specialized railcars to distributors.

- Lubricants (produces light machine oils, motor oils, and greases, adding viscosity stabilizers as required), usually shipped in bulk to an offsite packaging plant.

- Wax (paraffin), used in the packaging of frozen foods, among others. May be shipped in bulk to a site to prepare as packaged blocks.

- Sulfur (or sulfuric acid), byproducts of sulfur removal from petroleum which may have up to a couple percent sulfur as organic sulfur-containing compounds. Sulfur and sulfuric acid are useful industrial materials. Sulfuric acid is usually prepared and shipped as the acid precursor oleum.

- Bulk tar shipping for offsite unit packaging for use in tar-and-gravel roofing.

- Asphalt unit. Prepares bulk asphalt for shipment.

- Petroleum coke, used in specialty carbon products or as solid fuel.

- Petrochemicals or petrochemical feedstocks, which are often sent to petrochemical plants for further processing in a variety of ways. The petrochemicals may be olefins or their precursors, or various types of aromatic petrochemicals.

Siting/Locating of Petroleum Refineries

A party searching for a site to construct a refinery or a chemical plant needs to consider the following issues:

- The site has to be reasonably far from residential areas.

- Infrastructure should be available for supply of raw materials and shipment of products to markets.

- Energy to operate the plant should be available.

- Facilities should be available for waste disposal.

Refineries which use a large amount of steam and cooling water need to have an abundant source of water. Oil refineries therefore are often located nearby navigable rivers or on a sea shore, nearby a port. Such location also gives access to transportation by river or by sea. The advantages of transporting crude oil by pipeline are evident, and oil companies often transport a large volume of fuel to distribution terminals by pipeline. Pipeline may not be practical for products with small output, and rail cars, road tankers, and barges are used.

Petrochemical plants and solvent manufacturing (fine fractionating) plants need spaces for further processing of a large volume of refinery products for further processing, or to mix chemical additives with a product at source rather than at blending terminals.

Safety and Environmental Concerns

Fire-extinguishing operations after the Texas City refinery explosion.

The refining process releases a number of different chemicals into the atmosphere and a notable odor normally accompanies the presence of a refinery. Aside from air pollution impacts there are also wastewater concerns, risks of industrial accidents such as fire and explosion, and noise health effects due to industrial noise.

Many governments worldwide have mandated restrictions on contaminants that refineries release, and most refineries have installed the equipment needed to comply with the requirements of the pertinent environmental protection regulatory agencies. In the United States, there is strong pressure to prevent the development of new refineries, and no major refinery has been built in the country since Marathon'sGaryville, Louisiana facility in 1976. However, many existing refineries have been expanded during that time. Environmental restrictions and pressure to prevent construction of new refineries may have also contributed to rising fuel prices in the United States. Additionally, many refineries (more than 100 since the 1980s) have closed due to obsolescence and/or merger activity within the industry itself.

Environmental and safety concerns mean that oil refineries are sometimes located some distance away from major urban areas. Nevertheless, there are many instances where refinery operations are close to populated areas and pose health risks. In California's Contra Costa County and Solano County, a shoreline necklace of refineries, built in the early 20th century before this area was populated, and associated chemical plants are adjacent to urban areas in Richmond, Martinez, Pacheco, Concord, Pittsburg, Vallejo and Benicia, with occasional accidental events that require "shelter in place" orders to the adjacent populations.

NIOSH criteria for occupational exposure to refined petroleum solvents have been available since 1977.

Corrosion Problems and Prevention

Refinery of Slovnaft in Bratislava.

Petroleum refineries run as efficiently as possible to reduce costs. One major factor that decreases efficiency is corrosion of the metallic components found throughout refining process. Corrosion causes the failure of equipment items as well as dictating the maintenance schedule of the refinery, during which part or all of the refinery must be shut down. The corrosion-related direct costs in the U.S. petroleum industry as of 1996 was estimated as US$3.7 billion per year.

Corrosion occurs in various forms in the refining process, such as pitting corrosion from water droplets, embrittlement from hydrogen, and stress corrosion cracking from sulfide attack. From a materials standpoint, carbon steel is used for upwards of 80 per cent of refinery components, which is beneficial due to its low cost. Carbon steel is resistant to the most common forms of corrosion, particularly from hydrocarbon impurities at temperatures below 205 °C, but other corrosive chemicals and environments prevent its use everywhere. Common replacement materials are low alloy steels containing chromium and molybdenum, with stainless steels containing more chromium dealing with more corrosive environments. More expensive materials commonly used are nickel, titanium, and copper alloys. These are primarily saved for the most problematic areas where extremely high temperatures and/or very corrosive chemicals are present.

Corrosion is fought by a complex system of monitoring, preventative repairs and careful use of materials. Monitoring methods include both off-line checks taken during maintenance and on-line monitoring. Off-line checks measure corrosion after it has occurred, telling the engineer when equipment must be replaced based on the historical information he has collected. This is referred to as preventative management.

Oil refinery in Iran.

On-line systems are a more modern development, and are revolutionizing the way corrosion is approached. There are several types of on-line corrosion monitoring technologies such as linear polarization resistance, electrochemical noise and electrical resistance. On-Line monitoring has generally had slow reporting rates in the past (minutes or hours) and been limited by process conditions and sources of error but newer technologies can report rates up to twice per minute with much higher accuracy (referred to as real-time monitoring). This allows process engineers to treat corrosion as another process variable that can be optimized in the system. Immediate responses to process changes allow the control of corrosion mechanisms, so they can be minimized while also maximizing production output. In an ideal situation having on-line corrosion information that is accurate and real-time will allow conditions that cause high corrosion rates to be identified and reduced. This is known as predictive management.

Materials methods include selecting the proper material for the application. In areas of minimal corrosion, cheap materials are preferable, but when bad corrosion can occur, more expensive but longer lasting materials should be used. Other materials methods come in the form of protective barriers between corrosive substances and the equipment metals. These can be either a lining of refractory material such as standard Portland cement or other special acid-resistant cements that are shot onto the inner surface of the vessel. Also available are thin overlays of more expensive metals that protect cheaper metal against corrosion without requiring lots of material.

History

The world's first refinery opened at Ploie□ti, Romania, in 1856-1857. After being taken over by Nazi Germany, the Ploie□ti refineries were bombed in Operation Tidal Wave by the Allies during the Oil Campaign of World War II.

At one point, the refinery in Ras Tanura, Saudi Arabia owned by Saudi Aramco was claimed to be the largest oil refinery in the world. For most of the 20th century, the largest refinery was the Abadan Refinery in Iran. This refinery suffered extensive damage during the Iran–Iraq War. On the 31 December 2014, the world's largest refinery complex is the Jamnagar Refinery Complex, consisting of two refineries side by side operated by Reliance Industries Limited in Jamnagar, India with a combined production capacity of 1,240,000 barrels per day (197,000 m³/d). PDVSA's Paraguaná Refinery Complex in Paraguaná Peninsula, Venezuela with a capacity of 940,000 bbl/d (149,000 m³/d) and SK Energy's Ulsan in South Korea with 840,000 bbl/d (134,000 m³/d) are the second and third largest, respectively.

Oil Refining in the United States

In the 19th century, refineries in the U.S. processed crude oil primarily to recover the kerosene. There was no market for the more volatile fraction, including gasoline, which was considered waste and was often dumped directly into the nearest river. The invention of the automobile shifted the demand to gasoline and diesel, which remain the primary refined products today. Today, national and state legislation requires refineries to meet stringent air and water cleanliness standards. In fact, oil companies in the U.S. perceive obtaining a permit to build a modern refinery to be so difficult and costly that no new refineries were built (though many have been expanded) in the U.S. from 1976 until 2014, when the small Dakota Prairie Refinery in North Dakota is set to begin operation. More than half the refineries that existed in 1981 are now closed due to low utilization rates and accelerating mergers. As a result of these closures total US refinery capacity fell between 1981 and 1995, though the operating capacity stayed fairly constant in that time period at around 15,000,000 barrels per day (2,400,000 m³/d). Increases in facility size and improvements in efficiencies have offset much of the lost physical capacity of the industry. In 1982 (the earliest data provided), the United States operated 301 refineries with a combined capacity of 17.9 million barrels (2,850,000 m³) of crude oil each calendar day. In 2010, there were 149 operable U.S. refineries with a combined capacity of 17.6 million barrels (2,800,000 m³) per calendar day. By 2014 the number of refinery had reduced to 140 but the total capacity increased to 18.02 million barrels (2,865,000 m³) per calendar day. Indeed, in order to reduce operating costs and depreciation, refining is operated in less sites but of bigger capacity.

In 2009 through 2010, as revenue streams in the oil business dried up and profitability of oil refineries fell due to lower demand for product and high reserves of supply preceding the economic recession, oil companies began to close or sell the less profitable refineries.

ExxonMobil oil refinery in Baton Rouge, Louisiana (the fourth-largest in the United States)

Worldwide Oil Refining Capacity

According to the Oil and Gas Journal in the world a total of 636 refineries were operated on the 31 December 2014 for a total capacity of 87.75 million barrels (13,951,000 m³).

Amine Gas Treating

Amine gas treating, also known as amine scrubbing, gas sweetening and acid gas removal, refers to a group of processes that use aqueous solutions of various alkylamines (commonly referred to simply as amines) to remove hydrogen sulfide (H_2S) and carbon dioxide (CO_2) from gases. It is a common unit process used in refineries, and is also used in petrochemical plants, natural gas processing plants and other industries.

Processes within oil refineries or chemical processing plants that remove hydrogen sulfide are referred to as "sweetening" processes because the odor of the processed products is improved by the absence of hydrogen sulfide. An alternative to the use of amines involves membrane technology. However, membrane separation is less attractive due to the relatively high capital and operating costs as well as other technical factors.

Many different amines are used in gas treating:

- Diethanolamine (DEA)

- Monoethanolamine (MEA)

- Methyldiethanolamine (MDEA)

- Diisopropanolamine (DIPA)

- Aminoethoxyethanol (Diglycolamine) (DGA)

The most commonly used amines in industrial plants are the alkanolamines DEA, MEA, and MDEA. These amines are also used in many oil refineries to remove sour gases from liquid hydrocarbons such as liquified petroleum gas (LPG).

Description of a Typical Amine Treater

Gases containing H_2S or both H_2S and CO_2 are commonly referred to as sour gases or acid gases in the hydrocarbon processing industries.

The chemistry involved in the amine treating of such gases varies somewhat with the particular amine being used. For one of the more common amines, monoethanolamine (MEA) denoted as *RNH₂*, the chemistry may be expressed as:

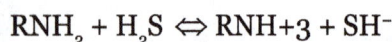

$$RNH_2 + H_2S \Leftrightarrow RNH+3 + SH^-$$

A typical amine gas treating process (the Girbotol process, as shown in the flow diagram below) includes an absorber unit and a regenerator unit as well as accessory equipment. In the absorber, the downflowing amine solution absorbs H_2S and CO_2 from the upflowing sour gas to produce a sweetened gas stream (i.e., a gas free of hydrogen sulfide and carbon dioxide) as a product and an amine solution rich in the absorbed acid gases. The resultant "rich" amine is then routed into the regenerator (a stripper with a reboiler) to produce regenerated or "lean" amine that is recycled for reuse in the absorber. The stripped overhead gas from the regenerator is concentrated H_2S and CO_2.

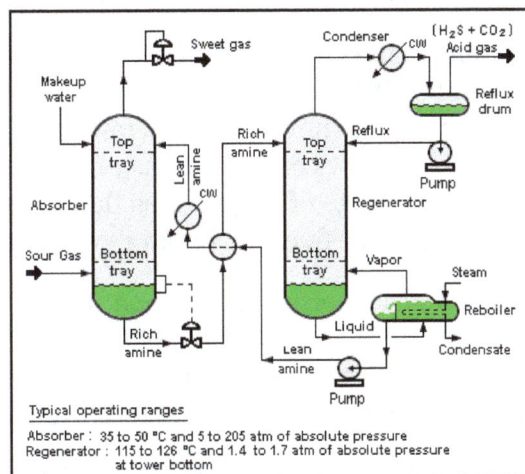

Process flow diagram of a typical amine treating process used in petroleum refineries, natural gas processing plants and other industrial facilities.

Alternative Processes

Alternative stripper configurations include matrix, internal exchange, flashing feed, and multipressure with split feed. Many of these configurations offer more energy efficiency for specific solvents or operating conditions. Vacuum operation favors solvents with low heats of absorption while operation at normal pressure favors solvents with high heats of absorption. Solvents with high heats of absorption require less energy for stripping from temperature swing at fixed capacity. The matrix stripper recovers 40% of CO_2 at a higher pressure and does not have inefficiencies associated with multipressure stripper. Energy and costs are reduced since the reboiler duty cycle is slightly less than normal pressure stripper. An Internal Exchange stripper has a smaller ratio of water vapor to CO_2 in the overheads stream, and therefore less steam is required. The multipressure configuration with split feed reduces the flow into the bottom section, which also reduces the equivalent work. Flashing feed requires less heat input because it uses the latent heat of water vapor to help strip some of the CO_2 in the rich stream entering the stripper at the bottom of the column. The multipressure configuration is more attractive for solvents with a higher heats of absorption.

Amines

The amine concentration in the absorbent aqueous solution is an important parameter in the design and operation of an amine gas treating process. Depending on which one of the following four amines the unit was designed to use and what gases it was designed to remove, these are some typical amine concentrations, expressed as weight percent of pure amine in the aqueous solution:

- Monoethanolamine: About 20 % for removing H_2S and CO_2, and about 32 % for removing only CO_2.

- Diethanolamine: About 20 to 25 % for removing H_2S and CO_2

- Methyldiethanolamine: About 30 to 55% % for removing H_2S and CO_2

- Diglycolamine: About 50 % for removing H_2S and CO_2

The choice of amine concentration in the circulating aqueous solution depends upon a number of factors and may be quite arbitrary. It is usually made simply on the basis of experience. The factors involved include whether the amine unit is treating raw natural gas or petroleum refinery by-product gases that contain relatively low concentrations of both H_2S and CO_2 or whether the unit is treating gases with a high percentage of CO_2 such as the offgas from the steam reforming process used in ammonia production or the flue gases from power plants.

Both H_2S and CO_2 are acid gases and hence corrosive to carbon steel. However, in an amine treating unit, CO_2 is the stronger acid of the two. H_2S forms a film of iron sulfide on the surface of the steel that acts to protect the steel. When treating gases with a high percentage of CO_2, corrosion inhibitors are often used and that permits the use of higher concentrations of amine in the circulating solution.

Another factor involved in choosing an amine concentration is the relative solubility of H_2S and CO_2 in the selected amine. The choice of the type of amine will affect the required circulation rate of amine solution, the energy consumption for the regeneration and the ability to selectively remove either H_2S alone or CO_2 alone if desired. For more information about selecting the amine concentration, the reader is referred to Kohl and Nielsen's book.

MEA and DEA

MEA and DEA are primary and secondary amines. They are very reactive and can effectively remove a high volume of gas removal due to a high reaction rate. However, due to stoichiometry, the loading capacity is limited to 0.5 mol CO_2 per mole of amine. MEA and DEA also require a large amount of energy to strip the CO_2 to during regeneration, which can be up to 70% of total operating costs. They are also more corrosive and chemically unstable compared to other amines.

Uses

In oil refineries, that stripped gas is mostly H_2S, much of which often comes from a sulfur-removing process called hydrodesulfurization. This H_2S-rich stripped gas stream is then usually routed into a Claus process to convert it into elemental sulfur. In fact, the vast majority of the 64,000,000 metric tons of sulfur produced worldwide in 2005 was byproduct sulfur from refineries and other hydrocarbon processing plants. Another sulfur-removing process is the WSA Process which recovers sulfur in any form as concentrated sulfuric acid. In some plants, more than one amine absorber unit may share a common regenerator unit. The current emphasis on removing CO_2 from the flue gases emitted by fossil fuel power plants has led to much interest in using amines for removing CO_2.

In the specific case of the industrial synthesis of ammonia, for the steam reforming process of hydrocarbons to produce gaseous hydrogen, amine treating is one of the commonly used processes for removing excess carbon dioxide in the final purification of the gaseous hydrogen.

In the biogas production it is sometimes necessary to remove carbon dioxide from the biogas to make it comparable with the natural. The removal of the sometimes high content of hydrogen sulfide is necessary to prevent corrosion of metallic parts after burning the bio gas.

Carbon Capture and Storage

Amines are used to remove CO_2 in various areas ranging from natural gas production to the food and beverage industry, and have been for over sixty years.

There are multiple classifications of amines, each of which has different characteristics relevant to CO_2 capture. For example, Monoethanolamine (MEA) reacts strongly with acid gases like CO_2 and has a fast reaction time and an ability to remove high percentages of CO_2, even at the low CO_2 concentrations. Typically, Monoethanolamine (MEA) can capture 85% to 90% of the CO_2 from the flue gas of a coal-fired plant, which is one of the most effective solvent to capture CO_2.

Challenges of carbon capture using amine include:

- Low pressure gas increases difficulty of transferring CO_2 from the gas into amine
- Oxygen content of the gas can cause amine degradation and acid formation
- CO_2 degradation of primary (and secondary) amines
- High energy consumption
- Very large facilities
- Finding suitable location for the removed CO_2

The partial pressure is the driving force to transfer CO_2 into the liquid phase. Under the low pressure, this transfer is hard to achieve without increasing the reboiler's heat duty, which will result in higher cost.

Primary and secondary amines, for example, MEA and DEA, will react with CO_2 and form degradation products. O_2 from the inlet gas will cause degradation as well. The degraded amine is no longer able to capture CO_2, which decreases the overall carbon capture efficiency.

Currently, variety of amine mixtures are being synthesized and tested to achieve a more desirable set of overall properties for use in CO_2 capture systems. One major focus is on lowering the energy required for solvent regeneration, which has a major impact on process costs. However, there are tradeoffs to consider. For example, the energy required for regeneration is typically related to the driving forces for achieving high capture capacities. Thus, reducing the regeneration energy can lower the driving force and thereby increase the amount of solvent and size of absorber needed to capture a given amount of CO_2, thus, increasing the capital cost.

Catalytic Reforming

Catalytic reforming is a chemical process used to convert petroleum refinerynaphthas distilled from crude oil (typically having low octane ratings) into high-octane liquid products called **refor-**

mates, which are premium blending stocks for high-octane gasoline. The process converts low-octane linear hydrocarbons (paraffins) into branched alkanes (isoparaffins) and cyclic naphthenes, which are then partially dehydrogenated to produce high-octane aromatic hydrocarbons. The dehydrogenation also produces significant amounts of byproduct hydrogen gas, which is fed into other refinery processes such as hydrocracking. A side reaction is hydrogenolysis, which produces light hydrocarbons of lower value, such as methane, ethane, propane and butanes. It is also the conversion of straight chains of alkane catalytically

In addition to a gasoline blending stock, reformate is the main source of aromatic bulk chemicals such as benzene, toluene, xylene and ethylbenzene which have diverse uses, most importantly as raw materials for conversion into plastics. However, the benzene content of reformate makes it carcinogenic, which has led to governmental regulations effectively requiring further processing to reduce its benzene content.

This process is quite different from and not to be confused with the catalytic steam reforming process used industrially to produce products such as hydrogen, ammonia, and methanol from natural gas, naphtha or other petroleum-derived feedstocks. Nor is this process to be confused with various other catalytic reforming processes that use methanol or biomass-derived feedstocks to produce hydrogen for fuel cells or other uses.

History

In the 1940s, Vladimir Haensel, a research chemist working for Universal Oil Products (UOP), developed a catalytic reforming process using a catalyst containing platinum. Haensel's process was subsequently commercialized by UOP in 1949 for producing a high octane gasoline from low octane naphthas and the UOP process become known as the Platforming process. The first Platforming unit was built in 1949 at the refinery of the Old Dutch Refining Company in Muskegon, Michigan.

In the years since then, many other versions of the process have been developed by some of the major oil companies and other organizations. Today, the large majority of gasoline produced worldwide is derived from the catalytic reforming process.

To name a few of the other catalytic reforming versions that were developed, all of which utilized a platinum and/or a rhenium catalyst:

- Rheniforming: Developed by Chevron Oil Company.

- Powerforming: Developed by Esso Oil Company, currently known as ExxonMobil.

- Magnaforming: Developed by Engelhard and Atlantic Richfield Oil Company.

- Ultraforming: Developed by Standard Oil of Indiana, now a part of the British Petroleum Company.

- Houdriforming: Developed by the Houdry Process Corporation.

- CCR Platforming: A Platforming version, designed for continuous catalyst regeneration, developed by UOP.

- Octanizing: A catalytic reforming version developed by Axens, a subsidiary of Institut fran-cais du petrole (IFP), designed for continuous catalyst regeneration.

Chemistry

Before describing the reaction chemistry of the catalytic reforming process as used in petroleum refineries, the typical naphthas used as catalytic reforming feedstocks will be discussed.

Typical Naphtha Feedstocks

A petroleum refinery includes many unit operations and unit processes. The first unit operation in a refinery is the continuous distillation of the petroleum crude oil being refined. The overhead liquid distillate is called naphtha and will become a major component of the refinery's gasoline (petrol) product after it is further processed through a catalytic hydrodesulfurizer to remove sulfur-containing hydrocarbons and a catalytic reformer to reform its hydrocarbon molecules into more complex molecules with a higher octane rating value. The naphtha is a mixture of very many different hydrocarbon compounds. It has an initial boiling point of about 35 °C and a final boiling point of about 200 °C, and it contains paraffin, naphthene (cyclic paraffins) and aromatic hydrocarbons ranging from those containing 4 carbon atoms to those containing about 10 or 11 carbon atoms.

The naphtha from the crude oil distillation is often further distilled to produce a "light" naphtha containing most (but not all) of the hydrocarbons with 6 or fewer carbon atoms and a "heavy" naphtha containing most (but not all) of the hydrocarbons with more than 6 carbon atoms. The heavy naphtha has an initial boiling point of about 140 to 150 °C and a final boiling point of about 190 to 205 °C. The naphthas derived from the distillation of crude oils are referred to as "straight-run" naphthas.

It is the straight-run heavy naphtha that is usually processed in a catalytic reformer because the light naphtha has molecules with 6 or fewer carbon atoms which, when reformed, tend to crack into butane and lower molecular weight hydrocarbons which are not useful as high-octane gasoline blending components. Also, the molecules with 6 carbon atoms tend to form aromatics which is undesirable because governmental environmental regulations in a number of countries limit the amount of aromatics (most particularly benzene) that gasoline may contain.

It should be noted that there are a great many petroleum crude oil sources worldwide and each crude oil has its own unique composition or "assay". Also, not all refineries process the same crude oils and each refinery produces its own straight-run naphthas with their own unique initial and final boiling points. In other words, naphtha is a generic term rather than a specific term.

The table just below lists some fairly typical straight-run heavy naphtha feedstocks, available for catalytic reforming, derived from various crude oils. It can be seen that they differ significantly in their content of paraffins, naphthenes and aromatics:

Typical Heavy Naphtha Feedstocks				
Crude oil name Location	Barrow Island Australia	Mutineer-Exeter Australia	CPC Blend Kazakhstan	Draugen North Sea
Initial boiling point, °C	149	140	149	150
Final boiling point, °C	204	190	204	180

Paraffins, liquid volume %	46	62	57	38
Naphthenes, liquid volume %	42	32	27	45
Aromatics, liquid volume %	12	6	16	17

Some refinery naphthas include olefinic hydrocarbons, such as naphthas derived from the fluid catalytic cracking and coking processes used in many refineries. Some refineries may also desulfurize and catalytically reform those naphthas. However, for the most part, catalytic reforming is mainly used on the straight-run heavy naphthas, such as those in the above table, derived from the distillation of crude oils.

The Reaction Chemistry

There are many chemical reactions that occur in the catalytic reforming process, all of which occur in the presence of a catalyst and a high partial pressure of hydrogen. Depending upon the type or version of catalytic reforming used as well as the desired reaction severity, the reaction conditions range from temperatures of about 495 to 525 °C and from pressures of about 5 to 45 atm.

The commonly used catalytic reforming catalysts contain noble metals such as platinum and/or rhenium, which are very susceptible to poisoning by sulfur and nitrogen compounds. Therefore, the naphtha feedstock to a catalytic reformer is always pre-processed in a hydrodesulfurization unit which removes both the sulfur and the nitrogen compounds. Most catalysts require both sulphur and nitrogen content to be lower than 1 ppm.

The four major catalytic reforming reactions are:

1: The dehydrogenation of naphthenes to convert them into aromatics as exemplified in the conversion methylcyclohexane (a naphthene) to toluene (an aromatic), as shown below:

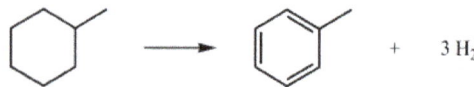

2: The isomerization of normal paraffins to isoparaffins as exemplified in the conversion of normal octane to 2,5-Dimethylhexane (an isoparaffin), as shown below:

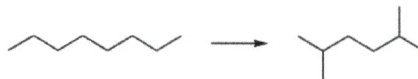

3: The dehydrogenation and aromatization of paraffins to aromatics (commonly called dehydrocyclization) as exemplified in the conversion of normal heptane to toluene, as shown below:

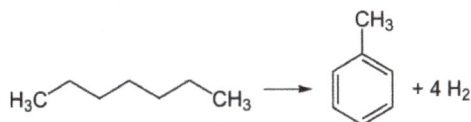

4: The hydrocracking of paraffins into smaller molecules as exemplified by the cracking of normal heptane into isopentane and ethane, as shown below:

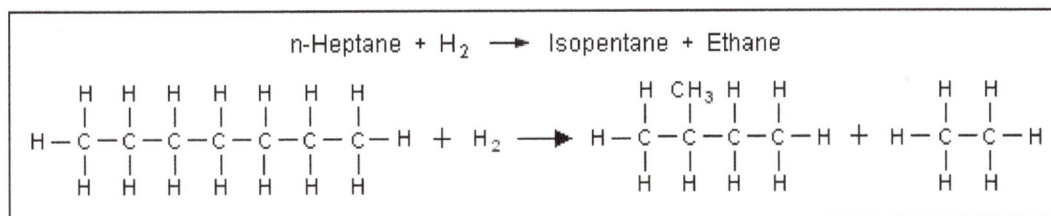

The hydrocracking of paraffins is the only one of the above four major reforming reactions that consumes hydrogen. The isomerization of normal paraffins does not consume or produce hydrogen. However, both the dehydrogenation of naphthenes and the dehydrocyclization of paraffins produce hydrogen. The overall net production of hydrogen in the catalytic reforming of petroleum naphthas ranges from about 50 to 200 cubic meters of hydrogen gas (at 0 °C and 1 atm) per cubic meter of liquid naphtha feedstock. In the United States customary units, that is equivalent to 300 to 1200 cubic feet of hydrogen gas (at 60 °F and 1 atm) per barrel of liquid naphtha feedstock. In many petroleum refineries, the net hydrogen produced in catalytic reforming supplies a significant part of the hydrogen used elsewhere in the refinery (for example, in hydrodesulfurization processes). The hydrogen is also necessary in order to hydrogenolyze any polymers that form on the catalyst.

In practice, the higher the content of naphtenes in the naphtha feedstock, the better will be the quality of the reformate and the higher the production of hydrogen. Crude oils containing the best naphtha for reforming are typically from Western Africa or the North Sea, such as Bonny light or Troll.

Process Description

The most commonly used type of catalytic reforming unit has three reactors, each with a fixed bed of catalyst, and all of the catalyst is regenerated in situ during routine catalyst regeneration shutdowns which occur approximately once each 6 to 24 months. Such a unit is referred to as a semi-regenerative catalytic reformer (SRR).

Some catalytic reforming units have an extra *spare* or *swing* reactor and each reactor can be individually isolated so that any one reactor can be undergoing in situ regeneration while the other reactors are in operation. When that reactor is regenerated, it replaces another reactor which, in turn, is isolated so that it can then be regenerated. Such units, referred to as *cyclic* catalytic reformers, are not very common. Cyclic catalytic reformers serve to extend the period between required shutdowns.

The latest and most modern type of catalytic reformers are called continuous catalyst regeneration (CCR) reformers. Such units are characterized by continuous in-situ regeneration of part of the catalyst in a special regenerator, and by continuous addition of the regenerated catalyst to the operating reactors. As of 2006, two CCR versions available: UOP's CCR Platformer process and Axens' Octanizing process. The installation and use of CCR units is rapidly increasing.

Many of the earliest catalytic reforming units (in the 1950s and 1960s) were non-regenerative in that they did not perform in situ catalyst regeneration. Instead, when needed, the aged catalyst was replaced by fresh catalyst and the aged catalyst was shipped to catalyst manufacturers to be either regenerated or to recover the platinum content of the aged catalyst. Very few, if any, catalytic reformers currently in operation are non-regenerative.

The process flow diagram below depicts a typical semi-regenerative catalytic reforming unit.

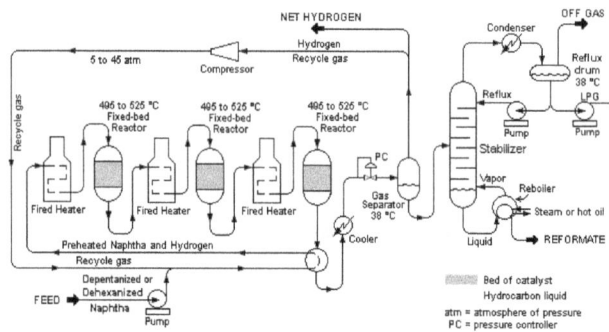

Schematic diagram of a typical semi-regenerative catalytic reformer unit in a petroleum refinery

The liquid feed (at the bottom left in the diagram) is pumped up to the reaction pressure (5–45 atm) and is joined by a stream of hydrogen-rich recycle gas. The resulting liquid–gas mixture is preheated by flowing through a heat exchanger. The preheated feed mixture is then totally vaporized and heated to the reaction temperature (495–520 °C) before the vaporized reactants enter the first reactor. As the vaporized reactants flow through the fixed bed of catalyst in the reactor, the major reaction is the dehydrogenation of naphthenes to aromatics (as described earlier herein) which is highly endothermic and results in a large temperature decrease between the inlet and outlet of the reactor. To maintain the required reaction temperature and the rate of reaction, the vaporized stream is reheated in the second fired heater before it flows through the second reactor. The temperature again decreases across the second reactor and the vaporized stream must again be reheated in the third fired heater before it flows through the third reactor. As the vaporized stream proceeds through the three reactors, the reaction rates decrease and the reactors therefore become larger. At the same time, the amount of reheat required between the reactors becomes smaller. Usually, three reactors are all that is required to provide the desired performance of the catalytic reforming unit.

Some installations use three separate fired heaters as shown in the schematic diagram and some installations use a single fired heater with three separate heating coils.

The hot reaction products from the third reactor are partially cooled by flowing through the heat exchanger where the feed to the first reactor is preheated and then flow through a water-cooled heat exchanger before flowing through the pressure controller (PC) into the gas separator.

Most of the hydrogen-rich gas from the gas separator vessel returns to the suction of the recycle hydrogen gas compressor and the net production of hydrogen-rich gas from the reforming reactions is exported for use in the other refinery processes that consume hydrogen (such as hydrodesulfurization units and/or a hydrocracker unit).

The liquid from the gas separator vessel is routed into a fractionating column commonly called a *stabilizer*. The overhead offgas product from the stabilizer contains the byproduct methane, ethane, propane and butane gases produced by the hydrocracking reactions as explained in the above discussion of the reaction chemistry of a catalytic reformer, and it may also contain some small amount of hydrogen. That offgas is routed to the refinery's central gas processing plant for removal and recovery of propane and butane. The residual gas after such processing becomes part of the refinery's fuel gas system.

The bottoms product from the stabilizer is the high-octane liquid reformate that will become a component of the refinery's product gasoline. Reformate can be blended directly in the gasoline pool but often it is separated in two or more streams. A common refining scheme consists in fractionating the reformate in two streams, light and heavy reformate. The light reformate has lower octane and can be used as isomerization feedstock if this unit is available. The heavy reformate is high in octane and low in benzene, hence it is an excellent blending component for the gasoline pool.

Benzene is often removed with a specific operation to reduce the content of benzene in the reformate as the finished gasoline has often an upper limit of benzene content (in the UE this is 1% volume). The benzene extracted can be marketed as feedstock for the chemical industry.

Catalysts and Mechanisms

Most catalytic reforming catalysts contain platinum or rhenium on a silica or silica-alumina support base, and some contain both platinum and rhenium. Fresh catalyst is chlorided (chlorinated) prior to use.

The noble metals (platinum and rhenium) are considered to be catalytic sites for the dehydrogenation reactions and the chlorinated alumina provides the acid sites needed for isomerization, cyclization and hydrocracking reactions. The biggest care has to be exercised during the chlorination. Indeed, if not chlorinated (or insufficiently chlorinated) the platinum and rhenium in the catalyst would be reduced almost immediately to metallic state by the hydrogen in the vapour phase. On the other an excessive chlorination could depress excessively the activity of the catalyst.

The activity (i.e., effectiveness) of the catalyst in a semi-regenerative catalytic reformer is reduced over time during operation by carbonaceous coke deposition and chloride loss. The activity of the catalyst can be periodically regenerated or restored by in situ high temperature oxidation of the coke followed by chlorination. As stated earlier herein, semi-regenerative catalytic reformers are regenerated about once per 6 to 24 months. The higher the severity of the reacting conditions (temperature), the higher is the octane of the produced reformate but also the shorter will be the duration of the cycle between two regenerations. Catalyst's cycle duration is also very dependent on the quality of the feedstock. However, independently of the crude oil used in the refinery, all catalysts require a maximum final boiling point of the naphtha feedstock of 180 °C.

Normally, the catalyst can be regenerated perhaps 3 or 4 times before it must be returned to the manufacturer for reclamation of the valuable platinum and/or rhenium content.

Cracking (Chemistry)

In petroleum geology and chemistry, cracking is the process whereby complex organicmolecules such as kerogens or long chain hydrocarbons are broken down into simpler molecules such as light hydrocarbons, by the breaking of carbon-carbon bonds in the precursors. The rate of cracking and the end products are strongly dependent on the temperature and presence of catalysts. Cracking is the breakdown of a large alkane into smaller, more useful alkanes and alkenes. Simply put, hydrocarbon cracking is the process of breaking a long-chain of hydrocarbons into short ones. This process might require high temperatures and high pressure.

More loosely, outside the field of petroleum chemistry, the term "cracking" is used to describe any type of splitting of molecules under the influence of heat, catalysts and solvents, such as in processes of destructive distillation or pyrolysis.

Fluid catalytic cracking produces a high yield of petrol and LPG, while hydrocracking is a major source of jet fuel, Diesel fuel, naphtha, and again yields LPG.

Refinery using the Shukhov cracking process, Baku, Soviet Union, 1934.

History and Patents

Among several variants of thermal cracking methods (variously known as the "Shukhov cracking process", "Burton cracking process", "Burton-Humphreys cracking process", and "Dubbs cracking process") Vladimir Shukhov, a Russian engineer, invented and patented the first in 1891 (Russian Empire, patent no. 12926, November 27, 1891). One installation was used to a limited extent in Russia, but development was not followed up. In the first decade of the 20th century the American engineers William Merriam Burton and Robert E. Humphreys independently developed and patented a similar process as U.S. patent 1,049,667 on June 8, 1908. Among its advantages was the fact that both the condenser and the boiler were continuously kept under pressure.

In its earlier versions however, it was a batch process, rather than continuous, and many patents were to follow in the USA and Europe, though not all were practical. In 1924, a delegation from the American Sinclair Oil Corporation visited Shukhov. Sinclair Oil apparently wished to suggest that the patent of Burton and Humphreys, in use by Standard Oil, was derived from Shukhov's patent for oil cracking, as described in the Russian patent. If that could be established, it could strengthen the hand of rival American companies wishing to invalidate the Burton-Humphreys patent. In the event Shukhov satisfied the Americans that in principle Burton's method closely resembled his 1891 patents, though his own interest in the matter was primarily to establish that "the Russian oil industry could easily build a cracking apparatus according to any of the described systems without being accused by the Americans of borrowing for free".

At that time, just a few years after the Russian Revolution, Russia was desperate to develop industry and earn foreign exchange, so their oil industry eventually did obtain much of their technology from foreign companies, largely American. At about that time however, fluid catalytic cracking was being explored and developed and soon replaced most of the purely thermal cracking processes in the fossil fuel processing industry. The replacement was however not complete; many types of cracking, including pure thermal cracking, still are in use, depending on the nature of the feedstock and the products required to satisfy market demands. Thermal cracking remains important however, for example in producing naphtha, gas oil, and coke, and more sophisticated forms of thermal cracking have been developed for various purposes. These include visbreaking, steam cracking, and coking.

Chemistry

A large number of chemical reactions take place during the cracking process, most of them based on free radicals. Computer simulations aimed at modeling what takes place during steam cracking have included hundreds or even thousands of reactions in their models. The main reactions that take place include:

Initiation

In these reactions a single molecule breaks apart into two free radicals. Only a small fraction of the feed molecules actually undergo initiation, but these reactions are necessary to produce the free radicals that drive the rest of the reactions. In steam cracking, initiation usually involves breaking a chemical bond between two carbon atoms, rather than the bond between a carbon and a hydrogen atom.

$$CH_3CH_3 \rightarrow 2\ CH_3\bullet$$

Hydrogen Abstraction

In these reactions a free radical removes a hydrogen atom from another molecule, turning the second molecule into a free radical.

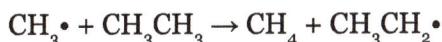

$$CH_3\bullet + CH_3CH_3 \rightarrow CH_4 + CH_3CH_2\bullet$$

Radical Decomposition

In these reactions a free radical breaks apart into two molecules, one an alkene, the other a free radical. This is the process that results in alkene products.

$$CH_3CH_2\bullet \rightarrow CH_2{=}CH_2 + H$$

Radical Addition

In these reactions, the reverse of radical decomposition reactions, a radical reacts with an alkene to form a single, larger free radical. These processes are involved in forming the aromatic products that result when heavier feedstocks are used.

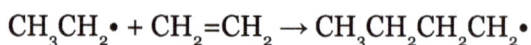

$$CH_3CH_2\bullet + CH_2=CH_2 \rightarrow CH_3CH_2CH_2CH_2\bullet$$

Termination

In these reactions two free radicals react with each other to produce products that are not free radicals. Two common forms of termination are *recombination*, where the two radicals combine to form one larger molecule, and *disproportionation*, where one radical transfers a hydrogen atom to the other, giving an alkene and an alkane.

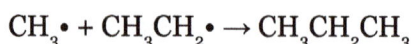

$$CH_3\bullet + CH_3CH_2\bullet \rightarrow CH_3CH_2CH_3$$

$$CH_3CH_2\bullet + CH_3CH_2\bullet \rightarrow CH_2=CH_2 + CH_3CH_3$$

Example: Cracking Butane

There are three places where a butane molecule (CH_3-CH_2-CH_2-CH_3) might be split. Each has a distinct likelihood:

- 48%: break at the CH_3-CH_2 bond.

$$CH_3* / *CH_2\text{-}CH_2\text{-}CH_3$$

  Ultimately this produces an alkane and an alkene: $CH_4 + CH_2=CH\text{-}CH_3$

- 38%: break at a CH_2-CH_2 bond.

$$CH_3\text{-}CH_2* / *CH_2\text{-}CH_3$$

  Ultimately this produces an alkane and an alkene of different types: CH_3-$CH_3 + CH_2=CH_2$

- 14%: break at a terminal C-H bond

$$H/CH_2\text{-}CH_2\text{-}CH_2\text{-}CH_3$$

  Ultimately this produces an alkene and hydrogen gas: $CH_2=CH\text{-}CH_2\text{-}CH_3 + H_2$

Cracking Methodologies

Thermal Methods

Thermal cracking was the first category of hydrocarbon cracking to be developed. Thermal cracking is an example of a reaction whose energetics are dominated by entropy ($\Delta S°$) rather than by enthalpy ($\Delta H°$) in the Gibbs Free Energy equation $\Delta G°=\Delta H°\text{-}T\Delta S°$. Although the bond dissociation energy D for a carbon-carbon single bond is relatively high (about 375 kJ/mol) and cracking is highly endothermic, the large positive entropy change resulting from the fragmentation of one large molecule into several smaller pieces, together with the extremely high temperature, makes TΔS° term larger than the ΔH° term, thereby favoring the cracking reaction.

Thermal Cracking

Modern high-pressure thermal cracking operates at absolute pressures of about 7,000 kPa. An overall process of disproportionation can be observed, where "light", hydrogen-rich products are formed at the expense of heavier molecules which condense and are depleted of hydrogen. The actual reaction is known as homolytic fission and produces alkenes, which are the basis for the economically important production of polymers.

Thermal cracking is currently used to "upgrade" very heavy fractions or to produce light fractions or distillates, burner fuel and/or petroleum coke. Two extremes of the thermal cracking in terms of product range are represented by the high-temperature process called "steam cracking" or pyrolysis (ca. 750 °C to 900 °C or higher) which produces valuable ethylene and other feedstocks for the petrochemical industry, and the milder-temperature delayed coking (ca. 500 °C) which can produce, under the right conditions, valuable needle coke, a highly crystalline petroleum coke used in the production of electrodes for the steel and aluminium industries.

William Merriam Burton developed one of the earliest thermal cracking processes in 1912 which operated at 700–750 °F (371–399 °C) and an absolute pressure of 90 psi (620 kPa) and was known as the Burton process. Shortly thereafter, in 1921, C.P. Dubbs, an employee of the Universal Oil Products Company, developed a somewhat more advanced thermal cracking process which operated at 750–860 °F (399–460 °C) and was known as the *Dubbs process*. The Dubbs process was used extensively by many refineries until the early 1940s when catalytic cracking came into use.

Steam Cracking

Steam cracking is a petrochemical process in which saturated hydrocarbons are broken down into smaller, often unsaturated, hydrocarbons. It is the principal industrial method for producing the lighter alkenes (or commonly olefins), including ethene (or ethylene) and propene (or propylene). Steam cracker units are facilities in which a feedstock such as naphtha, liquefied petroleum gas (LPG), ethane, propane or butane is thermally cracked through the use of steam in a bank of pyrolysis furnaces to produce lighter hydrocarbons. The products obtained depend on the composition of the feed, the hydrocarbon-to-steam ratio, and on the cracking temperature and furnace residence time.

In steam cracking, a gaseous or liquid hydrocarbon feed like naphtha, LPG or ethane is diluted with steam and briefly heated in a furnace without the presence of oxygen. Typically, the reaction temperature is very high, at around 850°C, but the reaction is only allowed to take place very briefly. In modern cracking furnaces, the residence time is reduced to milliseconds to improve yield, resulting in gas velocities up to the speed of sound. After the cracking temperature has been reached, the gas is quickly quenched to stop the reaction in a transfer line heat exchanger or inside a quenching header using quench oil.

The products produced in the reaction depend on the composition of the feed, the hydrocarbon to steam ratio and on the cracking temperature and furnace residence time. Light hydrocarbon feeds such as ethane, LPGs or light naphtha give product streams rich in the lighter alkenes, including ethylene, propylene, and butadiene. Heavier hydrocarbon (full range and heavy naphthas as well as other refinery products) feeds give some of these, but also give products rich in aromatic hydrocarbons and hydrocarbons suitable for inclusion in gasoline or fuel oil.

A higher cracking temperature (also referred to as severity) favors the production of ethene and benzene, whereas lower severity produces higher amounts of propene, C4-hydrocarbons and liquid products. The process also results in the slow deposition of coke, a form of carbon, on the reactor walls. This degrades the efficiency of the reactor, so reaction conditions are designed to minimize this. Nonetheless, a steam cracking furnace can usually only run for a few months at a time between de-cokings. Decokes require the furnace to be isolated from the process and then a flow of steam or a steam/air mixture is passed through the furnace coils. This converts the hard solid carbon layer to carbon monoxide and carbon dioxide. Once this reaction is complete, the furnace can be returned to service.

Catalytic Methods

The catalytic cracking process involves the presence of acidcatalysts (usually solid acids such as silica-alumina and zeolites) which promote a heterolytic (asymmetric) breakage of bonds yielding pairs of ions of opposite charges, usually a carbocation and the very unstable hydrideanion. Carbon-localized free radicals and cations are both highly unstable and undergo processes of chain rearrangement, C-C scission in position beta as in cracking, and intra- and intermolecular hydrogen transfer. In both types of processes, the corresponding reactive intermediates (radicals, ions) are permanently regenerated, and thus they proceed by a self-propagating chain mechanism. The chain of reactions is eventually terminated by radical or ion recombination.

Fluid Catalytic Cracking

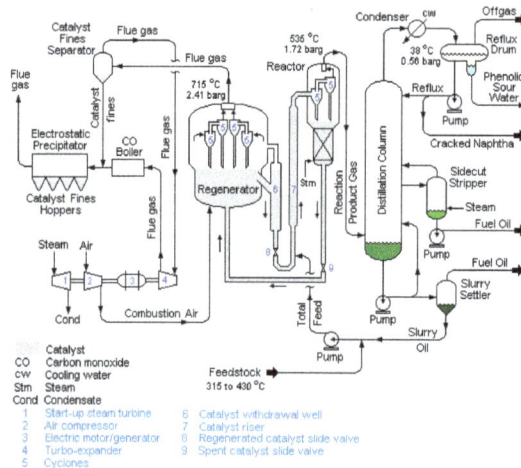

Schematic flow diagram of a fluid catalytic cracker

Fluid catalytic cracking is a commonly used process, and a modern oil refinery will typically include a *cat cracker*, particularly at refineries in the US, due to the high demand for gasoline. The process was first used around 1942 and employs a powdered catalyst. During WWII, the Allied Forces had plentiful supplies of the materials in contrast to the Axis Forces which suffered severe

shortages of gasoline and artificial rubber. Initial process implementations were based on low activity alumina catalyst and a reactor where the catalyst particles were suspended in a rising flow of
feed hydrocarbons in a fluidized bed.

Alumina-catalyzed cracking systems are still in use in high school and universitylaboratories in
experiments concerning alkanes and alkenes. The catalyst is usually obtained by crushing pumice
stones, which contain mainly aluminium oxide and silica into small, porous pieces. In the laboratory, aluminium oxide (or porous pot) must be heated.

In newer designs, cracking takes place using a very active zeolite-based catalyst in a short-contact
time vertical or upward-sloped pipe called the "riser". Pre-heated feed is sprayed into the base of
the riser via feed nozzles where it contacts extremely hot fluidized catalyst at 1,230 to 1,400 °F (666
to 760 °C). The hot catalyst vaporizes the feed and catalyzes the cracking reactions that break down
the high-molecular weight oil into lighter components including LPG, gasoline, and diesel. The cata-
lyst-hydrocarbon mixture flows upward through the riser for a few seconds, and then the mixture is
separated via cyclones. The catalyst-free hydrocarbons are routed to a main fractionator for separation into fuel gas, LPG, gasoline, naphtha, light cycle oils used in diesel and jet fuel, and heavy fuel oil.

During the trip up the riser, the cracking catalyst is "spent" by reactions which deposit coke on
the catalyst and greatly reduce activity and selectivity. The "spent" catalyst is disengaged from the
cracked hydrocarbon vapors and sent to a stripper where it contacts steam to remove hydrocarbons remaining in the catalyst pores. The "spent" catalyst then flows into a fluidized-bed regenerator where air (or in some cases air plus oxygen) is used to burn off the coke to restore catalyst
activity and also provide the necessary heat for the next reaction cycle, cracking being an endothermic reaction. The "regenerated" catalyst then flows to the base of the riser, repeating the cycle.

The gasoline produced in the FCC unit has an elevated octane rating but is less chemically stable
compared to other gasoline components due to its olefinic profile. Olefins in gasoline are responsible for the formation of polymeric deposits in storage tanks, fuel ducts and injectors. The FCC
LPG is an important source of C_3-C_4 olefins and isobutane that are essential feeds for the alkylation
process and the production of polymers such as polypropylene.

Hydrocracking

Hydrocracking is a catalytic cracking process assisted by the presence of added hydrogen gas.
Unlike a hydrotreater, where hydrogen is used to cleave C-S and C-N bonds, hydrocracking uses
hydrogen to break C-C bonds (hydrotreatment is conducted prior to hydrocracking to protect the
catalysts in a hydrocracking process).

The products of this process are saturated hydrocarbons; depending on the reaction conditions
(temperature, pressure, catalyst activity) these products range from ethane, LPG to heavier hydrocarbons consisting mostly of isoparaffins. Hydrocracking is normally facilitated by a bifunctional
catalyst that is capable of rearranging and breaking hydrocarbon chains as well as adding hydrogen to aromatics and olefins to produce naphthenes and alkanes.

The major products from hydrocracking are jet fuel and diesel, but low sulphur naphtha fractions
and LPG are also produced. All these products have a very low content of sulfur and other contaminants.

It is very common in Europe and Asia because those regions have high demand for diesel and kerosene. In the US, fluid catalytic cracking is more common because the demand for gasoline is higher.

The hydrocracking process depends on the nature of the feedstock and the relative rates of the two competing reactions, hydrogenation and cracking. Heavy aromatic feedstock is converted into lighter products under a wide range of very high pressures (1,000-2,000 psi) and fairly high temperatures (750°-1,500° F), in the presence of hydrogen and special catalysts.

The primary function of hydrogen is, thus: a) preventing the formation of polycyclic aromatic compounds if feedstock has a high paraffinic content. b) reduced tar formation c) reducing impurities d) preventing buildup of coke on the catalyst. e) converting sulfur and nitrogen compounds present in the feedstock to hydrogen sulfide and ammonia, and e) achieving high cetane number fuel.

Fluid Catalytic Cracking

A typical fluid catalytic cracking unit in a petroleum refinery.

Fluid catalytic cracking (FCC) is one of the most important conversion processes used in petroleum refineries. It is widely used to convert the high-boiling, high-molecular weight hydrocarbon fractions of petroleumcrude oils to more valuable gasoline, olefinic gases, and other products. Cracking of petroleum hydrocarbons was originally done by thermal cracking, which has been almost completely replaced by catalytic cracking because it produces more gasoline with a higher octane rating. It also produces byproduct gases that are more olefinic, and hence more valuable, than those produced by thermal cracking.

The feedstock to an FCC is usually that portion of the crude oil that has an initial boiling point of 340 °C or higher at atmospheric pressure and an average molecular weight ranging from about 200 to 600 or higher. This portion of crude oil is often referred to as heavy gas oil or vacuum gas oil (HVGO). The FCC process vaporizes and breaks the long-chain molecules of the high-boiling hydrocarbon liquids into much shorter molecules by contacting the feedstock, at high temperature and moderate pressure, with a fluidized powdered catalyst.

Economics

Oil refineries use fluid catalytic cracking to correct the imbalance between the market demand for gasoline and the excess of heavy, high boiling range products resulting from the distillation of crude oil.

As of 2006, FCC units were in operation at 400 petroleum refineries worldwide and about one-third of the crude oil refined in those refineries is processed in an FCC to produce high-octane gasoline and fuel oils. During 2007, the FCC units in the United States processed a total of 5,300,000 barrels (840,000 m³) per day of feedstock and FCC units worldwide processed about twice that amount.

FCC units are less common in Europe and Asia because those regions have high demand for diesel and kerosene, which can be satisfied with hydrocracking. In the US, fluid catalytic cracking is more common because the demand for gasoline is higher.

Flow Diagram and Process Description

The modern FCC units are all continuous processes which operate 24 hours a day for as long as 3 to 5 years between scheduled shutdowns for routine maintenance.

There are several different proprietary designs that have been developed for modern FCC units. Each design is available under a license that must be purchased from the design developer by any petroleum refining company desiring to construct and operate an FCC of a given design.

There are two different configurations for an FCC unit: the "stacked" type where the reactor and the catalyst regenerator are contained in a single vessel with the reactor above the catalyst regenerator and the "side-by-side" type where the reactor and catalyst regenerator are in two separate vessels. These are the major FCC designers and licensors:

Side-by-side configuration:

- CB&I
- ExxonMobil Research and Engineering (EMRE)
- Shell Global Solutions
- Axens / Stone & Webster Process Technology — currently owned by Technip
- Universal Oil Products (UOP) — currently fully owned subsidiary of Honeywell

Stacked configuration:

- Kellogg Brown & Root (KBR)

Each of the proprietary design licensors claims to have unique features and advantages. A complete discussion of the relative advantages of each of the processes is beyond the scope of this article.

Reactor and Regenerator

The reactor and regenerator are considered to be the heart of the fluid catalytic cracking unit. The schematic flow diagram of a typical modern FCC unit in Figure 1 below is based upon the "side-by-side" configuration. The preheated high-boiling petroleum feedstock (at about 315 to 430 °C) consisting of long-chain hydrocarbon molecules is combined with recycle slurry oil from the bottom of the distillation column and injected into the *catalyst riser* where it is vaporized and cracked into smaller molecules of vapor by contact and mixing with the very hot powdered catalyst from the regenerator. All of the cracking reactions take place in the catalyst riser within a period of 2–4 seconds. The hydrocarbon vapors "fluidize" the powdered catalyst and the mixture of hydrocarbon vapors and catalyst flows upward to enter the *reactor* at a temperature of about 535 °C and a pressure of about 1.72 bar.

The reactor is a vessel in which the cracked product vapors are: (a) separated from the so-called *spent catalyst* by flowing through a set of two-stage cyclones within the reactor and (b) the *spent catalyst* flows downward through a steam stripping section to remove any hydrocarbon vapors before the spent catalyst returns to the *catalyst regenerator*. The flow of spent catalyst to the regenerator is regulated by a slide valve in the spent catalyst line.

Since the cracking reactions produce some carbonaceous material (referred to as catalyst coke) that deposits on the catalyst and very quickly reduces the catalyst reactivity, the catalyst is regenerated by burning off the deposited coke with air blown into the regenerator. The regenerator operates at a temperature of about 715 °C and a pressure of about 2.41 barg, hence the regenerator operates at about 0.7 barg higher pressure than the reactor. The combustion of the coke is exothermic and it produces a large amount of heat that is partially absorbed by the regenerated catalyst and provides the heat required for the vaporization of the feedstock and the endothermic cracking reactions that take place in the catalyst riser. For that reason, FCC units are often referred to as being 'heat balanced'.

The hot catalyst (at about 715 °C) leaving the regenerator flows into a *catalyst withdrawal well* where any entrained combustion flue gases are allowed to escape and flow back into the upper part to the regenerator. The flow of regenerated catalyst to the feedstock injection point below the catalyst riser is regulated by a slide valve in the regenerated catalyst line. The hot flue gas exits the regenerator after passing through multiple sets of two-stage cyclones that remove entrained catalyst from the flue gas.

The amount of catalyst circulating between the regenerator and the reactor amounts to about 5 kg per kg of feedstock, which is equivalent to about 4.66 kg per litre of feedstock. Thus, an FCC unit processing 75,000 barrels per day (11,900 m^3/d) will circulate about 55,900 tonnes per day of catalyst.

Distillation Column

The reaction product vapors (at 535 °C and a pressure of 1.72 barg) flow from the top of the reactor to the bottom section of the distillation column (commonly referred to as the *main fractionator*) where they are distilled into the FCC end products of cracked naphtha, fuel oil, and offgas. After further processing for removal of sulfur compounds, the cracked naphtha becomes a high-octane component of the refinery's blended gasolines.

The main fractionator offgas is sent to what is called a *gas recovery unit* where it is separated into butanes and butylenes, propane and propylene, and lower molecular weight gases (hydrogen, methane, ethylene and ethane). Some FCC gas recovery units may also separate out some of the ethane and ethylene.

Although the schematic flow diagram above depicts the main fractionator as having only one side-cut stripper and one fuel oil product, many FCC main fractionators have two sidecut strippers and produce a light fuel oil and a heavy fuel oil. Likewise, many FCC main fractionators produce a light cracked naphtha and a heavy cracked naphtha. The terminology *light* and *heavy* in this context refers to the product boiling ranges, with light products having a lower boiling range than heavy products.

The bottom product oil from the main fractionator contains residual catalyst particles which were not completely removed by the cyclones in the top of the reactor. For that reason, the bottom product oil is referred to as a *slurry oil*. Part of that slurry oil is recycled back into the main fractionator above the entry point of the hot reaction product vapors so as to cool and partially condense the reaction product vapors as they enter the main fractionator. The remainder of the slurry oil is pumped through a slurry settler. The bottom oil from the slurry settler contains most of the slurry oil catalyst particles and is recycled back into the catalyst riser by combining it with the FCC feedstock oil. The so-called *clarified slurry oil* or decant oil is withdrawn from the top of slurry settler for use elsewhere in the refinery, as a heavy fuel oil blending component, or as carbon black feedstock.

Regenerator Flue Gas

Depending on the choice of FCC design, the combustion in the regenerator of the coke on the spent catalyst may or may not be complete combustion to carbon dioxide CO_2. The combustion air flow is controlled so as to provide the desired ratio of carbon monoxide (CO) to carbon dioxide for each specific FCC design.

In the design shown in Figure 1, the coke has only been partially combusted to CO_2. The combustion flue gas (containing CO and CO_2) at 715 °C and at a pressure of 2.41 barg is routed through a secondary catalyst separator containing *swirl tubes* designed to remove 70 to 90 percent of the particulates in the flue gas leaving the regenerator. This is required to prevent erosion damage to the blades in the turbo-expander that the flue gas is next routed through.

The expansion of flue gas through a turbo-expander provides sufficient power to drive the regenerator's combustion air compressor. The electrical motor-generator can consume or produce electrical power. If the expansion of the flue gas does not provide enough power to drive the air compressor, the electric motor/generator provides the needed additional power. If the flue gas expansion provides more power than needed to drive the air compressor, than the electric motor/generator converts the excess power into electric power and exports it to the refinery's electrical system.

The expanded flue gas is then routed through a steam-generating boiler (referred to as a *CO boiler*) where the carbon monoxide in the flue gas is burned as fuel to provide steam for use in the refinery as well as to comply with any applicable environmental regulatory limits on carbon monoxide emissions.

The flue gas is finally processed through an electrostatic precipitator (ESP) to remove residual particulate matter to comply with any applicable environmental regulations regarding particulate emissions. The ESP removes particulates in the size range of 2 to 20 µm from the flue gas. Particulate filter systems, known as Fourth Stage Separators (FSS) are sometimes required to meet particulate emission limits. These can replace the ESP when particulate emissions are the only concern.

The steam turbine in the flue gas processing system (shown in the above diagram) is used to drive the regenerator's combustion air compressor during start-ups of the FCC unit until there is sufficient combustion flue gas to take over that task..

Chemistry

Before delving into the chemistry involved in catalytic cracking, it will be helpful to briefly discuss the composition of petroleum crude oil.

Petroleum crude oil consists primarily of a mixture of hydrocarbons with small amounts of other organic compounds containing sulfur, nitrogen and oxygen. The crude oil also contains small amounts of metals such as copper, iron, nickel and vanadium.

Table 1	
Carbon	83-87%
Hydrogen	10-14%
Nitrogen	0.1-2%
Oxygen	0.1-1.5%
Sulfur	0.5-6%
Metals	< 0.1%

The elemental composition ranges of crude oil are summarized in Table 1 and the hydrocarbons in the crude oil can be classified into three types:

- Paraffins or alkanes: saturated straight-chain or branched hydrocarbons, without any ring structures

- Naphthenes or cycloalkanes: saturated hydrocarbons having one or more ring structures with one or more side-chain paraffins

- Aromatics: hydrocarbons having one or more unsaturated ring structures such as benzene or unsaturated polycyclic ring structures such as naphthalene or phenanthrene, any of which may also have one or more side-chain paraffins.

Olefins or alkenes, which are unsaturated straight-chain or branched hydrocarbons, do not occur naturally in crude oil.

In plain language, the fluid catalytic cracking process breaks large hydrocarbon molecules into smaller molecules by contacting them with powdered catalyst at a high temperature and moderate pressure which first vaporizes the hydrocarbons and then breaks them. The cracking reactions occur in the vapor phase and start immediately when the feedstock is vaporized in the catalyst riser.

Figure 2 is a very simplified schematic diagram that exemplifies how the process breaks high boiling, straight-chain alkane (paraffin) hydrocarbons into smaller straight-chain alkanes as well as branched-chain alkanes, branched alkenes (olefins) and cycloalkanes (naphthenes). The breaking of the large hydrocarbon molecules into smaller molecules is more technically referred to by organic chemists as *scission* of the carbon-to-carbon bonds.

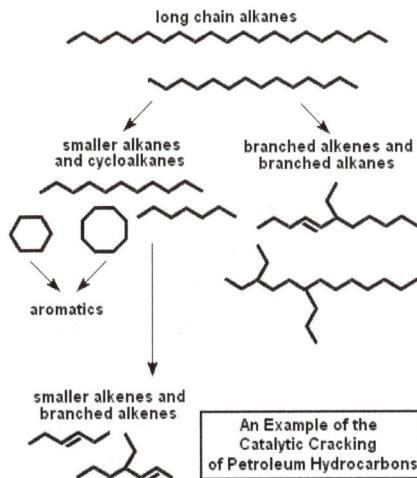

Figure 2: Diagrammatic example of the catalytic cracking of petroleum hydrocarbons

As depicted in Figure 2, some of the smaller alkanes are then broken and converted into even smaller alkenes and branched alkenes such as the gases ethylene, propylene, butylenes, and isobutylenes. Those olefinic gases are valuable for use as petrochemical feedstocks. The propylene, butylene and isobutylene are also valuable feedstocks for certain petroleum refining processes that convert them into high-octane gasoline blending components.

As also depicted in Figure 2, the cycloalkanes (naphthenes) formed by the initial breakup of the large molecules are further converted to aromatics such as benzene, toluene, and xylenes, which boil in the gasoline boiling range and have much higher octane ratings than alkanes.

In the cracking process carbon is also produced which gets deposited on the catalyst (catalyst coke). The carbon formation tendency or amount of carbon in a crude or FCC feed is measured with methods such as Micro Carbon Residue, Conradson Carbon Residue or Ramsbottom Carbon Residue.

By no means does Figure 2 include all the chemistry of the primary and secondary reactions taking place in the fluid catalytic process. There are a great many other reactions involved. However, a full discussion of the highly technical details of the various catalytic cracking reactions is beyond the scope of this article and can be found in the technical literature.

Catalysts

Modern FCC catalysts are fine powders with a bulk density of 0.80 to 0.96 g/cm³ and having a particle size distribution ranging from 10 to 150 μm and an average particle size of 60 to 100 μm. The design and operation of an FCC unit is largely dependent upon the chemical and physical properties of the catalyst. The desirable properties of an FCC catalyst are:

- Good stability to high temperature and to steam

- High activity

- Large pore sizes

- Good resistance to attrition

- Low coke production

A modern FCC catalyst has four major components: crystalline zeolite, matrix, binder, and filler. Zeolite is the primary active component and can range from about 15 to 50 weight percent of the catalyst. The zeolite used in FCC catalysts is referred to as faujasite or as *Type Y* and is composed of silica and alumina tetrahedra with each tetrahedron having either an aluminum or a silicon atom at the center and four oxygen atoms at the corners. It is a molecular sieve with a distinctive lattice structure that allows only a certain size range of hydrocarbon molecules to enter the lattice. In general, the zeolite does not allow molecules larger than 8 to 10 nm (i.e., 80 to 100 ångströms) to enter the lattice.

The catalytic sites in the zeolite are strong acids (equivalent to 90% sulfuric acid) and provide most of the catalytic activity. The acidic sites are provided by the alumina tetrahedra. The aluminum atom at the center of each alumina tetrahedra is at a +3 oxidation state surrounded by four oxygen atoms at the corners which are shared by the neighboring tetrahedra. Thus, the net charge of the alumina tetrahedra is -1 which is balanced by a sodiumion during the production of the catalyst. The sodium ion is later replaced by an ammonium ion, which is vaporized when the catalyst is subsequently dried, resulting in the formation of Lewis and Brønsted acidic sites. In some FCC catalysts, the Brønsted sites may be later replaced by rare earth metals such as cerium and lanthanum to provide alternative activity and stability levels. The acidic activity of the catalyst is deactivated by the nitrogen present in the feedstock. Typically the nitrogen content of the feedstock is between 1000 and 2000 ppm. Higher Nitrogen content results in significant depression of the conversion rate.

The matrix component of an FCC catalyst contains amorphous alumina which also provides catalytic activity sites and in larger pores that allows entry for larger molecules than does the zeolite. That enables the cracking of higher-boiling, larger feedstock molecules than are cracked by the zeolite.

The binder and filler components provide the physical strength and integrity of the catalyst. The binder is usually silica sol and the filler is usually a clay (kaolin).

Nickel, vanadium, iron, copper and other metal contaminants, present in FCC feedstocks in the parts per million range, all have detrimental effects on the catalyst activity and performance. Nickel and vanadium are particularly troublesome. There are a number of methods for mitigating the effects of the contaminant metals:

- Avoid feedstocks with high metals content: This seriously hampers a refinery's flexibility to process various crude oils or purchased FCC feedstocks.

- Feedstock feed pretreatment: Hydrodesulfurization of the FCC feedstock removes some of the metals and also reduces the sulfur content of the FCC products. However, this is quite a costly option.

- Increasing fresh catalyst addition: All FCC units withdraw some of the circulating *equilibrium catalyst* as spent catalyst and replaces it with fresh catalyst in order to maintain a desired level of activity. Increasing the rate of such exchange lowers the level of metals in the circulating *equilibrium catalyst*, but this is also quite a costly option.

- Demetallization: The commercial proprietary *Demet Process* removes nickel and vanadium from the withdrawn spent catalyst. The nickel and vanadium are converted to chlorides which are then washed out of the catalyst. After drying, the demetallized catalyst is recycled into the circulating catalyst. Removals of about 95 percent nickel removal and 67 to 85 percent vanadium have been reported. Despite that, the use of the Demet process has not become widespread, perhaps because of the high capital expenditure required.

- Metals passivation: Certain materials can be used as additives which can be impregnated into the catalyst or added to the FCC feedstock in the form of metal-organic compounds. Such materials react with the metal contaminants and passivate the contaminants by forming less harmful compounds that remain on the catalyst. For example, antimony and bismuth are effective in passivating nickel and tin is effective in passivating vanadium. A number of proprietary passivation processes are available and fairly widely used.

The major suppliers of FCC catalysts worldwide include Albemarle Corporation, W.R. Grace Company and BASF Catalysts (formerly Engelhard). The price for lanthanum oxide used in fluid catalytic cracking has risen from $5 per kilogram in early 2010 to $140 per kilogram in June 2011.

History

The first commercial use of catalytic cracking occurred in 1915 when Almer M. McAfee of Gulf Refining Company developed a batch process using aluminum chloride (a Friedel Crafts catalyst known since 1877) to catalytically crack heavy petroleum oils. However, the prohibitive cost of the catalyst prevented the widespread use of McAfee's process at that time.

In 1922, a French mechanical engineer named Eugene Jules Houdry and a French pharmacist named E.A. Prudhomme set up a laboratory near Paris to develop a catalytic process for converting lignite coal to gasoline. Supported by the French government, they built a small demonstration plant in 1929 that processed about 60 tons per day of lignite coal. The results indicated that the process was not economically viable and it was subsequently shut down.

Houdry had found that Fuller's earth, a clay mineral containing aluminosilicates, could convert oil derived from the lignite to gasoline. He then began to study the catalysis of petroleum oils and had some success in converting vaporized petroleum oil to gasoline. In 1930, the Vacuum Oil Company invited him to come to the United States and he moved his laboratory to Paulsboro, New Jersey.

In 1931, the Vacuum Oil Company merged with Standard Oil of New York (Socony) to form the Socony-Vacuum Oil Company. In 1933, a small Houdry unit processed 200 barrels per day ($32 \text{ m}^3/\text{d}$) of petroleum oil. Because of the economic depression of the early 1930s, Socony-Vacuum was no longer able to support Houdry's work and gave him permission to seek help elsewhere.

In 1933, Houdry and Socony-Vacuum joined with Sun Oil Company in developing the Houdry process. Three years later, in 1936, Socony-Vacuum converted an older thermal cracking unit in their Paulsboro refinery in New Jersey to a small demonstration unit using the Houdry process to catalytically crack 2,000 barrels per day (320 m³/d) of petroleum oil.

In 1937, Sun Oil began operation of a new Houdry unit processing 12,000 barrels per day (1,900 m³/d) in their Marcus Hook refinery in Pennsylvania. The Houdry process at that time used reactors with a fixed bed of catalyst and was a semi-batch operation involving multiple reactors with some of the reactors in operation while other reactors were in various stages of regenerating the catalyst. Motor-driven valves were used to switch the reactors between online operation and offline regeneration and a cycle timer managed the switching. Almost 50 percent of the cracked product was gasoline as compared with about 25 percent from the thermal cracking processes.

By 1938, when the Houdry process was publicly announced, Socony-Vacuum had eight additional units under construction. Licensing the process to other companies also began and by 1940 there were 14 Houdry units in operation processing 140,000 barrels per day (22,000 m³/d).

The next major step was to develop a continuous process rather than the semi-batch Houdry process. That step was implemented by advent of the moving-bed process known as the Thermofor Catalytic Cracking (TCC) process which used a bucket conveyor-elevator to move the catalyst from the regeneration kiln to the separate reactor section. A small semicommercial demonstration TCC unit was built in Socony-Vacuum's Paulsboro refinery in 1941 and operated successfully, producing 500 barrels per day (79 m³/d). Then a full-scale commercial TCC unit processing 10,000 barrels per day (1,600 m³/d) began operation in 1943 at the Beaumont, Texas refinery of Magnolia Oil Company, an affiliate of Socony-Vacuum. By the end of World War II in 1945, the processing capacity of the TCC units in operation was about 300,000 barrels per day (48,000 m³/d).

It is said that the Houdry and TCC units were a major factor in the winning of World War II by supplying the high-octane gasoline needed by the air forces of Great Britain and the United States for the more efficient higher compression ratio engines of the Spitfire and the Mustang.

In the years immediately after World War II, the Houdriflow process and the air-lift TCC process were developed as improved variations on the moving-bed theme. Just like Houdry's fixed-bed reactors, the moving-bed designs were prime examples of good engineering by developing a method of continuously moving the catalyst between the reactor and regeneration sections. The first air-lift TCC unit began operation in October 1950 at the Beaumont, Texas refinery.

This fluid catalytic cracking process had first been investigated in the 1920s by Standard Oil of New Jersey, but research on it was abandoned during the economic depression years of 1929 to 1939. In 1938, when the success of Houdry's process had become apparent, Standard Oil of New Jersey resumed the project as part of a consortium of that include five oil companies (Standard Oil of New Jersey, Standard Oil of Indiana, Anglo-Iranian Oil, Texas Oil and Dutch Shell), two engineering-construction companies (M.W. Kellogg and Universal Oil Products) and a German chemical company (I.G. Farben). The consortium was called Catalytic Research Associates (CRA) and its purpose was to develop a catalytic cracking process which would not impinge on Houdry's patents.

Chemical engineering professors Warren K. Lewis and Edwin R. Gilliland of the Massachusetts Institute of Technology (MIT) suggested to the CRA researchers that a low velocity gas flow through a powder might "lift" it enough to cause it to flow in a manner similar to a liquid. Focused on that idea of a fluidized catalyst, researchers Donald Campbell, Homer Martin, Eger Murphree and

Charles Tyson of the Standard Oil of New Jersey (now Exxon-Mobil Company) developed the first fluidized catalytic cracking unit. Their U.S. Patent No. 2,451,804, *A Method of and Apparatus for Contacting Solids and Gases*, describes their milestone invention. Based on their work, M. W. Kellogg Company constructed a large pilot plant in the Baton Rouge, Louisiana refinery of the Standard Oil of New Jersey. The pilot plant began operation in May 1940.

Based on the success of the pilot plant, the first commercial fluid catalytic cracking plant (known as the Model I FCC) began processing 13,000 barrels per day (2,100 m³/d) of petroleum oil in the Baton Rouge refinery on May 25, 1942, just four years after the CRA consortium was formed and in the midst of World War II. A little more than a month later, in July 1942, it was processing 17,000 barrels per day (2,700 m³/d). In 1963, that first Model I FCC unit was shut down after 21 years of operation and subsequently dismantled.

In the many decades since the Model I FCC unit began operation, the fixed bed Houdry units have all been shut down as have most of the moving bed units (such as the TCC units) while hundreds of FCC units have been built. During those decades, many improved FCC designs have evolved and cracking catalysts have been greatly improved, but the modern FCC units are essentially the same as that first Model I FCC unit.

Note: All of the refinery and company names in this history section (with the exception of Universal Oil Products) have changed over time by mergers and buyouts. Some have changed a number of times.

Gas Flare

Flare stack at the Shell Haven refinery in England.

A gas flare, alternatively known as a flare stack, is a gas combustion device used in industrial plants such as petroleum refineries, chemical plants, natural gas processing plants as well as at oil or gas production sites having oil wells, gas wells, offshore oil and gas rigs and landfills.

North Dakota Flaring of Gas

In industrial plants, flare stacks are primarily used for burning off flammable gas released by pressure relief valves during unplanned over-pressuring of plant equipment. During plant or partial plant startups and shutdowns, flare stacks are also often used for the planned combustion of gases over relatively short periods.

A great deal of gas flaring at many oil and gas production sites has to do with protection against the dangers of over-pressuring industrial plant equipment. When petroleum crude oil is extracted and produced from onshore or offshore oil wells, raw natural gas associated with the oil is brought to the surface as well. Especially in areas of the world lacking pipelines and other gas transportation infrastructure, vast amounts of such associated gas are commonly flared as waste or unusable gas. The flaring of associated gas may occur at the top of a vertical flare stack (as in the adjacent photo) or it may occur in a ground-level flare in an earthen pit. Preferably, associated gas is reinjected into the reservoir, which saves it for future use while maintaining higher well pressure and crude oil producibility.

Overall Flare System in Industrial Plants

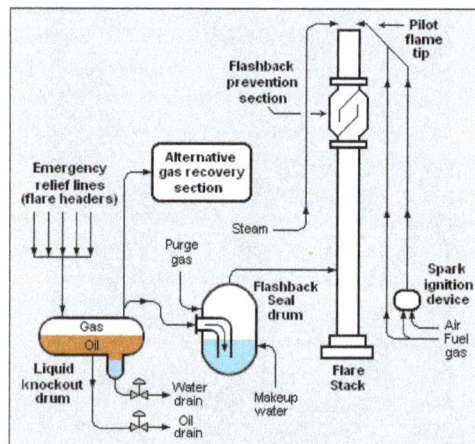

Schematic flow diagram of an overall vertical, elevated flare stack system in an industrial plant.

When industrial plant equipment items are over-pressured, the pressure relief valve is an essential safety device that automatically release gases and sometimes liquids. Those pressure relief valves are required by industrial design codes and standards as well as by law.

The released gases and liquids are routed through large piping systems called *flare headers* to a vertical elevated flare. The released gases are burned as they exit the flare stacks. The size and brightness of the resulting flame depends upon the flammable material's flow rate in joules per hour (or btu per hour).

Most industrial plant flares have a vapor-liquid separator (also known as a knockout drum) upstream of the flare to remove any large amounts of liquid that may accompany the relieved gases.

Steam is very often injected into the flame to reduce the formation of black smoke. When too much steam is added, a condition known as "over steaming" can occur resulting in reduced combustion efficiency and higher emissions. To keep the flare system functional, a small amount of gas is continuously burned, like a pilot light, so that the system is always ready for its primary purpose as an over-pressure safety system.

The adjacent flow diagram depicts the typical components of an overall industrial flare stack system:

- A knockout drum to remove any oil or water from the relieved gases.
- A water seal drum to prevent any flashback of the flame from the top of the flare stack.
- An alternative gas recovery system for use during partial plant startups and shutdowns as well as other times when required. The recovered gas is routed into the fuel gas system of the overall industrial plant.
- A steam injection system to provide an external momentum force used for efficient mixing of air with the relieved gas, which promotes smokeless burning.
- A pilot flame (with its ignition system) that burns all the time so that it is available to ignite relieved gases when needed.
- The flare stack, including a flashback prevention section at the upper part of the stack.

There is also a safe method to divert the flare gas which is insertion of Liquid U seal with Liquid Hold up vessel. The Liquid U seal is designed to take pressure up to permitted back pressure of the system. This helps to divert the flare gas to recovery system. In case of plant upset, pressure rises and liquid in the U seal will move into Liquid Hold up vessel. On normalization, the Liquid U seal will start diverting the gas again.

Impacts of Waste Flaring Associated Gas from Oil Drilling Sites and Other Facilities

Flaring of associated gas from an oil well site in Nigeria.

Flaring gases from an oil platform in the North Sea.

Improperly operated flares may emit methane and other volatile organic compounds as well as sulfur dioxide and other sulfur compounds, which are known to exacerbate asthma and other respiratory problems. Other emissions from improperly operated flares may include, aromatic hydrocarbons (benzene, toluene, xylenes) and benzapyrene, which are known to be carcinogenic.

Flaring can affect wildlife by attracting birds and insects to the flame. Approximately 7,500 migrating songbirds were attracted to and killed by the flare at the liquefied natural gas terminal in Saint John, New Brunswick, Canada on September 13, 2013. Similar incidents have occurred at flares on offshore oil and gas installations. Moths are known to be attracted to lights. A brochure published by the Secretariat of the Convention on Biological Diversity describing the Global Taxonomy Initiative describes a situation where "*a taxonomist working in a tropical forest noticed that a gas flare at an oil refinery was attracting and killing hundreds of these [hawk or sphinx] moths. Over the course of the months and years that the refinery was running a vast number of moths must have been killed, suggesting that plants could not be pollinated over a large area of forest*".

As of the end of 2011, 150×10^9 cubic meters (5.3×10^{12} cubic feet) of associated gas are flared annually. That is equivalent to about 25 per cent of the annual natural gas consumption in the United States or about 30 per cent of the annual gas consumption in the European Union. If it were to reach market, this quantity of gas (at a nominal value of \$5.62 per 1000 cubic feet) would be worth \$29.8 billion USD.

Also as of the end of 2011, 10 countries accounted for 72 per cent of the flaring, and twenty for 86 per cent. The top ten leading contributors to world gas flaring at the end of 2011, were (in declining order): Russia (27%), Nigeria (11%), Iran (8%), Iraq (7%), United States (5%), Algeria (4%), Kazakhstan (3%), Angola (3%), Saudi Arabia (3%) and Venezuela (3%).

That amount of flaring and burning of associated gas from oil drilling sites is a significant source of carbon dioxide (CO_2) emissions. Coupled with fossil fuel combustion and cement production, flaring's carbon dioxide emissions in 2010 have tripled (1300 ± 110 GtCO2) compared to the last recording (years 1750-1970, 420 ± 35 GtCO had been emitted.) 2400×10^6 tons of carbon dioxide are emitted annually in this way and it amounts to about 1.2 per cent of the worldwide emissions of carbon dioxide. That may seem to be insignificant, but in perspective it is more than half of the *Certified Emissions Reductions* (a type of carbon credits) that have been issued under the rules and mechanisms of the Kyoto Protocol as of June 2011.

Satellite data show that from 2005 to 2010, global gas flaring decreased by about 20%. The most significant reductions in terms of volume were made in Russia (down 40%) and Nigeria (down 29%).

Hydrodenitrogenation

Hydrodenitrogenation (HDN) is an industrial process for the removal of nitrogen from petroleum. Organonitrogen compounds, even though they occur at low levels, are undesirable because they poisoning with downstream catalysts. Furthermore, upon combustion, organonitrogen compounds generate NOx, a pollutant. HDN is effected as general hydroprocessing, which traditionally focuses on hydrodesulfurization (HDS) because sulfur compounds are even more problematic. To some extent, hydrodeoxygenation (HDO) is also effected.

Typical organonitrogen compounds in petroleum include quinolines and porphyrins and their derivatives. The total nitrogen content is typically less than 1% and the targeted levels are in the ppm range. As described in organic geochemistry, organonitrogen compounds are derivatives or degradation products of the compounds in the living matter that comprised the precursor to fossil fuels. In HDN, the organonitrogen compounds are treated at high temperatures with hydrogen in the presence of a catalyst, the net transformation being:

$$R_3N + 3\ H_2 \rightarrow 3\ RH + NH_3$$

The catalysts generally consist of cobalt and nickel as well as molybdenum disulfide or less often tungsten disulfidesupported on alumina]]. The precise composition of the catalyst, i.e. Co/Ni and Mo/W ratios, are tuned for particular feedstocks. A wide variety of catalyst compositions have been considered, including metal phosphides.

Hydrodesulfurization

Hydrodesulfurization (HDS) is a catalytic chemical process widely used to remove sulfur (S) from natural gas and from refined petroleum products, such as gasoline or petrol, jet fuel, kerosene, diesel fuel, and fuel oils. The purpose of removing the sulfur is to reduce the sulfur dioxide (SO 2) emissions that result from using those fuels in automotive vehicles, aircraft, railroad locomotives, ships, gas or oil burning power plants, residential and industrial furnaces, and other forms of fuel combustion.

Another important reason for removing sulfur from the naphtha streams within a petroleum refinery is that sulfur, even in extremely low concentrations, poisons the noble metal catalysts (platinum and rhenium) in the catalytic reforming units that are subsequently used to upgrade the octane rating of the naphtha streams.

The industrial hydrodesulfurization processes include facilities for the capture and removal of the resulting hydrogen sulfide (H2S) gas. In petroleum refineries, the hydrogen sulfide gas is then subsequently converted into byproduct elemental sulfur or sulfuric acid (H2SO4). In fact, the vast majority of the 64,000,000 metric tons of sulfur produced worldwide in 2005 was byproduct sulfur from refineries and other hydrocarbon processing plants.

An HDS unit in the petroleum refining industry is also often referred to as a hydrotreater.

History

Although some reactions involving catalytic hydrogenation of organic substances were already known, the property of finely divided nickel to catalyze the fixation of hydrogen on hydrocarbon (ethylene, benzene) double bonds was discovered by the French chemistPaul Sabatier in 1897. Through this work, he found that unsaturated hydrocarbons in the vapor phase could be converted into saturated hydrocarbons by using hydrogen and a catalytic metal, laying the foundation of the modern catalytic hydrogenation process.

Soon after Sabatier's work, a German chemist, Wilhelm Normann, found that catalytic hydrogenation could be used to convert unsaturated fatty acids or glycerides in the liquid phase into saturated ones. He was awarded a patent in Germany in 1902 and in Britain in 1903, which was the beginning of what is now a worldwide industry.

In the mid-1950s, the first noble metal catalytic reforming process (the Platformer process) was commercialized. At the same time, the catalytic hydrodesulfurization of the naphtha feed to such reformers was also commercialized. In the decades that followed, various proprietary catalytic hydrodesulfurization processes, such as the one depicted in the flow diagram below, have been commercialized. Currently, virtually all of the petroleum refineries worldwide have one or more HDS units.

By 2006, miniature microfluidic HDS units had been implemented for treating JP-8 jet fuel to produce clean feed stock for a fuel cellhydrogen reformer. By 2007, this had been integrated into an operating 5 kW fuel cell generation system.

Process Chemistry

Hydrogenation is a class of chemical reactions in which the net result is the addition of hydrogen (H). Hydrogenolysis is a type of hydrogenation and results in the cleavage of the C-X chemical bond, where C is a carbon atom and X is a sulfur (S), nitrogen (N) or oxygen (O) atom. The net result of a hydrogenolysis reaction is the formation of C-H and H-X chemical bonds. Thus, hydrodesulfurization is a hydrogenolysis reaction. Using ethanethiol (C2H5SH), a sulfur compound present in some petroleum products, as an example, the hydrodesulfurization reaction can be simply expressed as

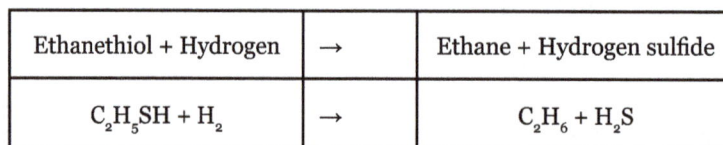

Ethanethiol + Hydrogen	$\rightarrow$	Ethane + Hydrogen sulfide
$C_2H_5SH + H_2$	$\rightarrow$	$C_2H_6 + H_2S$

For the mechanistic aspects of, and the catalysts used in this reaction see the section catalysts and mechanisms.

Process Description

In an industrial hydrodesulfurization unit, such as in a refinery, the hydrodesulfurization reaction takes place in a fixed-bed reactor at elevated temperatures ranging from 300 to 400 °C and elevated pressures ranging from 30 to 130 atmospheres of absolute pressure, typically in the presence of a catalyst consisting of an alumina base impregnated with cobalt and molybdenum (usually called a CoMo catalyst). Occasionally, a combination of nickel and molybdenum (called NiMo) is used, in addition to the CoMo catalyst, for specific difficult-to-treat feed stocks, such as those containing a high level of chemically bound nitrogen.

The image below is a schematic depiction of the equipment and the process flow streams in a typical refinery HDS unit.

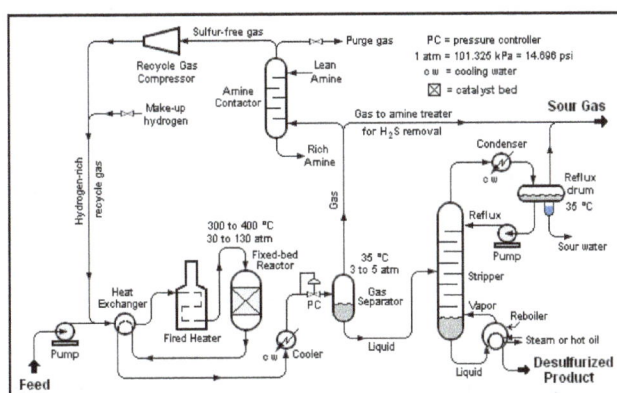

Schematic diagram of a typical Hydrodesulfurization (HDS) unit in a petroleum refinery

The liquid feed (at the bottom left in the diagram) is pumped up to the required elevated pressure and is joined by a stream of hydrogen-rich recycle gas. The resulting liquid-gas mixture is preheated by flowing through a heat exchanger. The preheated feed then flows through a fired heater where the feed mixture is totally vaporized and heated to the required elevated temperature before entering the reactor and flowing through a fixed-bed of catalyst where the hydrodesulfurization reaction takes place.

The hot reaction products are partially cooled by flowing through the heat exchanger where the reactor feed was preheated and then flows through a water-cooled heat exchanger before it flows through the pressure controller (PC) and undergoes a pressure reduction down to about 3 to 5 atmospheres. The resulting mixture of liquid and gas enters the gas separator vessel at about 35 °C and 3 to 5 atmospheres of absolute pressure.

Most of the hydrogen-rich gas from the gas separator vessel is recycle gas, which is routed through an amine contactor for removal of the reaction product H2S that it contains. The H2S-free hydrogen-rich gas is then recycled back for reuse in the reactor section. Any excess gas from the gas separator vessel joins the sour gas from the stripping of the reaction product liquid.

The liquid from the gas separator vessel is routed through a reboiled stripper distillation tower. The bottoms product from the stripper is the final desulfurized liquid product from hydrodesulfurization unit.

The overhead sour gas from the stripper contains hydrogen, methane, ethane, hydrogen sulfide, propane, and, perhaps, some butane and heavier components. That sour gas is sent to the refinery's central gas processing plant for removal of the hydrogen sulfide in the refinery's main amine gas treating unit and through a series of distillation towers for recovery of propane, butane and pentane or heavier components. The residual hydrogen, methane, ethane, and some propane is used as refinery fuel gas. The hydrogen sulfide removed and recovered by the amine gas treating unit is subsequently converted to elemental sulfur in a Claus process unit or to sulfuric acid in a wet sulfuric acid process or in the conventional Contact Process.

Note that the above description assumes that the HDS unit feed contains no olefins. If the feed does contain olefins (for example, the feed is a naphtha derived from a refinery fluid catalytic cracker (FCC) unit), then the overhead gas from the HDS stripper may also contain some ethene, propene, butenes and pentenes, or heavier components.

It should also be noted that the amine solution to and from the recycle gas contactor comes from and is returned to the refinery's main amine gas treating unit.

Sulfur Compounds in Refinery HDS Feedstocks

The refinery HDS feedstocks (naphtha, kerosene, diesel oil, and heavier oils) contain a wide range of organic sulfur compounds, including thiols, thiophenes, organic sulfides and disulfides, and many others. These organic sulfur compounds are products of the degradation of sulfur containing biological components, present during the natural formation of the fossil fuel, petroleum crude oil.

When the HDS process is used to desulfurize a refinery naphtha, it is necessary to remove the total sulfur down to the parts per million range or lower in order to prevent poisoning the noble metal catalysts in the subsequent catalytic reforming of the naphthas.

When the process is used for desulfurizing diesel oils, the latest environmental regulations in the United States and Europe, requiring what is referred to as ultra-low-sulfur diesel (ULSD), in turn requires that very deep hydrodesulfurization is needed. In the very early 2000s, the governmental regulatory limits for highway vehicle diesel was within the range of 300 to 500 ppm by weight of total sulfur. As of 2006, the total sulfur limit for highway diesel is in the range of 15 to 30 ppm by weight.

Thiophenes

A family of substrates that are particularly common in petroleum are the aromatic sulfur-containing heterocycles called thiophenes. Many kinds of thiophenes occur in petroleum ranging from thiophene itself to more condensed derivatives called benzothiophenes and dibenzothiophenes. Thiophene itself and its alkyl derivatives are easier to hydrogenolyse, whereas dibenzothiophene, especially its 4,6-disubstituted derivatives, are considered the most challenging substrates. Benzothiophenes are midway between the simple thiophenes and dibenzothiophenes in their susceptibility to HDS.

Catalysts and Mechanisms

The main HDS catalysts are based on molybdenum disulfide (MoS_2) together with smaller amounts of other metals. The nature of the sites of catalytic activity remains an active area of investigation, but it is generally assumed basal planes of the MoS_2 structure are not relevant to catalysis, rather the edges or rims of these sheet. At the edges of the MoS2 crystallites, the molybdenum centre can stabilize a coordinatively unsaturated site (CUS), also known as an anion vacancy. Substrates, such as thiophene, bind to this site and undergo a series of reactions that result in both C-S scission and C=C hydrogenation. Thus, the hydrogen serves multiple roles—generation of anion vacancy by removal of sulfide, hydrogenation, and hydrogenolysis. A simplified diagram for the cycle is shown:

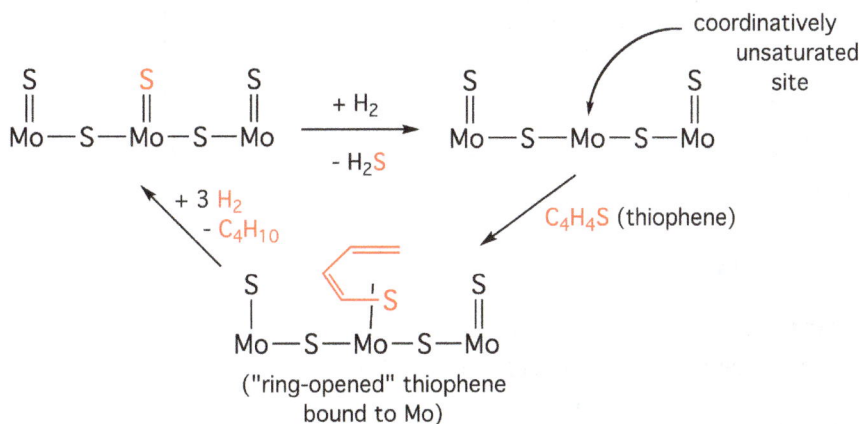

Simplified diagram of a HDS cycle for thiophene

Catalysts

Most metals catalyse HDS, but it is those at the middle of the transition metal series that are most active. Ruthenium disulfide appears to be the single most active catalyst, but binary combinations of cobalt and molybdenum are also highly active. Aside from the basic cobalt-modified MoS_2 catalyst, nickel and tungsten are also used, depending on the nature of the feed. For example, Ni-W catalysts are more effective for hydrodenitrogenation.

Supports

Metal sulfides are "supported" on materials with high surface areas. A typical support for HDS catalyst is γ-alumina. The support allows the more expensive catalyst to be more widely distributed, giving rise to a larger fraction of the MoS2 that is catalytically active. The interaction between the support and the catalyst is an area of intense interest, since the support is often not fully inert but participates in the catalysis.

Other Uses

The basic hydrogenolysis reaction has a number of uses other than hydrodesulfurization.

Hydrodenitrogenation

The hydrogenolysis reaction is also used to reduce the nitrogen content of a petroleum stream in a process referred to as hydrodenitrogenation (HDN). The process flow is the same as that for an HDS unit.

Using pyridine (C_5H_5N), a nitrogen compound present in some petroleum fractionation products, as an example, the hydrodenitrogenation reaction has been postulated as occurring in three steps:

Pyridine + Hydrogen	→	Piperdine + Hydrogen	→	Amylamine + Hydrogen	→	Pentane + Ammonia
$C_5H_5N + 5H_2$	→	$C_5H_{11}N + 2H_2$	→	$C_5H_{11}NH_2 + H_2$	→	$C_5H_{12} + NH_3$

and the overall reaction may be simply expressed as:

Pyridine + Hydrogen	→	Pentane + Ammonia
$C_5H_5N + 5H_2$	→	$C_5H_{12} + NH_3$

Many HDS units for desulfurizing naphthas within petroleum refineries are actually simultaneously denitrogenating to some extent as well.

Saturation of Olefins

The hydrogenolysis reaction may also be used to saturate or convert olefins (alkenes) into paraffins (alkanes). The process used is the same as for an HDS unit.

As an example, the saturation of the olefin pentene can be simply expressed as:

Pentene + Hydrogen	→	Pentane
$C_5H_{10} + H_2$	→	C_5H_{12}

Some hydrogenolysis units within a petroleum refinery or a petrochemical plant may be used solely for the saturation of olefins or they may be used for simultaneously desulfurizing as well as denitrogenating and saturating olefins to some extent.

Hydrogenation in the Food Industry

The food industry uses hydrogenation to completely or partially saturate the unsaturatedfatty acids in liquid vegetable fats and oils to convert them into solid or semi-solid fats, such as those in margarine and shortening.

References

- Gary, J.H. & Handwerk, G.E. (1984). Petroleum Refining Technology and Economics (2nd ed.). Marcel Dekker, Inc. ISBN 0-8247-7150-8.

- Gary, J.H.; Handwerk, G.E. (1984). Petroleum Refining Technology and Economics (2nd ed.). Marcel Dekker, Inc. ISBN 0-8247-7150-8.

- Wu, Ying; Carroll, John J. (5 July 2011). Carbon Dioxide Sequestration and Related Technologies. John Wiley & Sons. pp. 128–131. ISBN 978-0-470-93876-8.

- Gary, J.H.; Handwerk, G.E. (1984). Petroleum Refining Technology and Economics (2nd ed.). Marcel Dekker, Inc. ISBN 0-8247-7150-8.

- M. S. Vassiliou (2 March 2009). Historical Dictionary of the Petroleum Industry. Scarecrow Press. pp. 459–. ISBN 978-0-8108-6288-3.

- Newton Copp; Andrew Zanella (1993). Discovery, Innovation, and Risk: Case Studies in Science and Technology. MIT Press. pp. 172–. ISBN 978-0-262-53111-5.

- James H. Gary and Glenn E. Handwerk (2001). Petroleum Refining: Technology and Economics (4th ed.). CRC Press. ISBN 0-8247-0482-7.

- James. G. Speight (2006). The Chemistry and Technology of Petroleum (4th ed.). CRC Press. ISBN 0-8493-9067-2.

- David S.J. Jones and Peter P. Pujado (Editors) (2006). Handbook of Petroleum Processing (First ed.). Springer. ISBN 1-4020-2819-9.

- Alex C. Hoffmann; Lewis E. Stein (2002). Gas Cyclones and Swirl Tubes:Principles, Design and Operation (1st ed.). Springer. ISBN 3-540-43326-0.

- Wen-Ching Yang (2003). Handbook of Fluidization and Fluid Particle Systems. CRC Press. ISBN 0-8247-0259-X.

- Julius Scherzer (1990). Octane-enhancing Zeolitic FCC Catalysts: Scientific and Technical Aspects. CRC Press. ISBN 0-8247-8399-9.

- A. Kayode Coker (2007). Ludwig's Applied Process Design for Chemical And Petrochemical Plants, Volume 1 (4th ed.). Gulf Professional Publishing. pp. 732–737. ISBN 0-7506-7766-X.

- Milton R. Beychok (2005). Fundamentals of Stack Gas Dispersion (Fourth ed.). self-published. ISBN 0-9644588-0-2.

- Gary, J.H.; Handwerk, G.E. (1984). Petroleum Refining Technology and Economics (2nd ed.). Marcel Dekker, Inc. ISBN 0-8247-7150-8.

Formation Evaluation of Petroleum

Formation evaluation is the evaluation of a borehole to measure its ability to produce petroleum. Well logging, core sample, mud logging, gamma ray logging and formation evaluation neutron porosity are some of the chapters explained in this section. The text has been carefully written to provide an easy understanding of the formation evaluation of petroleum.

Formation Evaluation

In petroleum exploration and development, formation evaluation is used to determine the ability of a borehole to produce petroleum. Essentially, it is the process of "recognizing a commercial well when you drill one".

Modern rotary drilling usually uses a heavy mud as a lubricant and as a means of producing a confining pressure against the formation face in the borehole, preventing blowouts. Only in rare and catastrophic cases, do oil and gas wells *come in* with a fountain of gushing oil. In real life, that is a blowout—and usually also a financial and environmental disaster. But controlling blowouts has drawbacks—mud filtrate soaks into the formation around the borehole and a mud cake plasters the sides of the hole. These factors obscure the possible presence of oil or gas in even very porous formations. Further complicating the problem is the widespread occurrence of small amounts of petroleum in the rocks of many sedimentary provinces. In fact, if a sedimentary province is absolutely barren of traces of petroleum, it is not feasible to continue drilling there.

The formation evaluation problem is a matter of answering two questions:

1. What are the lower limits for porosity, permeability and upper limits for water saturation that permit profitable production from a particular formation or pay zone; in a particular geographic area; in a particular economic climate.

2. Do any of the formations in the well under consideration exceed these lower limits.

It is complicated by the impossibility of directly examining the formation. It is, in short, the problem of looking at the formation *indirectly*.

Formation Evaluation Tools

Tools to detect oil and gas have been evolving for over a century. The simplest and most direct tool is well cuttings examination. Some older oilmen ground the cuttings between their teeth and tasted to see if crude oil was present. Today, a wellsite geologist or mudlogger uses a low powered stereoscopic microscope to determine the lithology of the formation being drilled and to estimate porosity and possible oil staining. A portable ultraviolet light chamber or "Spook Box" is used to examine the cuttings for fluorescence. Fluorescence can be an indication of crude oil staining, or of the presence of fluorescent minerals. They can be differentiated by placing the cuttings in a solvent filled watchglass or dimple dish. The solvent is usually carbon tetrachlorethane. Crude oil dissolves and then redeposits as a fluorescent ring when the solvent evaporates. The written strip chart recording of these examinations is called a sample log or mudlog.

Well cuttings examination is a learned skill. During drilling, chips of rock, usually less than about 1/8 inch (6 mm) across, are cut from the bottom of the hole by the bit. Mud, jetting out of holes in the bit under high pressure, washes the cuttings away and up the hole. During their trip to the surface they may circulate around the turning drillpipe, mix with cuttings falling back down the hole, mix with fragments caving from the hole walls and mix with cuttings travelling faster and slower in the same upward direction. They then are screened out of the mudstream by the shale shaker and fall on a pile at its base. Determining the type of rock being drilled at any one time is a matter of knowing the 'lag time' between a chip being cut by the bit and the time it reaches the surface where it is then examined by the wellsite geologist (or mudlogger as they are sometimes called). A sample of the cuttings taken at the proper time will contain the current cuttings in a mixture of previously drilled material. Recognizing them can be very difficult at times, for example after a "bit trip" when a couple of miles of drill pipe has been extracted and returned to the hole in order to replace a dull bit. At such a time there is a flood of foreign material knocked from the borehole walls (cavings), making the mudloggers task all the more difficult.

Coring

One way to get more detailed samples of a formation is by coring. Two techniques commonly used at present. The first is the "whole core", a cylinder of rock, usually about 3" to 4" in diameter and up to 50 feet (15 m) to 60 feet (18 m) long. It is cut with a "core barrel", a hollow pipe tipped with a ring-shaped diamond chip-studded bit that can cut a plug and bring it to the surface. Often the plug breaks while drilling, usually in shales or fractures and the core barrel jams, slowly grinding the rocks in front of it to powder. This signals the driller to give up on getting a full length core and to pull up the pipe.

Taking a full core is an expensive operation that usually stops or slows drilling for at least the better part of a day. A full core can be invaluable for later reservoir evaluation. Once a section of well has been drilled, there is, of course, no way to core it without drilling another well.

Another, cheaper, technique for obtaining samples of the formation is "Sidewall Coring". One type of sidewall cores is percussion cores. In this method, a steel cylinder—a coring gun—has hollow-point steel bullets mounted along its sides and moored to the gun by short steel cables. The coring gun is lowered to the bottom of the interval of interest and the bullets are fired individually as the gun is pulled up the hole. The mooring cables ideally pull the hollow bullets and the enclosed plug of formation loose and the gun carries them to the surface. Advantages of this technique are low cost and the ability to sample the formation after it has been drilled. Disadvantages are possible non-recovery because of lost or misfired bullets and a slight uncertainty about the sample depth. Sidewall cores are often shot "on the run" without stopping at each core point because of the danger of differential sticking. Most service company personnel are skilled enough to minimize this problem, but it can be significant if depth accuracy is important.

A second method of sidewall coring is rotary sidewall cores. In this method, a circular-saw assembly is lowered to the zone of interest on a wireline, and the core is sawed out. Dozens of cores may be taken this way in one run. This method is roughly 20 times as expensive as percussion cores, but yields a much better sample.

A serious problem with cores is the change they undergo as they are brought to the surface. It might seem that cuttings and cores are very direct samples but the problem is whether the formation at depth will produce oil or gas. Sidewall cores are deformed and compacted and fractured by the bullet impact. Most full cores from any significant depth expand and fracture as they are brought to the surface and removed from the core barrel. Both types of core can be invaded or even flushed by mud, making the evaluation of formation fluids difficult. The formation analyst has to remember that all tools give indirect data.

Mud Logging

Mud logging (or Wellsite Geology) is a well logging process in which drilling mud and drill bit cuttings from the formation are evaluated during drilling and their properties recorded on a strip chart as a visual analytical tool and stratigraphic cross sectional representation of the well. The drilling mud which is analyzed for hydrocarbon gases, by use of a gas chromatograph, contains drill bit cuttings which are visually evaluated by a mudlogger and then described in the mud log. The total gas, chromatograph record, lithological sample, pore pressure, shale density,D-exponent, etc. (all lagged parameters because they are circulated up to the surface from the bit) are plotted along with surface parameters such as rate of penetration (ROP), Weight On Bit (WOB),rotation per minute etc. on the mudlog which serve as a tool for the mudlogger, drilling engineers, mud engineers, and other service personnel charged with drilling and producing the well.

Wireline Logging

The oil and gas industry uses wireline logging to obtain a continuous record of a formation's rock properties. Wireline logging can be defined as being "The acquisition and analysis of geophysical data performed as a function of well bore depth, together with the provision of related services." Note that "wireline logging" and "mud logging" are not the same, yet are closely linked through the integration of the data sets. The measurements are made referenced to "TAH" - True Along Hole depth: these and the associated analysis can then be used to infer further properties, such as hydrocarbon saturation and formation pressure, and to make further drilling and production decisions.

Wireline logging is performed by lowering a 'logging tool' - or a string of one or more instruments - on the end of a wireline into an oil well (or borehole) and recording petrophysical properties using a variety of sensors. Logging tools developed over the years measure the natural gamma ray, electrical, acoustic, stimulated radioactive responses, electromagnetic, nuclear magnetic resonance, pressure and other properties of the rocks and their contained fluids. For this article, they are broadly broken down by the main property that they respond to.

The data itself is recorded either at surface (real-time mode), or in the hole (memory mode) to an electronic data format and then either a printed record or electronic presentation called a "well log" is provided to the client, along with an electronic copy of the raw data. Well logging operations can either be performed during the drilling process, to provide real-time information about the formations being penetrated by the borehole, or once the well has reached Total Depth and the whole depth of the borehole can be logged.

Real-time data is recorded directly against measured cable depth. Memory data is recorded against time, and then depth data is simultaneously measured against time. The two data sets are then merged using the common time base to create an instrument response versus depth log. Memory recorded depth can also be corrected in exactly the same way as real-time corrections are made, so there should be no difference in the attainable TAH accuracy.

The measured cable depth can be derived from a number of different measurements, but is usually either recorded based on a calibrated wheel counter, or (more accurately) using magnetic marks which provide calibrated increments of cable length. The measurements made must then be corrected for elastic stretch and temperature.

There are many types of wireline logs and they can be categorized either by their function or by the technology that they use. "Open hole logs" are run before the oil or gas well is lined with pipe or cased. "Cased hole logs" are run after the well is lined with casing or production pipe.

Wireline logs can be divided into broad categories based on the physical properties measured.

Electric Logs

In 1928, the Schlumberger brothers in France developed the workhorse of all formation evaluation tools: the electric log. Electric logs have been improved to a high degree of precision and sophistication since that time, but the basic principle has not changed. Most underground formations contain water, often salt water, in their pores. The resistance to electric current of the total formation—rock and fluids—around the borehole is proportional to the sum of the volumetric proportions of mineral grains and conductive water-filled pore space. If the pores are partially filled with gas or oil, which are resistant to the passage of electric current, the bulk formation resistance is higher than for water filled pores. For the sake of a convenient comparison from measurement to measurement, the electrical logging tools measure the resistance of a cubic meter of formation. This measurement is called *resistivity*.

Modern resistivity logging tools fall into two categories, Laterolog and Induction, with various commercial names, depending on the company providing the logging services.

Laterolog tools send an electric current from an electrode on the sonde directly into the formation. The return electrodes are located either on surface or on the sonde itself. Complex arrays of electrodes on the sonde (guard electrodes) focus the current into the formation and prevent current lines from fanning out or flowing directly to the return electrode through the borehole fluid. Most tools vary the voltage at the main electrode in order to maintain a constant current intensity. This voltage is therefore proportional to the resistivity of the formation. Because current must flow from the sonde to the formation, these tools only work with conductive borehole fluid. Actually, since the resistivity of the mud is measured in series with the resistivity of the formation, laterolog tools give best results when mud resistivity is low with respect to formation resistivity, i.e., in salty mud.

Induction logs use an electric coil in the sonde to generate an alternating current loop in the formation by induction. This is the same physical principle as is used in electric transformers. The alternating current loop, in turn, induces a current in a receiving coil located elsewhere on the sonde. The amount of current in the receiving coil is proportional to the intensity of current loop, hence to the conductivity (reciprocal of resistivity) of the formation. Multiple transmitting and receiving coils are used to focus formation current loops both radially (depth of investigation) and axially (vertical resolution). Until the late 80's, the workhorse of induction logging has been the 6FF40 sonde which is made up of six coils with a nominal spacing of 40 inches (1,000 mm). Since the 90's all major logging companies use so-called array induction tools. These comprise a single transmitting coil and a large number of receiving coils. Radial and axial focusing is performed by software rather than by the physical layout of coils. Since the formation current flows in circular loops around the logging tool, mud resistivity is measured in parallel with formation resistivity. Induction tools therefore give best results when mud resistivity is high with respect to formation resistivity, i.e., fresh mud or non-conductive fluid. In oil-base mud, which is non conductive, induction logging is the only option available.

Until the late 1950s electric logs, mud logs and sample logs comprised most of the oilman's armamentarium. Logging tools to measure porosity and permeability began to be used at that time. The first was the microlog. This was a miniature electric log with two sets of electrodes. One measured the formation resistivity about 1/2" deep and the other about 1"-2" deep. The purpose of this seemingly pointless measurement was to detect permeability. Permeable sections of a borehole wall develop a thick layer of mudcake during drilling. Mud liquids, called filtrate, soak into the formation, leaving the mud solids behind to -ideally- seal the wall and stop the filtrate "invasion" or soaking. The short depth electrode of the microlog sees mudcake in permeable sections. The deeper 1" electrode sees filtrate invaded formation. In nonpermeable sections both tools read alike and the traces fall on top of each other on the stripchart log. In permeable sections they separate.

Also in the late 1950s porosity measuring logs were being developed. The two main types are: nuclear porosity logs and sonic logs.

Porosity Logs

The two main nuclear porosity logs are the Density and the Neutron log.

Density logging tools contain a caesium-137gamma ray source which irradiates the formation with 662 keV gamma rays. These gamma rays interact with electrons in the formation through Compton scattering and lose energy. Once the energy of the gamma ray has fallen below 100 keV, photolectric absorption dominates: gamma rays are eventually absorbed by the formation. The amount of energy loss by Compton scattering is related to the number electrons per unit volume of formation. Since for most elements of interest (below $Z = 20$) the ratio of atomic weight, A, to atomic number, Z, is close to 2, gamma ray energy loss is related to the amount of matter per unit volume, i.e., formation density.

A gamma ray detector located some distance from the source, detects surviving gamma rays and sorts them into several energy windows. The number of high-energy gamma rays is controlled by compton scattering, hence by formation density. The number of low-energy gamma rays is controlled by photoelectric absorption, which is directly related to the average atomic number, Z, of the formation, hence to lithology. Modern density logging tools include two or three detectors, which allow compensation for some borehole effects, in particular for the presence of mud cake between the tool and the formation.

Since there is a large contrast between the density of the minerals in the formation and the density of pore fluids, porosity can easily be derived from measured formation bulk density if both mineral and fluid densities are known.

Neutron porosity logging tools contain an americium-berylliumneutron source, which irradiates the formation with neutrons. These neutrons lose energy through elastic collisions with nuclei in the formation. Once their energy has decreased to thermal level, they diffuse randomly away from the source and are ultimately absorbed by a nucleus. Hydrogen atoms have essentially the same mass as the neutron; therefore hydrogen is the main contributor to the slowing down of neutrons. A detector at some distance from the source records the number of neutron reaching this point. Neutrons that have been slowed down to thermal level have a high probability of being absorbed by the formation before reaching the detector. The neutron counting rate is therefore inversely related to the amount of hydrogen in the formation. Since hydrogen is mostly present in pore fluids (water, hydrocarbons) the count rate can be converted into apparent porosity. Modern neutron logging tools usually include two detectors to compensate for some borehole effects. Porosity is derived from the ratio of count rates at these two detectors rather than from count rates at a single detector.

The combination of neutron and density logs takes advantage of the fact that lithology has opposite effects on these two porosity measurements. The average of neutron and density porosity values is usually close to the true porosity, regardless of lithology. Another advantage of this combination is the "gas effect." Gas, being less dense than liquids, translates into a density-derived porosity that is too high. Gas, on the other hand, has much less hydrogen per unit volume than liquids: neutron-derived porosity, which is based on the amount of hydrogen, is too low. If both logs are displayed on compatible scales, they overlay each other in liquid-filled clean formations and are widely separated in gas-filled formations.

Sonic logs use a pinger and microphone arrangement to measure the velocity of sound in the formation from one end of the sonde to the other. For a given type of rock, acoustic velocity varies indirectly with porosity. If the velocity of sound through solid rock is taken as a measurement of 0% porosity, a slower velocity is an indication of a higher porosity that is usually filled with formation water with a slower sonic velocity.

Both sonic and density-neutron logs give porosity as their primary information. Sonic logs read farther away from the borehole so they are more useful where sections of the borehole are caved. Because they read deeper, they also tend to average more formation than the density-neutron logs do. Modern sonic configurations with pingers and microphones at both ends of the log, combined with computer analysis, minimize the averaging somewhat. Averaging is an advantage when the formation is being evaluated for seismic parameters, a different area of formation evaluation. A special log, the Long Spaced Sonic, is sometimes used for this purpose. Seismic signals (a single undulation of a sound wave in the earth) average together tens to hundreds of feet of formation, so an averaged sonic log is more directly comparable to a seismic waveform.

Density-neutron logs read the formation within about four to seven inches (178 mm) of the borehole wall. This is an advantage in resolving thin beds. It is a disadvantage when the hole is badly caved. Corrections can be made automatically if the cave is no more than a few inches deep. A caliper arm on the sonde measures the profile of the borehole and a correction is calculated and incorporated in the porosity reading. However, if the cave is much more than four inches deep, the density-neutron log is reading little more than drilling mud.

Lithology Logs - SP and Gamma Ray

There are two other tools, the SP log and the Gamma Ray log, one or both of which are almost always used in wireline logging. Their output is usually presented along with the electric and porosity logs described above. They are indispensable as additional guides to the nature of the rock around the borehole.

The SP log, known variously as a "Spontaneous Potential", "Self Potential" or "Shale Potential" log is a voltmeter measurement of the voltage or electrical potential difference between the mud in the hole at a particular depth and a copper ground stake driven into the surface of the earth a short distance from the borehole. A salinity difference between the drilling mud and the formation water acts as a natural battery and will cause several voltage effects. This "battery" causes a movement of charged ions between the hole and the formation water where there is enough permeability in the rock. The most important voltage is set up as a permeable formation permits ion movement, reducing the voltage between the formation water and the mud. Sections of the borehole where this occurs then have a voltage difference with other nonpermeable sections where ion movement is restricted. Vertical ion movement in the mud column occurs much more slowly because the mud is not circulating while the drill pipe is out of the hole. The copper surface stake provides a reference point against which the SP voltage is measured for each part of the borehole. There can also be several other minor voltages, due for example to mud filtrate streaming into the formation under the effect of an overbalanced mud system. This flow carries ions and is a voltage generating current. These other voltages are secondary in importance to the voltage resulting from the salinity contrast between mud and formation water.

The nuances of the SP log are still being researched. In theory, almost all porous rocks contain water. Some pores are completely filled with water. Others have a thin layer of water molecules wetting the surface of the rock, with gas or oil filling the rest of the pore. In sandstones and porous limestones there is a continuous layer of water throughout the formation. If there is even a little permeability to water, ions can move through the rock and decrease the voltage difference with the mud nearby. Shales do not allow water or ion movement. Although they may have a large water content, it is bound to the surface of the flat clay crystals comprising the shale. Thus mud opposite shale sections maintains its voltage difference with the surrounding rock. As the SP logging tool is drawn up the hole it measures the voltage difference between the reference stake and the mud opposite shale and sandstone or limestone sections. The resulting log curve reflects the permeability of the rocks and, indirectly, their lithology. SP curves degrade over time, as the ions diffuse up and down the mud column. It also can suffer from stray voltages caused by other logging tools that are run with it. Older, simpler logs often have better SP curves than more modern logs for this reason. With experience in an area, a good SP curve can even allow a skilled interpreter to infer sedimentary environments such as deltas, point bars or offshore tidal deposits.

The gamma ray log is a measurement of naturally occurring gamma radiation from the borehole walls. Sandstones are usually nonradioactive quartz and limestones are nonradioactive calcite. Shales however, are naturally radioactive due to potassium isotopes in clays, and adsorbed uranium and thorium. Thus the presence or absence of gamma rays in a borehole is an indication of the amount of shale or clay in the surrounding formation. The gamma ray log is useful in holes drilled with air or with oil based muds, as these wells have no SP voltage. Even in water-based muds, the gamma ray and SP logs are often run together. They comprise a check on each other and can indicate unusual shale sections which may either not be radioactive, or may have an abnormal ionic chemistry. The gamma ray log is also useful to detect coal beds, which, depending on the local geology, can have either low radiation levels, or high radiation levels due to adsorption of uranium. In addition, the gamma ray log will work inside a steel casing, making it essential when a cased well must be evaluated.

Interpreting the Tools

The immediate questions that have to be answered in deciding to complete a well or to plug and abandon (P&A) it are:

- Do any zones in the well contain producible hydrocarbons?

- How much?

- How much, if any, water will be produced with them?

The elementary approach to answering these questions uses the Archie Equation.

Well Logging

Well logging, also known as borehole logging is the practice of making a detailed record (a *well log*) of the geologic formations penetrated by a borehole. The log may be based either on visual inspection of samples brought to the surface (*geological* logs) or on physical measurements made by instruments lowered into the hole (*geophysical* logs). Some types of geophysical well logs can be done during any phase of a well's history: drilling, completing, producing, or abandoning. Well logging is performed in boreholes drilled for the oil and gas, groundwater, mineral and geothermal exploration, as well as part of environmental and geotechnical studies.

Wireline Logging

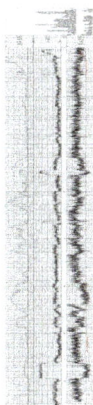

Wireline log consisting of caliper, density and resistivity logs

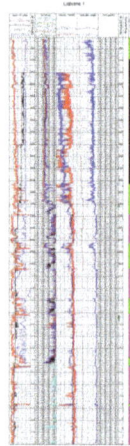

Wireline log consisting of a complete set of logs

The oil and gas industry uses wireline logging to obtain a continuous record of a formation's rock properties. Wireline logging can be defined as being "The acquisition and analysis of geophysical data performed as a function of well bore depth, together with the provision of related services." Note that "wireline logging" and "mud logging" are not the same, yet are closely linked through the integration of the data sets. The measurements are made referenced to "TAH" - True Along Hole depth: these and the associated analysis can then be used to infer further properties, such as hydrocarbon saturation and formation pressure, and to make further drilling and production decisions.

Wireline logging is performed by lowering a 'logging tool' - or a string of one or more instruments - on the end of a wireline into an oil well (or borehole) and recording petrophysical properties using a variety of sensors. Logging tools developed over the years measure the natural gamma ray, electrical, acoustic, stimulated radioactive responses, electromagnetic, nuclear magnetic resonance, pressure and other properties of the rocks and their contained fluids. For this article, they are broadly broken down by the main property that they respond to.

The data itself is recorded either at surface (real-time mode), or in the hole (memory mode) to an electronic data format and then either a printed record or electronic presentation called a "well log" is provided to the client, along with an electronic copy of the raw data. Well logging operations can either be performed during the drilling process, to provide real-time information about the formations being penetrated by the borehole, or once the well has reached Total Depth and the whole depth of the borehole can be logged.

Real-time data is recorded directly against measured cable depth. Memory data is recorded against time, and then depth data is simultaneously measured against time. The two data sets are then merged using the common time base to create an instrument response versus depth log. Memory recorded depth can also be corrected in exactly the same way as real-time corrections are made, so there should be no difference in the attainable TAH accuracy.

The measured cable depth can be derived from a number of different measurements, but is usually either recorded based on a calibrated wheel counter, or (more accurately) using magnetic marks which provide calibrated increments of cable length. The measurements made must then be corrected for elastic stretch and temperature.

There are many types of wireline logs and they can be categorized either by their function or by the technology that they use. "Open hole logs" are run before the oil or gas well is lined with pipe or cased. "Cased hole logs" are run after the well is lined with casing or production pipe.

Wireline logs can be divided into broad categories based on the physical properties measured.

History

Conrad and Marcel Schlumberger, who founded Schlumberger Limited in 1926, are considered the inventors of electric well logging. Conrad developed the Schlumberger array, which was a technique for prospecting for metalore deposits, and the brothers adapted that surface technique to subsurface applications. On September 5, 1927, a crew working for Schlumberger lowered an electric sonde or tool down a well in Pechelbronn, Alsace, France creating the first well log. In modern terms, the first log was a resistivity log that could be described as 3.5-meter upside-down lateral log.

In 1931, Henri George Doll and G. Dechatre, working for Schlumberger, discovered that the galvanometer wiggled even when no current was being passed through the logging cables down in the well. This led to the discovery of the spontaneous potential (SP) which was as important as the ability to measure resistivity. The SP effect was produced naturally by the borehole mud at the boundaries of permeable beds. By simultaneously recording SP and resistivity, loggers could distinguish between permeable oil-bearing beds and impermeable nonproducing beds.

In 1940, Schlumberger invented the spontaneous potentialdipmeter; this instrument allowed the calculation of the dip and direction of the dip of a layer. The basic dipmeter was later enhanced by the resistivity dipmeter (1947) and the continuous resistivity dipmeter (1952).

Oil-based mud (OBM) was first used in Rangely Field, Colorado in 1948. Normal electric logs require a conductive or water-based mud, but OBMs are nonconductive. The solution to this problem was the induction log, developed in the late 1940s.

The introduction of the transistor and integrated circuits in the 1960s made electric logs vastly more reliable. Computerization allowed much faster log processing, and dramatically expanded log data-gathering capacity. The 1970s brought more logs and computers. These included combo type logs where resistivity logs and porosity logs were recorded in one pass in the borehole.

The two types of porosity logs (acoustic logs and nuclear logs) date originally from the 1940s. Sonic logs grew out of technology developed during World War II. Nuclear logging has supplemented acoustic logging, but acoustic or sonic logs are still run on some combination logging tools.

Nuclear logging was initially developed to measure the natural gamma radiation emitted by underground formations. However, the industry quickly moved to logs that actively bombard rocks with nuclear particles. The gamma ray log, measuring the natural radioactivity, was introduced by Well Surveys Inc. in 1939, and the WSI neutron log came in 1941. The gamma ray log is particularly useful as shale beds which often provide a relatively low permeability cap over hydrocarbon reservoirs usually display a higher level of gamma radiation. These logs were important because they can be used in cased wells (wells with production casing). WSI quickly became part of Lane-Wells.

During World War II, the US Government gave a near wartime monopoly on open-hole logging to Schlumberger, and a monopoly on cased-hole logging to Lane-Wells. Nuclear logs continued to evolve after the war.

The nuclear magnetic resonance log was developed in 1958 by Borg Warner. Initially, the NMR log was a scientific success but an engineering failure. However, the development of a continuous NMR logging tool by Numar (now a subsidiary of Halliburton) is a promising new technology.

Many modern oil and gas wells are drilled directionally. At first, loggers had to run their tools somehow attached to the drill pipe if the well was not vertical. Modern techniques now permit continuous information at the surface. This is known as logging while drilling (LWD) or measure-ment-while-drilling (MWD). MWD logs use mud pulse technology to transmit data from the tools on the bottom of the drillstring to the processors at the surface.

Electrical Logs

Resistivity Log

Resistivity logging measures the subsurface electrical resistivity, which is the ability to impede the flow of electric current. This helps to differentiate between formations filled with salty waters (good conductors of electricity) and those filled with hydrocarbons (poor conductors of electric-ity). Resistivity and porosity measurements are used to calculate water saturation. Resistivity is expressed in ohms or ohms\meter, and is frequently charted on a logarithm scale versus depth because of the large range of resistivity. The distance from the borehole penetrated by the current varies with the tool, from a few centimeters to one meter.

Borehole Imaging

The term "borehole imaging" refers to those logging and data-processing methods that are used to produce centimeter-scale images of the borehole wall and the rocks that make it up. The context is, therefore, that of open hole, but some of the tools are closely related to their cased-hole equiv-alents. Borehole imaging has been one of the most rapidly advancing technologies in wireline well logging. The applications range from detailed reservoir description through reservoir performance to enhanced hydrocarbon recovery. Specific applications are fracture identification, analysis of small-scale sedimentological features, evaluation of net pay in thinly bedded formations, and the identification of breakouts (irregularities in the borehole wall that are aligned with the minimum horizontal stress and appear where stresses around the wellbore exceed the compressive strength of the rock). The subject area can be classified into four parts:

1. Optical imaging

2. Acoustic imaging

3. Electrical imaging

4. Methods that draw on both acoustic and electrical imaging techniques using the same logging tool

Porosity Logs

Porosity logs measure the fraction or percentage of pore volume in a volume of rock. Most porosity logs use either acoustic or nuclear technology. Acoustic logs measure characteristics of sound waves propagated through the well-bore environment. Nuclear logs utilize nuclear reactions that take place in the downhole logging instrument or in the formation. Nuclear logs include density logs and neutron logs, as well as gamma ray logs which are used for correlation. The basic principle behind the use of nuclear technology is that a neutron source placed near the formation whose porosity is being measured will result in neutrons being scattered by the hydrogen atoms, largely those present in the formation fluid. Since there is little difference in the neutrons scattered by hydrocarbons or water, the porosity measured gives a figure close to the true physical porosity whereas the figure obtained from electrical resistivity measurements is that due to the conductive formation fluid. The difference between neutron porosity and electrical porosity measurements therefore indicates the presence of hydrocarbons in the formation fluid.

Density

The density log measures the bulk density of a formation by bombarding it with a radioactive source and measuring the resulting gamma ray count after the effects of Compton Scattering and Photoelectric absorption. This bulk density can then be used to determine porosity.

Neutron Porosity

The neutron porosity log works by bombarding a formation with high energy epithermal neutrons that lose energy through elastic scattering to near thermal levels before being absorbed by the nuclei of the formation atoms. Depending on the particular type of neutron logging tool, either the gamma ray of capture, scattered thermal neutrons or scattered, higher energy epithermal neutrons are detected. The neutron porosity log is predominantly sensitive to the quantity of hydrogen atoms in a particular formation, which generally corresponds to rock porosity.

Boron is known to cause anomalously low neutron tool count rates due to it having a high capture cross section for thermal neutron absorption. An increase in hydrogen concentration in clay minerals has a similar effect on the count rate.

Sonic

A sonic log provides a formation interval transit time, which typically varies lithology and rock texture but particularly porosity. The logging tool consists of a piezoelectric transmitter and receiver and the time taken to for the sound wave to travel the fixed distance between the two is recorded as an *interval transit time*.

Lithology Logs

Gamma Ray

A log of the natural radioactivity of the formation along the borehole, measured in API units, particularly useful for distinguishing between sands and shales in a siliclastic environment. This is because sandstones are usually nonradioactive quartz, whereas shales are naturally radioactive due to potassium isotopes in clays, and adsorbed uranium and thorium.

In some rocks, and in particular in carbonate rocks, the contribution from uranium can be large and erratic, and can cause the carbonate to be mistaken for a shale. In this case, The carbonate gamma ray is a better indicator of shaliness. the carbonate gamma ray log is a gamma ray log from which the uranium contribution has been subtracted.

Self/Spontaneous Potential

The Spontaneous Potential (SP) log measures the natural or spontaneous potential difference between the borehole and the surface, without any applied current. It was one of the first wireline logs to be developed, found when a single potential electrode was lowered into a well and a potential was measured relative to a fixed reference electrode at the surface.

The most useful component of this potential difference is the electrochemical potential because it can cause a significant deflection in the SP response opposite permeable beds. The magnitude of this deflection depends mainly on the salinity contrast between the drilling mud and the formation water, and the clay content of the permeable bed. Therefore, the SP log is commonly used to detect permeable beds and to estimate clay content and formation water salinity.

Miscellaneous

Caliper

A tool that measures the diameter of the borehole, using either 2 or 4 arms. It can be used to detect regions where the borehole walls are compromised and the well logs may be less reliable.

Nuclear Magnetic Resonance

Nuclear magnetic resonance (NMR) logging uses the NMR response of a formation to directly determine its porosity and permeability, providing a continuous record along the length of the borehole. The chief application of the NMR tool is to determine moveable fluid volume (BVM) of a rock. This is the pore space excluding clay bound water (CBW) and irreducible water (BVI). Neither of these are moveable in the NMR sense, so these volumes are not easily observed on older logs. On modern tools, both CBW and BVI can often be seen in the signal response after transforming the relaxation curve to the porosity domain. Note that some of the moveable fluids (BVM) in the NMR sense are not actually moveable in the oilfield sense of the word. Residual oil and gas, heavy oil, and bitumen may appear moveable to the NMR precession measurement, but these will not necessarily flow into a well bore.

Spectral Noise Logging

Spectral noise logging (SNL) is an acoustic noise measuring technique used in oil and gas wells for well integrity analysis, identification of production and injection intervals and hydrodynamic characterisation of the reservoir. SNL records acoustic noise generated by fluid or gas flow through the reservoir or leaks in downhole well components.

Noise logging tools have been used in the petroleum industry for several decades. As far back as

1955, an acoustic detector was proposed for use in well integrity analysis to identify casing holes. Over many years, downhole noise logging tools proved effective in inflow and injectivity profiling of operating wells, leak detection, location of cross-flows behind casing, and even in determining reservoir fluid compositions. Robinson (1974) described how noise logging can be used to determine effective reservoir thickness.

Logging while Drilling

In the 1970s, a new approach to wireline logging was introduced in the form of logging while drilling (LWD). This technique provides similar well information to conventional wireline logging but instead of sensors being lowered into the well at the end of wireline cable, the sensors are integrated into the drill string and the measurements are made in real-time, whilst the well is being drilled. This allows drilling engineers and geologists to quickly obtain information such as porosity, resistivity, hole direction and weight-on-bit and they can use this information to make immediate decisions about the future of the well and the direction of drilling.

In LWD, measured data is transmitted to the surface in real time via pressure pulses in the well's mud fluid column. This mud telemetry method provides a bandwidth of less than 10 bits per second, although, as drilling through rock is a fairly slow process, data compression techniques mean that this is an ample bandwidth for real-time delivery of information. A higher sample rate of data is recorded into memory and retrieved when the drillstring is withdrawn at bit changes. High-definition downhole and subsurface information is available through networked or wired drillpipe that deliver memory quality data in real time.

Corrosion Well Logging

Throughout the life of the wells, integrity controles of the steel and cemented column (casing and tubing) are performed using calipers and thickness gauges. These advanced technical methods use non destructive technologies as ultrasonic, electromagnetic and magnetic transducers.

Memory Log

This method of data acquisition involves recording the sensor data into a down hole memory, rather than transmitting "Real Time" to surface. There are some advantages and disadvantages to this memory option.

- The tools can be conveyed into wells where the trajectory is deviated or extended beyond the reach of conventional Electric Wireline cables. This can involve a combination of weight to strength ratio of the electric cable over this extended reach. In such cases the memory tools can be conveyed on Pipe or Coil Tubing.

- The type of sensors are limited in comparison to those used on Electric Line, and tend to be focussed on the cased hole,production stage of the well. Although there are now developed some memory "Open Hole" compact formation evaluation tool combinations. These tools can be deployed and carried downhole concealed internally in drill pipe to protect them from damage while running in the hole, and then "Pumped" out the end at depth to initiate logging. Other basic open hole formation evaluation memory tools are available for use in "Commodity" markets on slickline to reduce costs and operating time.

- In cased hole operation there is normally a "Slick Line" intervention unit. This uses a solid mechanical wire (0.072 - 0.125 inches in OD), to manipulate or otherwise carry out operations in the well bore completion system. Memory operations are often carried out on this Slickline conveyance in preference to mobilizing a full service Electric Wireline unit.

- Since the results are not known until returned to surface, any realtime well dynamic changes cannot be monitored real time. This limits the ability to modify or change the well down hole production conditions accurately during the memory logging by changing the surface production rates. Something that is often done in Electric Line operations.

- Failure during recording is not known until the memory tools are retrieved. This loss of data can be a major issue on large offshore (expensive) locations. On land locations (e.g. South Texas, US) where there is what is called a "Commodity" Oil service sector, where logging often is without the rig infrastructure. this is less problematic, and logs are often run again without issue.

Coring

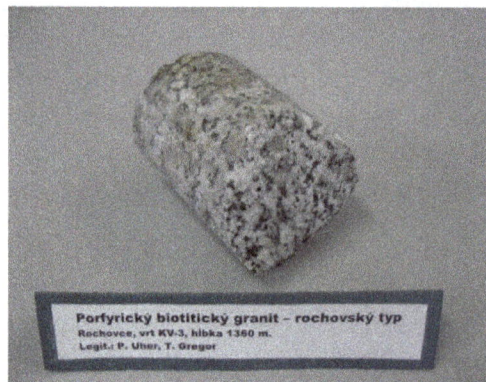

An example of a granite core

Coring is the process of obtaining an actual sample of a rock formation from the borehole. There are two main types of coring: 'full coring', in which a sample of rock is obtained using a specialised drill-bit as the borehole is first penetrating the formation and 'sidewall coring', in which multiple samples are obtained from the side of the borehole after it has penetrated through a formation. The main advantage of sidewall coring over full coring are that it is cheaper (drilling doesn't have to be stopped) and multiple samples can be easily acquired, with the main disadvantages being that there can be uncertainty in the depth at which the sample was acquired and the tool can fail to acquire the sample.

Mudlogging

Mud logs are well logs prepared by describing rock or soil cuttings brought to the surface by mud circulating in the borehole. In the oil industry they are usually prepared by a mud logging company contracted by the operating company. One parameter a typical mud log displays is the formation gas (gas units or ppm). "The gas recorder usually is scaled in terms of arbitrary gas units, which are defined differently by the various gas-detector manufactures. In practice, significance is placed only on relative changes in the gas concentrations detected." The current oil industry standard mud log normally includes real-time drilling parameters such as rate of penetration (ROP), lithology, gas hydrocarbons, flow line temperature (temperature of the drilling fluid) and chlorides but may also include mud weight, estimated pore pressure and corrected d-exponent (corrected drilling exponent) for a pressure pack log. Other information that is normally notated on a mud log include directional data (deviation surveys), weight on bit, rotary speed, pump pressure, pump rate, viscosity, drill bit info, casing shoe depths, formation tops, mud pump info, to name just a few.

Information Use

In the oil industry, the well and mud logs are usually transferred in 'real time' to the operating company, which uses these logs to make operational decisions about the well, to correlate formation depths with surrounding wells, and to make interpretations about the quantity and quality of hydrocarbons present. Specialists involved in well log interpretation are called log analysts.

Core Sample

A core sample is a cylindrical section of (usually) a naturally occurring substance. Most core samples are obtained by drilling with special drills into the substance, for example sediment or rock, with a hollow steel tube called a core drill. The hole made for the core sample is called the "core bowling". A variety of core samplers exist to sample different media under different conditions. More continue to be invented on a regular basis. In the coring process, the sample is pushed more or less intact into the tube. Removed from the tube in the laboratory, it is inspected and analyzed by different techniques and equipment depending on the type of data desired.

Core samples can be taken to test the properties of manmade materials, such as concrete, ceramics, some metals and alloys, especially the softer ones. Core samples can also be taken of living things, including human beings, especially of a person's bones for microscopic examination to help diagnose diseases.

Methods

Granitic rock core from Stillwater igneous complex, Montana (from a spoil pile).

The composition of the subject materials can vary from almost liquid to the strongest materials found in nature or technology, and the location of the subject materials can vary from on the laboratory bench to over 10 km from the surface of the Earth in a borehole. The range of equipment and techniques applied to the task is correspondingly great. Core samples are most often taken with their long axis oriented roughly parallel to the axis of a borehole, or parallel to the gravity field for the gravity-driven tools. However it is also possible to take core samples from the wall of an existing borehole. Taking samples from an exposure, albeit an overhanging rock face or on a different planet, is almost trivial. (The Mars Exploration Rovers carry a Rock Abrasion Tool, which is logically equivalent to the "rotary sidewall core" tool described below.)

Some common techniques include:

- *gravity coring*, in which the core sampler is dropped into the sample, usually the bed of a water body, but essentially the same technique can also be done on soft materials on land. The penetration forces, if recorded, give information about the strength of different depths in the material, which may be the only information required, with samples as an incidental benefit. This technique is common in both civil engineering site investigations (where the techniques tend towards into pile driving) and geological studies of recent aquatic deposits. The low strength of the materials penetrated means that cores have to be relatively small.

- *vibracoring*, in which the sampler is vibrated to allow penetration into thixotropic media. Again, the physical strength of the subject material limits the size of core that can be retrieved.

- *drilling*exploration diamond drilling where a rotating annular tool backed up by a cylindrical core sample storage device is pressed against the subject materials to cut out a cylinder of the subject material. A mechanism is normally needed to retain the cylindrical sample in the coring tool. Depending on circumstances, particularly the consistency and composition of the subject materials, different arrangements may be needed within the core tools to support and protect the sample on its way to surface; it is often also necessary to control or reduce the contact between the drilling fluid and the core sample, to reduce changes from the coring process. The mechanical forces imposed on the core sample by the tool frequently lead to fracture of the core and loss of less-competent intervals, which can greatly complicate interpretation of the core. Cores can routinely be cut as small as a few millimeters in diameter (in wood, for dendrochronology) up to over 150 millimeters in diameter (routine in oil exploration). The lengths of samples can range from less than a meter (again, in wood, for dendrochronology) up to around 200 metres in one run, though 27m to 54m is more usual (in oil exploration), and many runs can be made in succession if "quick look" analysis in the field suggests that the zone of interest is continuing.

- *percussion sidewall coring* coring uses robust cylindrical "bullets" explosively propelled into the wall of a borehole to retrieve a (relatively) small, short core sample. These tend to be heavily shattered, rendering porosity/ permeability measurements dubious, but are often sufficient for lithological and micropalaeontological study. Many samples can be attempted in a single run of the tools, which are typically configured with 20 to 30 "bullets" and propulsive charges along the length of a tool. Several tools can often be ganged together for a single run. The success rates for firing a particular bullet, it penetrating the borehole wall, the retention system recovering the bullet from the borehole wall, and the sample being retained in the bullet are all relatively low, so it is not uncommon for only half the samples attempted to be successful. This is an important consideration in planning sample programmes.

- *rotary sidewall coring* where a miniaturised automated rotary drilling tool is applied to the side of the borehole to cut a sample similar in size to a percussion sidewall core (described above). These tend to suffer less deformation than percussion cores. However the core-cutting process takes longer and jams are common in the ancillary equipment which retrieves the sample from the drill bit and stores it within the tool body. This complicates the planning of a coring programme.

Management of Cores and Data

Although often neglected, core samples always degrade to some degree in the process of cutting the core, handling it, and studying it. Non-destructive techniques are increasingly common (e.g. the use of MRI scanning to characterize grains, pore fluids, pore spaces (porosity) and their interactions (constituting part of permeability) but such expensive subtlety is likely wasted on a core that has been shaken on an unsprung lorry for 300 km of dirt road. What happens to cores between the retrieval equipment and the final laboratory (or archive) is an often neglected part of record keeping and core management.

Cut Bakken Core samples

Coring has come to be recognized as an important source of data, and more attention and care is being put on preventing damage to the core during various stages of it transportation and analysis. The usual way to do this is to freeze the core completely using liquid nitrogen, which is cheaply sourced. In some cases, special polymers are also used to preserve and seat/cushion the core from damage.

Equally, a core sample which cannot be related to its context (where it was before it became a core sample) has lost much of its benefit. The identification of the borehole, and the position and orientation ("way up") of the core in the borehole is critical, even if the borehole is in a tree trunk

- dendrochronologists always try to include a bark surface in their samples so that the date of most-recent growth of the tree can be unambiguously determined.

If these data become separated from core samples, it is generally impossible to regain that data. The cost of a coring operation can vary from a few currency units (for a hand-caught core from a soft soil section) to tens of millions of currency units (for sidewall cores from a remote-area off-shore borehole many kilometres deep). Inadequate recording of such basic data has ruined the utility of both types of core.

Different disciplines have different local conventions of recording these data, and the user should familiarize themselves with their area's conventions. For example, in the oil industry, orientation of the core is typically recorded by marking the core with two longitudinal colour streaks, with the red one on the right when the core is being retrieved and marked at surface. Cores cut for mineral mining may have their own, different, conventions. Civil engineering or soil studies may have their own, different, conventions as their materials are often not competent enough to make permanent marks on.

Bakken Core samples under a black light

It is becoming increasingly common to retain core samples in cylindrical packaging which forms part of the core-cutting equipment, and to make the marks of record on these "inner barrels" in the field prior to further processing and analysis in the laboratory. Sometimes core is shipped from the field to the laboratory in as long a length as it comes out of the ground; other times it is cut into standard lengths (5m or 1m or 3 ft) for shipping, then reassembled in the laboratory. Some of the "inner barrel" systems are capable of being reversed on the core sample, so that in the laboratory the sample goes "wrong way up" when the core is reassembled. This can complicate interpretation.

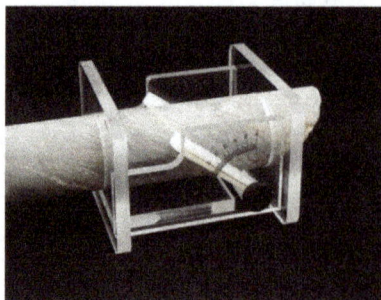

Goniometers are used to measure angles of fractures and other features in a core sample relative to its standard orientation.

If the borehole has petrophysical measurements made of the wall rocks, and these measurements are repeated along the length of the core then the two data sets correlated, one will almost universally find that the depth "of record" for a particular piece of core differs between the two methods of measurement. Which set of measurements to believe then becomes a matter of policy for the client (in an industrial setting) or of great controversy (in a context without an overriding authority). Recording that there are discrepancies, for whatever reason, retains the possibility of correcting an incorrect decision at a later date ; destroying the "incorrect" depth data makes it impossible to correct a mistake later. Any system for retaining and archiving data and core samples needs to be designed so that dissenting opinion like this can be retained.

If core samples from a campaign are competent, it is common practice to "slab" them - cut the sample into two or more samples longitudinally - quite early in laboratory processing so that one set of samples can be archived early in the analysis sequence as a protection against errors in processing. "Slabbing" the core into a 2/3 and a 1/3 set is common. It is also common for one set to be retained by the main customer while the second set goes to the government (who often impose a condition for such donation as a condition of exploration/ exploitation licensing). "Slabbing" also has the benefit of preparing a flat, smooth surface for examination and testing of profile permeability, which is very much easier to work with than the typically rough, curved surface of core samples when they're fresh from the coring equipment. Photography of raw and "slabbed" core surfaces is routine, often under both natural and ultra-violet light.

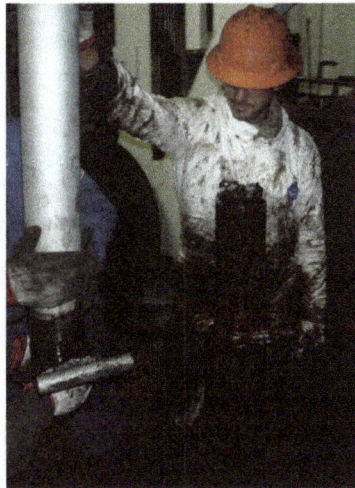

Pulling the Bakken Core out of the core barrell

A unit of length occasionally used in the literature on seabed cores is *cmbsf*, an abbreviation for centimeters below sea floor.

History of Coring

The technique of coring long predates attempts to drill into the Earth's mantle by the Deep Sea Drilling Program. The value to oceanic and other geologic history of obtaining cores over a wide area of sea floors soon became apparent. Core sampling by many scientific and exploratory organizations expanded rapidly. To date hundreds of thousands of core samples have been collected from floors of all the planet's oceans and many of its inland waters. Access to many of these samples is facilitated by the Index to Marine & Lacustrine Geological Samples,

"A collaboration between twenty institutions and agencies that operate geological repositories."

The above agency keeps a record of the samples held in the repositories of its member organizations. Data includes

Bakken Core

"Lithography, texture, age, principal investigator, province, weathering/metamorphism, glass remarks and descriptive comments"

Lower Bakken-3 Forks Transition

Informational Value of Core Samples

Coring began as a method of sampling surroundings of ore deposits and oil exploration. It soon expanded to oceans, lakes, ice, mud, soil and wood. Cores on very old trees give information about their growth rings without destroying the tree.

Cores indicate variations of climate, species and sedimentary composition during geologic history. The dynamic phenomena of the Earth's surface are for the most part cyclical in a number of ways, especially temperature and rainfall.

There are many ways to date a core. Once dated, it gives valuable information about changes of climate and terrain. For example, cores in the ocean floor, soil and ice have altered the view of the geologic history of the Pleistocene entirely.

Mud Logging

Mud logging is the creation of a detailed record (well log) of a borehole by examining the cuttings of rock brought to the surface by the circulating drilling medium (most commonly drilling mud). Mud logging is usually performed by a third-party mud logging company. This provides well owners and producers with information about the lithology and fluid content of the borehole while drilling. Historically it is the earliest type of well log. Under some circumstances compressed air is employed as a circulating fluid, rather than mud. Although most commonly used in petroleum exploration, mud logging is also sometimes used when drilling water wells and in other mineral exploration, where drilling fluid is the circulating medium used to lift cuttings out of the hole. In hydrocarbon exploration, hydrocarbon surface gas detectors record the level of natural gas brought up in the mud. A mobile laboratory is situated by the mud logging company near the drilling rig or on deck of an offshore drilling rig, or on a drill ship.

The Service Provided

Mud logging technicians in an oil field drilling operation determine positions of hydrocarbons with respect to depth, identify downhole lithology, monitor natural gas entering the drilling mud stream, and draw well logs for use by oil company geologist. Rock cuttings circulated to the surface in drilling mud are sampled and analyzed.

The mud logging company is normally contracted by the oil company (or operator). They then organize this information in the form of a graphic log, showing the data charted on a representation of the wellbore.

Well-site geologist mudlogging

The oil company representative (Company Man or "CoMan") together with the tool pusher, and well-site geologist (WSG) provides mud loggers their instruction. The mud logging company is contracted specifically as to when to start well-logging activity and what services to provide. Mud logging may begin on the first day of drilling, known as the "spud in" date but is more likely at some later time (and depth) determined by the oil industry geologist's research. The mud logger may also possess logs from wells drilled in the surrounding area. This information (known as "offset data") can provide valuable clues as to the characteristics of the particular geo-strata that the rig crew is about to drill through.

Mud loggers connect various sensors to the drilling apparatus and install specialized equipment to monitor or "log" drill activity. This can be physically and mentally challenging, especially when having to be done during drilling activity. Much of the equipment will require precise calibration or alignment by the mud logger to provide accurate readings.

Mud logging technicians observe and interpret the indicators in the mud returns during the drilling process, and at regular intervals log properties such as drilling rate, mud weight, flowline temperature, oil indicators, pump pressure, pump rate, lithology (rock type) of the drilled cuttings, and other data. Mud logging requires a good deal of diligence and attention. Sampling the drilled cuttings must be performed at predetermined intervals, and can be difficult during rapid drilling.

Another important task of the mud logger is to monitor gas levels (and types) and notify other personnel on the rig when gas levels may be reaching dangerous levels, so appropriate steps can be taken to avoid a dangerous well blowout condition. Because of the lag time between drilling and the time required for the mud and cuttings to return to the surface, a modern augmentation has come into use: Measurement while drilling. The MWD technician, often a separate service company employee, logs data in a similar manner but the data is different in source and content. Most of the data logged by an MWD technician comes from expensive and complex, sometimes electronic, tools that are downhole installed at or near the drill bit.

Scope

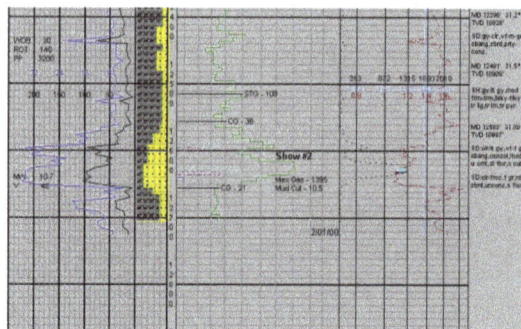

1" (5 foot average) mud log showing heavy (hydrocarbons) (large area of yellow)

Mud logging includes observation and microscopic examination of drill cuttings (formation rock chips), and evaluation of gas hydrocarbon and its constituents, basic chemical and mechanical parameters of drilling fluid or drilling mud (such as chlorides and temperature), as well as compiling other information about the drilling parameters. Then data is plotted on a graphic log called a mud log. Example1, Example2.

Other real-time drilling parameters that may be compiled include, but are not limited to; rate of penetration (ROP) of the bit (sometimes called the drill rate), pump rate (quantity of fluid being pumped), pump pressure, weight on bit, drill string weight, rotary speed, rotary torque, RPM (Revolutions Per Minute), SPM (Strokes Per Minute) mud volumes, mud weight and mud viscosity. This information is usually obtained by attaching *monitoring devices* to the drilling rig's equipment with a few exceptions such as the mud weight and mud viscosity which are measured by the derrickhand or the mud engineer.

Rate of drilling is affected by the pressure of the column of mud in the borehole and its relative counterbalance to the internal pore pressures of the encountered rock. A rock pressure greater than the mud fluid will tend to cause rock fragments to spall as it is cut and can increase the drilling rate. "D-exponents" are mathematical trend lines which estimate this internal pressure. Thus both visual evidence of spalling and mathematical plotting assist in formulating recommendations for optimum drilling mud densities for both safety (blowout prevention) and economics. (Faster drilling is generally preferred.)

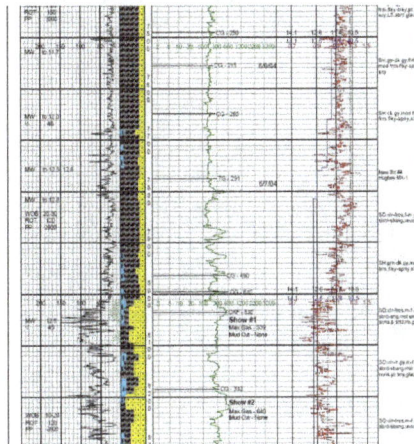

1" (every foot) mud log showing corrected d-Exponent trending into pressure above the sand

Mud logging is often written as a single word "mudlogging". The finished product can be called a "mud log" or "mudlog". The occupational description is "mud logger" or "mudlogger". In most cases, the two word usage seems to be more common. The mud log provides a reliable *time log* of drilled formations.

Details of the Mud Log

- The *rate of penetration* in (Figure 1 & 2) is represented by the black line on the left side of the log. The farther to the left that the line goes, the faster the rate of penetration. On this mud log, ROP is measured in feet per hour but on some older, hand drawn mud logs, it is measured in minutes per foot.

- The *porosity* in (Figure 1) is represented by the blue line farthest to the left of the log. It indicates the pore space within the rock structure. An analogy would be the holes in a sponge. The oil and gas resides within this pore space. Notice how far to the left the porosity goes where all the sand (in yellow) is. This indicates that the sand has good porosity. Porosity is not a direct or physical measurement of the pore space but rather an extrapolation from other drilling parameters and therefore not always reliable.

- The *lithology* in (Figure 1 & 2) is represented by the cyan, gray/black and yellow blocks of color. Cyan = lime, gray/black = shale and yellow = sand. More yellow represents more sand identified at that depth. The lithology is measured as percentage of the total sample, as visually inspected under a microscope, normally at 10x magnification (Figure 3). These are but a fraction of the different types of formations that might be encountered. (Color coding is not necessarily standardized among different mud logging companies, though the symbol representation for each are very similar.) In (Figure 3) you can see a sample of cuttings under a microscope at 10x magnification after they have been washed off. Some of the larger shale and lime fragments are separated from this sample by running it through sieves and must be considered when estimating percentages. Also, this image view is only a fragment of the total sample and some of the sand at the bottom of the tray can not be seen and must also be considered in the total estimation. With that in mind this sample would be considered to be about 90% shale, 5% sand and 5% lime (In 5% increments).

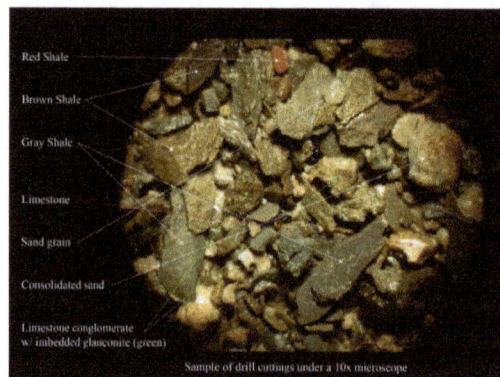

(Figure 3)Sample of drill cuttings of shale while drilling an oil well in Louisiana. For reference, the sand grain and red shale are approximately 2 mm. in dia.

- The *gas* in (Figure 1 & 2) is represented by the green line and is measured in units of ppm (parts per million) as the quantity of total gas, but does not represent the actual quantity of oil or gas the reservoir contains. In (Figure 1) the squared-off dash-dot lines just to the right of the sand (in yellow) and left of the gas (in green) represents the heavier hydrocarbons detected. Cyan = C_2 (ethane), purple = C_3 (propane) and blue = C_4 (butane). Detecting and analyzing these heavy gases help to determine the type of oil or gas the formation contains.

Gamma Ray Logging

Gamma ray logging is a method of measuring naturally occurring gamma radiation to characterize the rock or sediment in a borehole or drill hole. It is a wireline logging method used in mining, mineral exploration, water-well drilling, for formation evaluation in oil and gas well drilling and for other related purposes. Different types of rock emit different amounts and different spectra of natural gamma radiation. In particular, shales usually emit more gamma rays than other sedimentary rocks, such as sandstone, gypsum, salt, coal, dolomite, or limestone because radioactive potassium is a common component in their clay content, and because the cation exchange capacity of clay causes them to absorb uranium and thorium. This difference in radioactivity between shales and sandstones/carbonate rocks allows the gamma tool to distinguish between shales and non-shales.

The gamma ray log, like other types of well logging, is done by lowering an instrument down the drill hole and recording gamma radiation variation with depth. In the United States, the device most commonly records measurements at 1/2-foot intervals. Gamma radiation is usually recorded in API units, a measurement originated by the petroleum industry. Gamma logs are attenuated by diameter of the borehole because of the properties of the fluid filling the borehole, but because gamma logs are most often used in a qualitative way, corrections are usually not necessary.

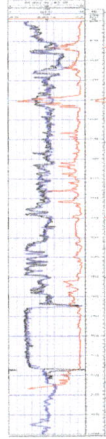

Example gamma ray log. Blue and black lines indicate the measured gamma rays. Sand section of interest is located at bottom of log where the log moves to the left.

Three elements and their decay chains are responsible for the radiation emitted by rock: potassium, thorium and uranium. Shales often contain potassium as part of their clay content, and tend to absorb uranium and thorium as well. A common gamma-ray log records the total radiation and cannot distinguish between the radioactive elements, while a spectral gamma ray log can.

For standard GR logs, the value measured is calculated from thorium in ppm, Uranium in ppm and potassium in percent. GR API = 8 × Uranium concentration in ppm + 4 × thorium concentration in ppm + 16 × potassium concentration in percent. Due to the weight of uranium concentration in the calculation, anomalous concentrations of uranium can cause clean sand reservoirs to appear shaley. Spectral Gamma ray is used to provide an individual reading for each element so anomalies in concentration can be found and interpreted.

An advantage of the gamma log over some other types of well logs is that it works through the steel and cement walls of cased boreholes. Although concrete and steel absorb some of the gamma radiation, enough travels through the steel and cement to allow qualitative determinations.

Sometimes non-shales also have elevated levels of gamma radiation. Sandstone can contain uranium mineralization, potassium feldspar, clay filling, or rock fragments that cause it to have higher-than usual gamma readings. Coal and dolomite may contain absorbed uranium. Evaporite deposits may contain potassium minerals such as carnallite. When this is the case, spectral gamma ray logging can be done to identify these anomalies.

Spectral Logging

The technique of measuring the spectrum, or number and energy, of gamma rays emitted as natural radioactivity by the formation. There are three sources of natural radioactivity in the Earth: 40K, 232Th and 238U, or potassium, thorium and uranium. These radioactive isotopes emit gamma rays that have characteristic energy levels. The quantity and energy of these gamma rays can be measured in a scintillation detector. A log of natural gamma ray spectroscopy is usually presented as a total gamma ray log and the weight fraction of potassium (%), thorium (ppm) and uranium (ppm). The primary standards for the weight fractions are formations with known quantities of the three isotopes. Natural gamma ray spectroscopy logs were introduced in the early 1970s, although they had been studied from the 1950s.

The characteristic gamma ray line that is associated with each component:

- Potassium : Gamma ray energy 1.46 MeV

- Thorium series: Gamma ray energy 2.61 MeV

- Uranium-Radium series: Gamma ray energy 1.76 MeV

Another example of the use of spectral gamma ray logs is to identify specific clay types, like Kaolinite or Illite. This can be used for environmental interpretation as Kaolinite forms from Feldspars in tropic soils by leaching of Potassium; and low Potassium readings may thus indicate paleosols. The identification of clay types is also useful for calculating the effective porosity of reservoir rock.

Use in Mineral Exploration

Gamma ray logs are also used in mineral exploration, especially exploration for phosphates, uranium, and potassium salts.

Spontaneous Potential Logging

The spontaneous potential log, commonly called the self potential log or SP log, is a passive measurement taken by oil industrywell loggers to characterise rock formation properties. The log works by measuring small electric potentials (measured in millivolts) between depths in the borehole and a grounded electrode at the surface. Conductive bore hole fluids are necessary to create a SP response, so the SP log cannot be used in nonconductive drilling muds (e.g. oil-based mud) or air filled holes.

The change in voltage through the well bore is caused by a buildup of charge on the well bore walls. Clays and shales (which are composed predominantly of clays) will generate one charge and permeable formations such as sandstone will generate an opposite one. Spontaneous potentials occur when two aqueous solutions with different ionic concentrations are placed in contact through a porous, semi-permeable membrane. In nature, ions tend to migrate from high to low ionic concentrations. In the case of SP logging, the two aqueous solutions are the well bore fluid (drilling mud) and the formation water (connate water). The potential opposite shales is called the baseline, and typically shifts only slowly over the depth of the borehole.

The relative salinity of the mud and the formation water will determine the which way the SP curve will deflect opposite a permeable formation. Generally if the ionic concentration of the well bore fluid is less than the formation fluid then the SP reading will be more negative (usually plotted as a deflection to the left). If the formation fluid has an ionic concentration less than the well bore fluid, the voltage deflection will be positive (usually plotted as an excursion to the right). The amplitudes of the line made by the changing SP will vary from formation to formation and will not give a definitive answer to how permeable or the porosity of the formation that it is logging.

The presence of hydrocarbons (e.g. oil, natural gas, condensate) will reduce the response on an SP log because the interstitial water contact with the well bore fluid is reduced. This phenomena is called hydrocarbon suppression and can be used to diagnose rocks for commercial potential. The SP curve is usually 'flat' opposite shale formations because there is no ion exchange due to the low permeability, low porosity properties (tight)thus creating a baseline. Tight rocks other than shale (e.g. tight sandstones, tight carbonates) will also result in poor or no response on the SP curve because of no ion exchange.

The SP tool is one of the simplest tools and is generally run as standard when logging a hole, along with the gamma ray. SP data can be used to find:

- Depths of permeable formations
- The boundaries of these formations
- Correlation of formations when compared with data from other analogue wells
- Values for the formation-water resistivity

The SP curve can be influenced by various factors both in the formation and introduced into the wellbore by the drilling process. These factors can cause the SP curve to be muted or even inverted depending on the situation.

- Formation bed thickness
- Resistivities in the formation bed and the adjacent formations
- Resistivity and make up of the drilling mud
- Wellbore diameter
- The depth of invasion by the drilling mud into the formation

Mud invasion into the permeable formation can cause the deflections in the SP curve to be rounded off and to reduce the amplitude of thin beds.

A smaller wellbore will cause, like a mud filtrate invasion, the deflections on the SP curve to be rounded off and decrease the amplitude opposite thin beds, while a larger diameter wellbore has the opposite effect. If the salinity of the mud filtrate is greater than formation water the SP currents will flow in opposite direction.In that case SP deflection will be positive towards to the right. Positive deflections are observed for fresh water bearing formations.

Formation Evaluation Neutron Porosity

In the field of formation evaluation, porosity is one of the key measurements to quantify oil and gas reserves. Neutron porosity measurement employs a neutron source to measure the hydrogen index in a reservoir, which is directly related to porosity. The Hydrogen Index (HI) of a material is defined as the ratio of the concentration of hydrogen atoms per cm³ in the material, to that of pure water at 75 °F. As hydrogen atoms are present in both water and oil filled reservoirs, measurement of the amount allows estimation of the amount of liquid-filled porosity.

Physics

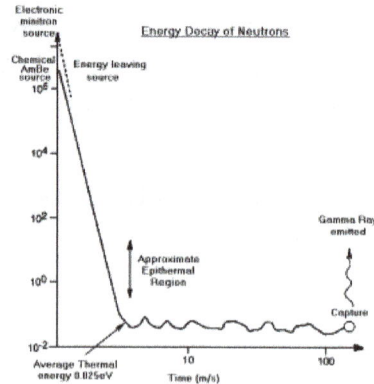

Fig1: Neutron Energy Decay

Neutrons are typically emitted by a radioactive source such as Americium Beryllium (Am-Be) or Plutonium Beryllium (Pu-Be), or generated by electronic neutron generators such as minitron. Fast neutrons are emitted by these sources with energy ranges from 4 MeV to 14 MeV, and inelastically interact with matter. Once slowed down to 2 MeV, they start to scatter elastically and slow down further until the neutrons reach a thermal energy level of about 0.025 eV. When thermal neutrons are then absorbed, gamma rays are emitted. A suitable detector, positioned at a certain distance from the source, can measure either epithermal neutron population, thermal neutron population, or the gamma rays emitted after the absorption.

Mechanics of elastic collisions predict that the maximum energy transfer occurs during collisions of two particles of equal mass. Therefore, a hydrogen atom (H) will cause a neutron to slow down the most, as they are of roughly equal mass. As hydrogen is fundamentally associated to the amount of water and/or oil present in the pore space, measurement of neutron population within the investigated volume is directly linked to porosity.

Correction

Determination of porosity is one of the most important uses of neutron porosity log. Correction parameters for lithology, borehole parameters, and others are necessary for accurate porosity determination as follow:

1. Borehole size

2. Borehole salinity

3. Borehole temperature and pressure

4. Mud cake

5. Mud weight

6. Formation salinity

7. Tool standoff from borehole wall

Interpretation

Subject to various assumptions and corrections, values of apparent porosity can be derived from any neutron log. One can not underestimate the slow down of neutrons by other elements even if they are less effective. Certain effects, such as lithology, clay content, and amount and type of hydrocarbons, can be recognized and corrected for only if additional porosity information is available, for example from sonic and/or density log. Any interpretation of a neutron log alone should be undertaken with a realization of the uncertainties involved.

Effect of Light Hydrocarbon and Gas

The quantitative response of neutron tool to gas or light hydrocarbon depends primarily on hydrogen index and "excavation effect". The hydrogen index can be estimated from the composition and density of the hydrocarbons

Given a fixed volume, gas has considerably lower hydrogen concentration. When pore spaces in the rock are excavated and replaced with gas, the formation has smaller neutron-slowing characteristic, hence the terms "Excavation Effect". If this effect is ignored, a neutron log will show a low porosity value. This characteristic allows a neutron porosity log to be used with other porosity logs (such as a density log) to detect gas zones and identify gas-liquid contacts.

Measurement Technique

Neutron tools are based on the measurement of a neutron cloud of different energy levels within the investigated volume. Epithermal-neutron tools measure epithermal neutron density with energy levels between 100eV and 0.1eV in the formation. Thermal-neutron tools only measure the population of neutrons with a thermal energy level, and Neutron-gamma tools measure the intensity of gamma flux generated by thermal neutron capture. The tools usually have two detectors (or more) with different spacings from the source to produce ratio of count rates, which theoretically reduce borehole effects.

A Helium-3 (He-3) filled proportional counter is the most common epithermal and thermal neutron detector. Helium has a high neutron capture cross section and produces the following reaction when interacting with a neutron.

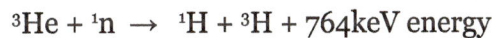

$$^{3}He + {^{1}n} \rightarrow {^{1}H} + {^{3}H} + 764keV \text{ energy}$$

To boost the charge produced by the interaction between Helium and a Neutron, a high voltage is applied to the anode of the counter. A high operating voltage is chosen to give enough gain for counting purposes. Most Helium-3 counters use a quench gas to stabilize high voltage performance and prevent run-away.

References

- Sengel, E.W. "Bill" (1981). Handbook on well logging. Oklahoma City, Oklahoma: Institute for Energy Development. p. 168 p. ISBN 0-89419-112-8.

- Etnyre, L.M. (1989). Finding Oil and Gas from Well Logs. Kluwer Academic Publishers. p. 249 p. ISBN 978-0442223090.

- Darling, Toby (2005). Well Logging and Formation Evaluation. Oxford, UK: Elsevier. p. 5 p. ISBN 0-7506-7883-6.

- P.W. Purcell "Chapter 16 Mud Logging" pp. 347-354 in L.W. Leroy, D.O. Leroy, and J.W. Raese, editors, 1977, Subsurface Geology, Colorado School of Mines, Golden, 941 pp. ISBN 0-918062-00-4

Environmental Concerns of Petroleum

The environmental concerns caused by petroleum are high, for it is poisonous for every form of life. It immensely damages the ecosystem and has resulted in climate change. The other aspects of the environmental concerns are oil spill, the environmental impact of the oil shale industry and the environmental impact of hydraulic fracturing.

Environmental Impact of the Petroleum Industry

The environmental impact of petroleum is often negative because it is toxic to almost all forms of life and its extraction fuels climate change. Petroleum, commonly referred to as oil, is closely linked to virtually all aspects of present society, especially for transportation and heating for both homes and for commercial and industrial activities.

A beach after an oil spill.

Issues

Toxicity

Crude oil is a mixture of many different kinds of organic compounds, many of which are highly toxic and cancer causing (carcinogenic). Oil is "acutely lethal" to fish - that is, it kills fish quickly, at a concentration of 4000 parts per million (ppm) (0.4%). Crude oil and petroleum distillates cause birth defects.

Benzene is present in both crude oil and gasoline and is known to cause leukaemia in humans. The compound is also known to lower the white blood cell count in humans, which would leave people exposed to it more susceptible to infections. "Studies have linked benzene exposure in the mere parts per billion (ppb) range to terminal leukemia, Hodgkin's lymphoma, and other blood and immune system diseases within 5-15 years of exposure."

Exhaust

Petroleum diesel exhaust from a truck

When oil or petroleum distillates are burned, usually the combustion is not complete. This means that incompletely burned compounds are created in addition to just water and carbon dioxide. The other compounds are often toxic to life. Examples are carbon monoxide and methanol. Also, fine particulates of soot blacken humans' and other animals' lungs and cause heart problems or death. Soot is cancer causing (carcinogenic).

Acid Rain

Trees killed by acid rain, an unwanted side effect of burning petroleum

High temperatures created by the combustion of petroleum cause nitrogen gas in the surrounding air to oxidize, creating nitrous oxides. Nitrous oxides, along with sulfur dioxide from the sulfur in the oil, combine with water in the atmosphere to create acid rain. Acid rain causes many problems such as dead trees and acidified lakes with dead fish. Coral reefs in the world's oceans are killed by acidic water caused by acid rain.

Acid rain leads to increased corrosion of machinery and structures (large amounts of capital), and to the slow destruction of archaeological structures like the marble ruins in Rome and Greece.

Climate Change

Humans burning large amounts of petroleum create large amounts of CO_2 (carbon dioxide) gas that traps heat in the Earth's atmosphere.

Oil Spills

An oil spill is the release of a liquidpetroleumhydrocarbon into the environment, especially marine areas, due to human activity, and is a form of pollution. The term is usually applied to marine oil spills, where oil is released into the ocean or coastal waters, but spills may also occur on land. Oil spills may be due to releases of crude oil from tankers, pipelines, railcars, offshore platforms, drilling rigs and wells, as well as spills of refined petroleum products (such as gasoline, diesel) and their by-products, heavier fuels used by large ships such as bunker fuel, or the spill of any oily refuse or waste oil.

Major oil spills include the Kuwaiti oil fires, Kuwaiti oil lakes, Lakeview Gusher, Gulf War oil spill, and the Deepwater Horizon oil spill. Spilt oil penetrates into the structure of the plumage of birds and the fur of mammals, reducing its insulating ability, and making them more vulnerable to temperature fluctuations and much less buoyant in the water. Cleanup and recovery from an oil spill is difficult and depends upon many factors, including the type of oil spilled, the temperature of the water (affecting evaporation and biodegradation), and the types of shorelines and beaches involved. Spills may take weeks, months or even years to clean up.

Volatile Organic Compounds

Volatile organic compounds (VOCs) are gases or vapours emitted by various solids and liquids, many of which have short- and long-term adverse effects on human health and the environment. VOCs from petroleum are toxic and foul the air, and some like benzene are extremely toxic, carcinogenic and cause DNA damage. Benzene often makes up about 1% of crude oil and gasoline. Benzene is present in automobile exhaust. More important for vapors from spills of diesel and crude oil are aliphatic, volatile compounds. Although "less toxic" than compounds like benzene, their overwhelming abundance can still cause health concerns even when benzene levels in the air are relatively low. The compounds are sometimes collectively measured as "total petroleum hydrocarbons" or "TPH." Petroleum hydrocarbons such as gasoline, diesel, or jet fuel intruding into indoor spaces from underground storage tanks or brownfields threaten safety (e.g., explosive potential) and causes adverse health effects from inhalation.

Waste Oil

Waste oil is used oil containing not only breakdown products but also impurities from use. Some examples of waste oil are used oils such as hydraulic oil, transmission oil, brake fluids, motor oil, crankcase oil, gear box oil and synthetic oil. Many of the same problems associated with natural petroleum exist with waste oil. When waste oil from vehicles drips out engines over streets and roads, the oil travels into the water table bringing with it such toxins as benzene. This poisons both soil and drinking water. Runoff from storms carries waste oil into rivers and oceans, poisoning them as well.

Waste oil in the form of motor oil

Mitigation

Conservation and Phasing Out

- Creating laws to completely phase out the use of petroleum (Sweden's 15-year plan)

- Making use of petroleum more efficiently via better technology

Substitution of Other Energy Sources

- Using "cleaner" energy sources such as natural gas and biodiesel, especially in critical areas like cities where there are people.

Use of Biomass Instead of Petroleum

- It is suggested that cellulose from fibrous plant material, such as hemp, can be used to produce alternatives to many oil-based products.

- Plastics can be created from cellulose instead of from oil.

- Lubricants like motor oil and grease can be made from plants and animal fat.

Safety Measures

- Decreasing the risk of spills

- False floors at gasoline stations to catch gasoline and oil drips from making it into the water table

- Double-hulled tanker ships

Environmental Impact of Hydraulic Fracturing

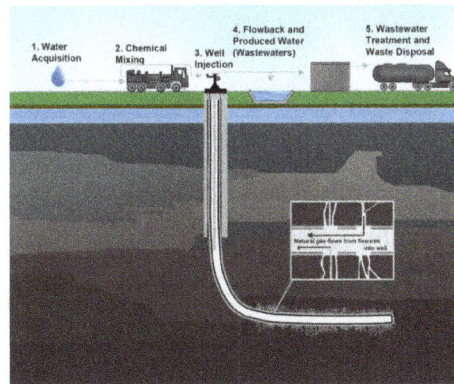

Illustration of hydraulic fracturing and related activities

The environmental impact **of** hydraulic fracturing affects land use and water consumption, methane emissions, air emissions, water contamination, noise pollution, and health. Water and air pollution are the biggest risks to human health from hydraulic fracturing. Research is underway to determine if human health has been affected, and rigorous adherence to regulation and safety procedures is required to avoid harm. Noise from hydraulic fracturing and associated transport can also affect residents and local wildlife.

Hydraulic fracturing fluids include proppants and other substances, which may include toxic chemicals. In the United States, such additives may be treated as trade secrets by companies who use them. Lack of knowledge about specific chemicals has complicated efforts to develop risk management policies and to study health effects. In other jurisdictions, such as the United Kingdom, these chemicals must be made public and their applications are required to be nonhazardous.

Water usage by hydraulic fracturing can be a problem in areas that experience water shortage. Surface water may be contaminated through spillage and improperly built and maintained waste pits, in jurisdictions where these are permitted. Further, ground water can be contaminated if fluid is able to escape during fracking. Produced water, the water that returns to the surface after fracking, is managed by underground injection, municipal and commercialwastewater treatment, and reuse in future wells. There is potential for methane to leak into ground water and the air, though escape of methane is a bigger problem in older wells than in those built under more recent legislation.

Hydraulic fracturing causes induced seismicity called microseismic events or microearthquakes. The magnitude of these events is too small to be detected at the surface, being of magnitude M-3 to M-1 usually. However, fluid disposal wells (which are often used in the USA to dispose of polluted waste from several industries) have been responsible for earthquakes up to 5.6M in Oklahoma and other states.

Governments worldwide are developing regulatory frameworks to assess and manage environmental and associated health risks, working under pressure from industry on the one hand, and from anti-fracking groups on the other. In some countries like France a precautionary approach has been favored and hydraulic fracturing has been banned. Some countries such as the United States have adopted the approach of identifying risks before regulating. The United Kingdom's regulatory framework is based on the conclusion that the risks associated with hydraulic fracturing are manageable if carried out under effective regulation and if operational best practices are implemented.

Air Emissions

A report for the European Union on the potential risks was produced in 2012. Potential risks are "methane emissions from the wells, diesel fumes and other hazardous pollutants, ozone precursors or odours from hydraulic fracturing equipment, such as compressors, pumps, and valves". Also gases and hydraulic fracturing fluids dissolved in flowback water pose air emissions risks.

"In the UK, all oil and gas operators must minimise the release of gases as a condition of their licence from the Department of Energy and Climate Change (DECC). Natural gas may only be vented for safety reasons."

Also transportation of necessary water volume for hydraulic fracturing, if done by trucks, can cause emissions. Piped water supplies can reduce the number of truck movements necessary.

A report from the Pennsylvania Dept of Environmental Protection indicated that there is little potential for radiation exposure from oil and gas operations.

Climate Change

Whether natural gas produced by hydraulic fracturing causes higher well-to-burner emissions than gas produced from conventional wells is a matter of contention. Some studies have found that hydraulic fracturing has higher emissions due to methane released during completing wells as some gas returns to the surface, together with the fracturing fluids. Depending on their treatment, the well-to-burner emissions are 3.5%–12% higher than fore conventional gas.

A debate has arisen particularly around a study by professor Robert W. Howarth finding shale gas significantly worse for global warming than oil or coal. Other researchers have criticized Howarth's analysis, including Cathles *et al.*, whose estimates were substantially lower." A 2012 industry funded report co-authored by researchers at the United States Department of Energy's National Renewable Energy Laboratory found emissions from shale gas, when burned for electricity, were "very similar" to those from so-called "conventional well" natural gas, and less than half the emissions of coal.

Several studies which have estimated lifecycle methane leakage from shale gas development and production have found a wide range of leakage rates, from less than 1% of total production to 10%. According to the Environmental Protection Agency's Greenhouse Gas Inventory a methane leakage rate is about 1.4%. The American Gas Association, an industry trade group, calculated a 1.2% leakage rate. The most comprehensive study of methane leakage from shale gas to date, initiated by the Environmental Defense Fund and released in the Proceedings of the National Academy of Sciences on September 16, 2013, finds that fugitive emissions in key stages of the natural gas production process are significantly lower than estimates in the EPA's national emissions inventory. The study reports direct measurements from 190 onshore natural gas sites, all hydraulically fractured, across the country and estimates a leakage rate of 0.42% for gas production.

Water Consumption

Massive hydraulic fracturing typical of shale wells uses between 1.2 and 3.5 million US gallons (4,500 and 13,200 m³) of water per well, with large projects using up to 5 million US gallons (19,000 m³). Additional water is used when wells are refractured. An average well requires 3 to 8 million US gallons (11,000 to 30,000 m³) of water over its lifetime. According to the Oxford Institute for Energy Studies, greater volumes of fracturing fluids are required in Europe, where the shale depths average 1.5 times greater than in the U.S. Whilst the published amounts may seem large, they are small in comparison with the overall water usage in most areas. A study in Texas, which is a water shortage area, indicates "Water use for shale gas is <1% of statewide water withdrawals; however, local impacts vary with water availability and competing demands."

A report by the Royal Society and the Royal Academy of Engineering shows the usage expected for hydraulic fracturing a well is approximately the amount needed to run a 1,000 MW coal-fired power plant for 12 hours. A 2011 report from the Tyndall Centre estimates that to support a 9 billion cubic metres per annum ($320{\times}10^9$ cu ft/a) gas production industry, between 1.25 to 1.65 million cubic metres ($44{\times}10^6$ to $58{\times}10^6$ cu ft) would be needed annually, which amounts to 0.01% of the total water abstraction nationally.

Concern has been raised over the increasing quantities of water for hydraulic fracturing in areas that experience water stress. Use of water for hydraulic fracturing can divert water from stream flow, water supplies for municipalities and industries such as power generation, as well as recreation and aquatic life. The large volumes of water required for most common hydraulic fracturing methods have raised concerns for arid regions, such as the Karoo in South Africa, and in drought-prone Texas, in North America. It may also require water overland piping from distant sources.

A 2014 life cycle analysis of natural gas electricity by the National Renewable Energy Laboratory concluded that electricity generated by natural gas from massive hydraulically fractured wells consumed between 249 gallons per megawatt-hour (gal/MWhr) (Marcellus trend) and 272 gal/MWhr (Barnett Shale). The water consumption for the gas from massive hydraulic fractured wells was from 52 to 75 gal/MWhr greater (26 percent to 38 percent greater) than the 197 gal/MWhr consumed for electricity from conventional onshore natural gas.

Some producers have developed hydraulic fracturing techniques that could reduce the need for water. Using carbon dioxide, liquid propane or other gases instead of water have been proposed to reduce water consumption. After it is used, the propane returns to its gaseous state and can be collected and reused. In addition to water savings, gas fracturing reportedly produces less damage to rock formations that can impede production. Recycled flowback water can be reused in hydraulic fracturing. It lowers the total amount of water used and reduces the need to dispose of wastewater after use. The technique is relatively expensive, however, since the water must be treated before each reuse and it can shorten the life of some types of equipment.

Water Contamination

Injected Fluid

In the United States, hydraulic fracturing fluids include proppants, radionuclide tracers, and other chemicals, many of which are toxic. The type of chemicals used in hydraulic fracturing and their properties vary. While most of them are common and generally harmless, some chemicals are carcinogenic. Out of 2,500 products used as hydraulic fracturing additives in the United States, 652 contained one or more of 29 chemical compounds which are either known or possible human carcinogens, regulated under the Safe Drinking Water Act for their risks to human health, or listed as hazardous air pollutants under the Clean Air Act. Another 2011 study identified 632 chemicals used in United States natural gas operations, of which only 353 are well-described in the scientific literature. The Ground Water Protection Council has launched FracFocus.org, an online voluntary disclosure database for hydraulic fracturing fluids funded by oil and gas trade groups and the Department of Energy.

The European Union regulatory regime requires full disclosure of all additives. According to the EU groundwater directive of 2006, "in order to protect the environment as a whole, and human health in particular, detrimental concentrations of harmful pollutants in groundwater must be avoided, prevented or reduced." In the United Kingdom, only chemicals that are "non hazardous in their application" are licensed by the Environment Agency.

Some of the water used in hydraulic fracturing is recovered at the surface as flowback or later production brine. The water left in place is called residual treatment water. According to Engelder and Cathles, this residual treatment water becomes permanently sequestered in the shale and cannot seep into and contaminate ground water.

Flowback

Less than half of injected water is recovered as flowback or later production brine, and in many cases recovery is <30%. As the fracturing fluid flows back through the well, it consists of spent fluids and may contain dissolved constituents such as minerals and brine waters. In some cases, depending on the geology of the formation, it may contain uranium, radium, radon and thorium. Estimates of the amount of injected fluid returning to the surface range from 15-20% to 30–70%.

Approaches to managing these fluids, commonly known as produced water, include underground injection, municipal and commercialwastewater treatment and discharge, self-contained systems at well sites or fields, and recycling to fracture future wells. The vacuum multi-effect membrane distillation system as a more effective treatment system has been proposed for treatment of flowback. However, the quantity of waste water needing treatment and the improper configuration of sewage plants have become an issue in some regions of the United States. Part of the wastewater from hydraulic fracturing operations is processed there by public sewage treatment plants, which are not equipped to remove radioactive material and are not required to test for it.

Surface Spills

Surface spills related to the hydraulic fracturing occur mainly because of equipment failure or engineering misjudgments.

Volatile chemicals held in waste water evaporation ponds can evaporate into the atmosphere, or overflow. The runoff can also end up in groundwater systems. Groundwater may become contaminated by trucks carrying hydraulic fracturing chemicals and wastewater if they are involved in accidents on the way to hydraulic fracturing sites or disposal destinations.

In the evolving European Union legislation, it is required that "Member States should ensure that the installation is constructed in a way that prevents possible surface leaks and spills to soil, water or air." Evaporation and open ponds are not permitted. Regulations call for all pollution pathways to be identified and mitigated. The use of chemical proof drilling pads to contain chemical spills is required. In the UK, total gas security is required, and venting of methane is only permitted in an emergency.

Methane

In September 2014, a study from the US 'Proceedings of the National Academy of Sciences' released a report that indicated that methane contamination can be correlated to distance from a well in wells that were known to leak. This however was not caused by the hydraulic fracturing process, but by poor cementation of casings.

Groundwater methane contamination has adverse effect on water quality and in extreme cases may lead to potential explosion. A scientific study conducted by researchers of Duke University found high correlations of gas well drilling activities, including hydraulic fracturing, and methane pollution of the drinking water. According to the 2011 study of the MIT Energy Initiative, "there is evidence of natural gas (methane) migration into freshwater zones in some areas, most likely as a result of substandard well completion practices i.e. poor quality cementing job or bad casing, by a few operators." A 2013 Duke study suggested that either faulty construction (defective cement seals in the upper part of wells, and faulty steel linings within deeper layers) combined with a peculiarity of local geology may be allowing methane to seep into waters; the latter cause may also release injected fluids to the aquifer. Abandoned gas and oil wells also provide conduits to the surface in areas like Pennsylvania, where these are common.

Some drinking water aquifers naturally contain methane, and drawing down the water level in the aquifer may cause an increase of methane in the drinking water, unrelated to oil or gas drilling. Tests can distinguish between the biogenic methane created by bacteria at shallow depths, and the thermogenic methane, which forms under conditions of high pressure and temperature deeper underground. Most oil and gas development produces the deeper-sourced thermogenic methane. Although methane that occurs naturally in shallow aquifers is usually biogenic, some drinking-water aquifers contain naturally occurring thermogenic methane, or mixed biogenic-thermogenic methane.

A study by Cabot Oil and Gas examined the Duke study using a larger sample size, found that methane concentrations were related to topography, with the highest readings found in low-lying areas, rather than related to distance from gas production areas. Using a more precise isotopic analysis, they showed that the methane found in the water wells came from both the formations where hydraulic fracturing occurred, and from the shallower formations. The Colorado Oil & Gas Conservation Commission investigates complaints from water well owners, and has found some wells to contain biogenic methane unrelated to oil and gas wells, but others that have thermogenic methane due to oil and gas wells with leaking well casing. A review published in February 2012 found no direct evidence that hydraulic fracturing actual injection phase resulted in contamination of ground water, and suggests that reported problems occur due to leaks in its fluid or waste storage apparatus; the review says that methane in water wells in some areas probably comes from natural resources.

Another 2013 review found that hydraulic fracturing technologies are not free from risk of contaminating groundwater, and described the controversy over whether the methane that has been detected in private groundwater wells near hydraulic fracturing sites has been caused by drilling or by natural processes.

Radionuclides

There are naturally occurring radioactive materials (NORM), for example radium, radon,uranium, and thorium, in shale deposits. Brine co-produced and brought to the surface along with the oil and gas sometimes contains naturally occurring radioactive materials; brine from many shale gas wells, contains these radioactive materials. When NORM is concentrated or exposed by human activities, such as hydraulic fracturing, EPA classifies it as TENORM (technologically enhanced naturally occurring radioactive material).

The U.S. Environmental Protection Agency and regulators in North Dakota consider radioactive material in flowback a potential hazard to workers at hydraulic fracturing drilling and waste disposal sites and those living or working nearby if the correct procedures are not followed. A report from the Pennsylvania Department of Environmental Protection indicated that there is little potential for radiation exposure from oil and gas operations.

Land Usage

In the UK, the likely well spacing visualised by the Dec 2013 DECC Strategic Environmental Assessment report indicated that well pad spacings of 5 km were likely in crowded areas, with up to 3 hectares (7.4 acres) per well pad. Each pad could have 24 separate wells. This amounts to 0.16% of land area. A study published in 2015 on the Fayetteville Shale found that a mature gas field impacted about 2% of the land area and substantially increased edge habitat creation. Average land impact per well was 3 hectares (about 7 acres)

Seismicity

Hydraulic fracturing causes induced seismicity called microseismic events or microearthquakes. These microseismic events are often used to map the horizontal and vertical extent of the fracturing. The magnitude of these events is usually too small to be detected at the surface, although the biggest micro-earthquakes may have the magnitude of about -1.5 (M_w).

Induced Seismicity from Hydraulic Fracturing

As of late 2014, there have been three instances of hydraulic fracturing, through induced seismicity, triggering quakes large enough to be felt by people: one each in the United States, Canada, and England. In England, two earthquakes that occurred in April and May 2011 of a magnitude of respectively 1.5 and 2.3 on the Richter scale were felt by local populations. The United Kingdom Department of Energy and Climate Change said the "observed seismicity in April and May 2011 was induced by the hydraulic fracture treatments at Preese Hall", in the North of England.

The National Research Council (part of the National Academy of Sciences) has also observed that hydraulic fracturing, when used in shale gas recovery, does not pose a serious risk of causing earthquakes that can be felt.

Induced Seismicity from Water Disposal Wells

According to the USGS only a small fraction of roughly 30,000 waste fluid disposal wells for oil and gas operations in the United States have induced earthquakes that are large enough to be of concern to the public. Although the magnitudes of these quakes has been small, the USGS says that there is no guarantee that larger quakes will not occur. In addition, the frequency of the quakes has been increasing. In 2009, there were 50 earthquakes greater than magnitude 3.0 in the area spanning Alabama and Montana, and there were 87 quakes in 2010. In 2011 there were 134 earthquakes in the same area, a sixfold increase over 20th century levels. There are also concerns that quakes may damage underground gas, oil, and water lines and wells that were not designed to withstand earthquakes.

Several earthquakes in 2011, including a 4.0 magnitude quake on New Year's Eve that hit Youngstown, Ohio, are likely linked to a disposal of hydraulic fracturing wastewater, according to seismologists at Columbia University. A similar series of small earthquakes occurred in 2012 in Texas. Earthquakes are not common occurrences in either area.

A 2012 US Geological Survey study reported that a "remarkable" increase in the rate of $M \geq 3$ earthquakes in the US midcontinent "is currently in progress", having started in 2001 and culminating in a 6-fold increase over 20th century levels in 2011. The overall increase was tied to earthquake increases in a few specific areas: the Raton Basin of southern Colorado (site of coalbed methane activity), and gas-producing areas in central and southern Oklahoma, and central Arkansas. While analysis suggested that the increase is "almost certainly man-made", the USGS noted: "USGS's studies suggest that the actual hydraulic fracturing process is only very rarely the direct cause of felt earthquakes." The increased earthquakes were said to be most likely caused by increased injection of gas-well wastewater into disposal wells. The injection of waste water from oil and gas operations, including from hydraulic fracturing, into saltwater disposal wells may cause bigger low-magnitude tremors, being registered up to 3.3 (M_w).

In 2013, Researchers from Columbia University and the University of Oklahoma demonstrated that in the midwestern United States, some areas with increased human-induced seismicity are susceptible to additional earthquakes triggered by the seismic waves from remote earthquakes. They recommended increased seismic monitoring near fluid injection sites to determine which areas are vulnerable to remote triggering and when injection activity should be ceased.

A British Columbia Oil and Gas Commission investigation concluded that a series of 38 earthquakes (magnitudes ranging from 2.2 to 3.8 on the Richter scale) occurring in the Horn River Basin area between 2009 and 2011 were caused by fluid injection during hydraulic fracturing in proximity to pre-existing faults. The tremors were small enough that only one of them was reported felt by people; there were no reports of injury or property damage.

Noise

Each well pad (in average 10 wells per pad) needs during preparatory and hydraulic fracturing process about 800 to 2,500 days of activity, which may affect residents. In addition, noise is created by transport related to the hydraulic fracturing activities.

The UK Onshore Oil and Gas (UKOOG) is the industry representative body, and it has published a charter that shows how noise concerns will be mitigated, using sound insulation, and heavily silenced rigs where this is needed.

Safety Issues

In July 2013, the United States Federal Railroad Administration listed oil contamination by hydraulic fracturing chemicals as "a possible cause" of corrosion in oil tank cars.

Health Risks

There is worldwide concern over the possible adverse public health implications of hydraulic fracturing activity. A 2013 review on shale gas production in the United States stated, "with increasing numbers of drilling sites, more people are at risk from accidents and exposure to harmful substances used at fractured wells." A 2011 hazard assessment found that most of the chemicals used for hydraulic fracturing and drilling have immediate health effects, and many may have long-term health effects.

In June 2014 Public Health England published a review of the potential public health impacts of exposures to chemical and radioactive pollutants as a result of shale gas extraction in the UK, based on the examination of literature and data from countries where hydraulic fracturing already occurs. The executive summary of the report stated: "An assessment of the currently available evidence indicates that the potential risks to public health from exposure to the emissions associated with shale gas extraction will be low if the operations are properly run and regulated. Most evidence suggests that contamination of groundwater, if it occurs, is most likely to be caused by leakage through the vertical borehole. Contamination of groundwater from the underground hydraulic fracturing process itself (ie the fracturing of the shale) is unlikely. However, surface spills of hydraulic fracturing fluids or wastewater may affect groundwater, and emissions to air also have the potential to impact on health. Where potential risks have been identified in the literature, the reported problems are typically a result of operational failure and a poor regulatory environment."[iii]

A 2013 review focusing on Marcellus shale gas hydraulic fracturing and the New York City water supply stated, "Although potential benefits of Marcellus natural gas exploitation are large for transition to a clean energy economy, at present the regulatory framework in New York State is inadequate to prevent potentially irreversible threats to the local environment and New York City water supply. Major investments in state and federal regulatory enforcement will be required to avoid these environmental consequences, and a ban on drilling within the NYC water supply watersheds is appropriate, even if more highly regulated Marcellus gas production is eventually permitted elsewhere in New York State." In 2014, New York State banned hydraulic fracturing entirely, citing health risks.

A 2012 report prepared for the European Union Directorate-General for the Environment identified risks to humans from air pollution and ground water contamination posed by hydraulic fracturing. This led to a series of recommendations in 2014 to mitigate these concerns.

A 2012 guidance for pediatric nurses in the US, said that hydraulic fracturing had a potential negative impact on public health, and that pediatric nurses should be prepared to gather information on such topics so as to advocate for improved community health.

Policy and Science

There are two main approaches to regulation that derive from policy debates about how to manage risk and a corresponding debate about how to assess risk.

The two main schools of regulation are science-based assessment of risk and the taking of measures to prevent harm from those risks through an approach like hazard analysis, and the precautionary principle, where action is taken before risks are well-identified. The relevance and reliability of risk assessments in communities where hydraulic fracturing occurs has also been debated amongst environmental groups, health scientists, and industry leaders. The risks, to some, are overplayed and the current research is insufficient in showing the link between hydraulic fracturing and adverse health effects, while to others the risks are obvious and risk assessment is underfunded.

Different regulatory approaches have thus emerged. In France and Vermont for instance, a precautionary approach has been favored and hydraulic fracturing has been banned based on two principles: the precautionary principle and the prevention principle. Nevertheless, some States such as the U.S. have adopted a risk assessment approach, which had led to many regulatory debates over the issue of hydraulic fracturing and its risks.

In the UK, the regulatory framework is largely being shaped by a report commissioned by the UK Government in 2012, whose purpose was to identify the problems around hydraulic fracturing and to advise the country's regulatory agencies. Jointly published by the Royal Society and the Royal Academy of Engineering, under the chairmanship of Professor Robert Mair, the report features ten recommendations covering issues such as groundwater contamination, well integrity, seismic risk, gas leakages, water management, environmental risks, best practice for risk management, and also includes advice for regulators and research councils. The report was notable for stating that the risks associated with hydraulic fracturing are manageable if carried out under effective regulation and if operational best practices are implemented.

A 2013 review concluded that, in the US, confidentiality requirements dictated by legal investigations have impeded peer-reviewed research into environmental impacts.

Oil Spill

Oil slick from the Montara oil spill in the Timor Sea, September 2009

An oil spill is the release of a liquidpetroleumhydrocarbon into the environment, especially marine areas, due to human activity, and is a form of pollution. The term is usually applied to marine oil spills, where oil is released into the ocean or coastal waters, but spills may also occur on land. Oil spills may be due to releases of crude oil from tankers, offshore platforms, drilling rigs and wells, as well as spills of refined petroleum products (such as gasoline, diesel) and their by-products, heavier fuels used by large ships such as bunker fuel, or the spill of any oily refuse or waste oil.

Oil spills penetrate into the structure of the plumage of birds and the fur of mammals, reducing its insulating ability, and making them more vulnerable to temperature fluctuations and much less buoyant in the water. Cleanup and recovery from an oil spill is difficult and depends upon many factors, including the type of oil spilled, the temperature of the water (affecting evaporation and biodegradation), and the types of shorelines and beaches involved. Spills may take weeks, months or even years to clean up.

Oil spills can have disastrous consequences for society; economically, environmentally, and socially. As a result, oil spill accidents have initiated intense media attention and political uproar, bringing many together in a political struggle concerning government response to oil spills and what actions can best prevent them from happening.

Largest Oil Spills

Crude oil and refined fuel spills from tanker ship accidents have damaged vulnerable ecosystems in Alaska, the Gulf of Mexico, the Galapagos Islands, France, the Sundarbans, Ogoniland, and many other places. The quantity of oil spilled during accidents has ranged from a few hundred tons to several hundred thousand tons (e.g., Deepwater Horizon Oil Spill, Atlantic Empress, Amoco Cadiz), but volume is a limited barometer of damage or impact. Smaller spills have already proven to have a great impact on ecosystems, such as the Exxon Valdez oil spill because of the remoteness of the site or the difficulty of an emergency environmental response.

Oil spills at sea are generally much more damaging than those on land, since they can spread for hundreds of nautical miles in a thin oil slick which can cover beaches with a thin coating of oil. These can kill seabirds, mammals, shellfish and other organisms they coat. Oil spills on land are more readily containable if a makeshift earth dam can be rapidly bulldozed around the spill site before most of the oil escapes, and land animals can avoid the oil more easily.

Largest oil spills						
Spill / Tanker	Location	Date	Tonnes of crude oil (thousands)	Barrels (thousands)	US Gallons (thousands)	References
Kuwaiti oil fires	Kuwait	January 16, 1991 - November 6, 1991	136,000	1,000,000	42,000,000	
Kuwaiti oil lakes	Kuwait	January 1991 - November 1991	3,409-6,818	25,000-50,000	1,050,000-2,100,000	
Lakeview Gusher	United States, Kern County, California	March 14, 1910 – September 1911	1,200	9,000	378,000	

Largest oil spills						
Spill / Tanker	Location	Date	Tonnes of crude oil (thousands)	Barrels (thousands)	US Gallons (thousands)	References
Gulf War oil spill[d]	Kuwait, Iraq, and the Persian Gulf	January 19, 1991 - January 28, 1991	818–1,091	6,000–8,000	252,000–336,000	
Deepwater Horizon	United States, Gulf of Mexico	April 20, 2010 – July 15, 2010	560-585	4,100-4,900	172,000-180,800	
Ixtoc I	Mexico, Gulf of Mexico	June 3, 1979 – March 23, 1980	454–480	3,329–3,520	139,818–147,840	
Atlantic Empress / Aegean Captain	Trinidad and Tobago	July 19, 1979	287	2,105	88,396	
Fergana Valley	Uzbekistan	March 2, 1992	285	2,090	87,780	
Nowruz Field Platform	Iran, Persian Gulf	February 4, 1983	260	1,907	80,080	
ABT Summer	Angola, 700 nmi (1,300 km; 810 mi) offshore	May 28, 1991	260	1,907	80,080	
Castillo de Bellver	South Africa, Saldanha Bay	August 6, 1983	252	1,848	77,616	
Amoco Cadiz	France, Brittany	March 16, 1978	223	1,635	68,684	

1. One metric ton (tonne) of crude oil is roughly equal to 308 US gallons or 7.33 barrels approx.; 1 oil barrel (bbl) is equal to 35 imperial or 42 US gallons. Approximate conversion factors.

2. Estimates for the amount of oil burned in the Kuwaiti oil fires range from 500,000,000 barrels (79,000,000 m³) to nearly 2,000,000,000 barrels (320,000,000 m³). Between 605 and 732 wells were set ablaze, while many others were severely damaged and gushed uncontrolled for several months. It took over ten months to bring all of the wells under control. The fires alone were estimated to consume approximately 6,000,000 barrels (950,000 m³) of oil per day at their peak.

3. Oil spilled from sabotaged fields in Kuwait during the 1991 Persian Gulf War pooled in approximately 300 oil lakes, estimated by the Kuwaiti Oil Minister to contain approximately 25,000,000 to 50,000,000 barrels (7,900,000 m³) of oil. According to the U.S. Geological Survey, this figure does not include the amount of oil absorbed by the ground, forming a layer of "tarcrete" over approximately five percent of the surface of Kuwait, fifty times the area occupied by the oil lakes.

4.   Estimates for the Gulf War oil spill range from 4,000,000 to 11,000,000 barrels (1,700,000 m³). The figure of 6,000,000 to 8,000,000 barrels (1,300,000 m³) is the range adopted by the U.S. Environmental Protection Agency and the United Nations in the immediate aftermath of the war, 1991–1993, and is still current, as cited by NOAA and The New York Times in 2010. This amount only includes oil discharged directly into the Persian Gulf by the retreating Iraqi forces from January 19 to 28, 1991. However, according to the U.N. report, oil from other sources not included in the official estimates continued to pour into the Persian Gulf through June, 1991. The amount of this oil was estimated to be at least several hundred thousand barrels, and may have factored into the estimates above 8,000,000 barrels (1,300,000 m³).

Human Impact

An oil spill represents an immediate fire hazard. The Kuwaiti oil fires produced air pollution that caused respiratory distress. The Deepwater Horizon explosion killed eleven oil rig workers. The fire resulting from the Lac-Mégantic derailment killed 47 and destroyed half of the town's centre.

Spilled oil can also contaminate drinking water supplies. For example, in 2013 two different oil spills contaminated water supplies for 300,000 in Miri, Malaysia; 80,000 people in Coca, Ecuador,. In 2000, springs were contaminated by an oil spill in Clark County, Kentucky.

Contamination can have an economic impact on tourism and marine resource extraction industries. For example, the Deepwater Horizon oil spill impacted beach tourism and fishing along the Gulf Coast, and the responsible parties were required to compensate economic victims.

Environmental Effects

A surf scoter covered in oil as a result of the 2007 San Francisco Bay oil spill

In general, spilled oil can affect animals and plants in two ways: direct from the oil and from the response or cleanup process. There is no clear relationship between the amount of oil in the aquatic environment and the likely impact on biodiversity. A smaller spill at the wrong time/wrong season and in a sensitive environment may prove much more harmful than a larger spill at another time of the year in another or even the same environment. Oil penetrates into the structure of the plumage of birds and the fur of mammals, reducing its insulating ability, and making them more vulnerable to temperature fluctuations and much less buoyant in the water.

A bird covered in oil from the Black Sea oil spill

Animals who rely on scent to find their babies or mothers cannot due to the strong scent of the oil. This causes a baby to be rejected and abandoned, leaving the babies to starve and eventually die. Oil can impair a bird's ability to fly, preventing it from foraging or escaping from predators. As they preen, birds may ingest the oil coating their feathers, irritating the digestive tract, altering liver function, and causing kidney damage. Together with their diminished foraging capacity, this can rapidly result in dehydration and metabolic imbalance. Some birds exposed to petroleum also experience changes in their hormonal balance, including changes in their luteinizing protein. The majority of birds affected by oil spills die from complications without human intervention. Some studies have suggested that less than one percent of oil-soaked birds survive, even after cleaning, although the survival rate can also exceed ninety percent, as in the case of the Treasure oil spill.

Heavily furred marine mammals exposed to oil spills are affected in similar ways. Oil coats the fur of sea otters and seals, reducing its insulating effect, and leading to fluctuations in body temperature and hypothermia. Oil can also blind an animal, leaving it defenseless. The ingestion of oil causes dehydration and impairs the digestive process. Animals can be poisoned, and may die from oil entering the lungs or liver.

There are three kinds of oil-consuming bacteria. Sulfate-reducing bacteria (SRB) and acid-producing bacteria are anaerobic, while general aerobic bacteria (GAB) are aerobic. These bacteria occur naturally and will act to remove oil from an ecosystem, and their biomass will tend to replace other populations in the food chain.

Sources and Rate of Occurrence

A VLCC tanker can carry 2 million barrels (320,000 m³) of crude oil. This is about eight times the amount spilled in the widely known *Exxon Valdez* incident. In this spill, the ship ran aground and dumped 260,000 barrels (41,000 m³) of oil into the ocean in March 1989. Despite efforts of scientists, managers, and volunteers over 400,000 seabirds, about 1,000 sea otters, and immense numbers of fish were killed. Considering the volume of oil carried by sea, however, tanker owners' organisations often argue that the industry's safety record is excellent, with only a tiny fraction of a percentage of oil cargoes carried ever being spilled. The International Association of Independent Tanker Owners has observed that "accidental oil spills this decade have been at record low levels—one third of the previous decade and one tenth of the 1970s—at a time when oil transported has more than doubled since the mid 1980s."

Oil tankers are only one source of oil spills. According to the United States Coast Guard, 35.7% of the volume of oil spilled in the United States from 1991 to 2004 came from tank vessels (ships/

barges), 27.6% from facilities and other non-vessels, 19.9% from non-tank vessels, and 9.3% from pipelines; 7.4% from mystery spills. On the other hand, only 5% of the actual spills came from oil tankers, while 51.8% came from other kinds of vessels.

The International Tanker Owners Pollution Federation has tracked 9,351 accidental spills that have occurred since 1974. According to this study, most spills result from routine operations such as loading cargo, discharging cargo, and taking on fuel oil. 91% of the operational oil spills are small, resulting in less than 7 metric tons per spill. On the other hand, spills resulting from accidents like collisions, groundings, hull failures, and explosions are much larger, with 84% of these involving losses of over 700 metric tons.

Cleanup and Recovery

A U.S. Air Force Reserve plane sprays Corexit dispersant over the Deepwater Horizon oil spill in the Gulf of Mexico.

A US Navy oil spill response team drills with a "Harbour Buster high-speed oil containment system".

Cleanup and recovery from an oil spill is difficult and depends upon many factors, including the type of oil spilled, the temperature of the water (affecting evaporation and biodegradation), and the types of shorelines and beaches involved.

Methods for cleaning up include:

- Bioremediation: use of microorganisms or biological agents to break down or remove oil; such as the bacteria Alcanivorax or Methylocella Silvestris.

- Bioremediation Accelerator: Oleophilic, hydrophobic chemical, containing no bacteria, which chemically and physically bonds to both soluble and insoluble hydrocarbons. The bioremediation accelerator acts as a herding agent in water and on the surface, floating molecules to the surface of the water, including solubles such as phenols and BTEX, forming gel-like agglomerations. Undetectable levels of hydrocarbons can be obtained in produced water and manageable water columns. By overspraying sheen with bioremediation accelerator, sheen is eliminated within minutes. Whether applied on land or on water, the nutrient-rich emulsion creates a bloom of local, indigenous, pre-existing, hydrocarbon-consuming bacteria. Those specific bacteria break down the hydrocarbons into water and carbon dioxide, with EPA tests showing 98% of alkanes biodegraded in 28 days; and aromatics being biodegraded 200 times faster than in nature they also sometimes use the hydrofireboom to clean the oil up by taking it away from most of the oil and burning it.

- Controlled burning can effectively reduce the amount of oil in water, if done properly. But it can only be done in low wind, and can cause air pollution.

Volunteers cleaning up the aftermath of the Prestige oil spill

- Dispersants can be used to dissipate oil slicks. A dispersant is either a non-surface active polymer or a surface-active substance added to a suspension, usually a colloid, to improve the separation of particles and to prevent settling or clumping. They may rapidly disperse large amounts of certain oil types from the sea surface by transferring it into the water column. They will cause the oil slick to break up and form water-soluble micelles that are rapidly diluted. The oil is then effectively spread throughout a larger volume of water than the surface from where the oil was dispersed. They can also delay the formation of persistent oil-in-water emulsions. However, laboratory experiments showed that dispersants increased toxic hydrocarbon levels in fish by a factor of up to 100 and may kill fish eggs. Dispersed oil droplets infiltrate into deeper water and can lethally contaminate coral. Research indicates that some dispersants are toxic to corals. A 2012 study found that Corexit dispersant had increased the toxicity of oil by up to 52 times.

- Watch and wait: in some cases, natural attenuation of oil may be most appropriate, due to the invasive nature of facilitated methods of remediation, particularly in ecologically sensitive areas such as wetlands.

- Dredging: for oils dispersed with detergents and other oils denser than water.

- Skimming: Requires calm waters at all times during the process.

- Solidifying: Solidifiers are composed of tiny, floating, dry ice pellets, and hydrophobicpolymers that both adsorb and absorb. They clean up oil spills by changing the physical state of spilled oil from liquid to a solid, semi-solid or a rubber-like material that floats on water. Solidifiers are insoluble in water, therefore the removal of the solidified oil is easy and the oil will not leach out. Solidifiers have been proven to be relatively non-toxic to aquatic and wild life and have been proven to suppress harmful vapors commonly associated with hydrocarbons such as Benzene, Xylene, Methyl Ethyl, Acetone and Naphtha. The reaction time for solidification of oil is controlled by the surface area or size of the polymer or dry pellets as well as the viscosity and thickness of the oil layer. Some solidifier product manufactures claim the solidified oil can be thawed and used if frozen with dry ice or disposed of in landfills, recycled as an additive in asphalt or rubber products, or burned as a low ash fuel. A solidifier called C.I.Agent (manufactured by C.I.Agent Solutions of Louisville, Kentucky) is being used by BP in granular form, as well as in Marine and Sheen Booms at Dauphin Island and Fort Morgan, Alabama, to aid in the Deepwater Horizon oil spill cleanup.

- Vacuum and centrifuge: oil can be sucked up along with the water, and then a centrifuge can be used to separate the oil from the water - allowing a tanker to be filled with near pure oil. Usually, the water is returned to the sea, making the process more efficient, but allowing small amounts of oil to go back as well. This issue has hampered the use of centrifuges due to a United States regulation limiting the amount of oil in water returned to the sea.

- Beach Raking: coagulated oil that is left on the beach can be picked up by machinery.

Bags of oily waste from the Exxon Valdez oil spill

Equipment used includes:

- Booms: large floating barriers that round up oil and lift the oil off the water

- Skimmers: skim the oil

- Sorbents: large absorbents that absorb oil

- Chemical and biological agents: helps to break down the oil

- Vacuums: remove oil from beaches and water surface

- Shovels and other road equipment: typically used to clean up oil on beaches

Prevention

- Secondary containment - methods to prevent releases of oil or hydrocarbons into environment.

- Oil Spill Prevention Containment and Countermeasures (SPCC) program by the United States Environmental Protection Agency.

- Double-hulling - build double hulls into vessels, which reduces the risk and severity of a spill in case of a collision or grounding. Existing single-hull vessels can also be rebuilt to have a double hull.

- Thick-hulled railroad transport tanks

Spill response procedures should include elements such as;

- A listing of appropriate protective clothing, safety equipment, and cleanup materials required

for spill cleanup (gloves, respirators, etc.) and an explanation of their proper use;

- Appropriate evacuation zones and procedures;

- Availability of fire suppression equipment;

- Disposal containers for spill cleanup materials; and

- The first aid procedures that might be required.

Environmental Sensitivity Index (ESI) Mapping

Environmental Sensitivity Index (ESI) maps are used to identify sensitive shoreline resources prior to an oil spill event in order to set priorities for protection and plan cleanup strategies. By planning spill response ahead of time, the impact on the environment can be minimized or prevented. Environmental sensitivity index maps are basically made up of information within the following three categories: shoreline type, and biological and human-use resources.

Shoreline Type

Shoreline type is classified by rank depending on how easy the target site would be to clean up, how long the oil would persist, and how sensitive the shoreline is. The floating oil slicks put the shoreline at particular risk when they eventually come ashore, covering the substrate with oil. The differing substrates between shoreline types vary in their response to oiling, and influence the type of cleanup that will be required to effectively decontaminate the shoreline. In 1995, the US National Oceanic and Atmospheric Administration extended ESI maps to lakes, rivers, and estuary shoreline types. The exposure the shoreline has to wave energy and tides, substrate type, and slope of the shoreline are also taken into account—in addition to biological productivity and sensitivity. The productivity of the shoreline habitat is also taken into account when determining ESI ranking. Mangroves and marshes tend to have higher ESI rankings due to the potentially long-lasting and damaging effects of both the oil contamination and cleanup actions. Impermeable and exposed surfaces with high wave action are ranked lower due to the reflecting waves keeping oil from coming onshore, and the speed at which natural processes will remove the oil.

Biological Resources

Habitats of plants and animals that may be at risk from oil spills are referred to as "elements" and are divided by functional group. Further classification divides each element into species groups with similar life histories and behaviors relative to their vulnerability to oil spills. There are eight element groups: Birds, Reptiles, Amphibians, Fish, Invertebrates, Habitats and Plants, Wetlands, and Marine Mammals and Terrestrial Mammals. Element groups are further divided into subgroups, for example, the 'marine mammals' element group is divided into dolphins, manatees, pinnipeds (seals, sea lions & walruses), polar bears, sea otters and whales. Problems taken into consideration when ranking biological resources include the observance of a large number of individuals in a small area, whether special life stages occur ashore (nesting or molting), and whether there are species present that are threatened, endangered or rare.

Human-use Resources

Human use resources are divided into four major classifications; archaeological importance or cultural resource site, high-use recreational areas or shoreline access points, important protected management areas, or resource origins. Some examples include airports, diving sites, popular beach sites, marinas, natural reserves or marine sanctuaries.

Estimating the Volume of a Spill

By observing the thickness of the film of oil and its appearance on the surface of the water, it is possible to estimate the quantity of oil spilled. If the surface area of the spill is also known, the total volume of the oil can be calculated.

Appearance	Film thickness			Quantity spread	
	inches	mm	nm	gal/ sq mi	L/ha
Barely visible	0.0000015	0.0000380	38	25	0.370
Silvery sheen	0.0000030	0.0000760	76	50	0.730
First trace of color	0.0000060	0.0001500	150	100	1.500
Bright bands of color	0.0000120	0.0003000	300	200	2.900
Colors begin to dull	0.00004	0.0010000	1000	666	9.700
Colors are much darker	0.0000800	0.0020000	2000	1332	19.500

Oil spill model systems are used by industry and government to assist in planning and emergency decision making. Of critical importance for the skill of the oil spill model prediction is the adequate description of the wind and current fields. There is a worldwide oil spill modelling (WOSM) program. Tracking the scope of an oil spill may also involve verifying that hydrocarbons collected during an ongoing spill are derived from the active spill or some other source. This can involve sophisticated analytical chemistry focused on finger printing an oil source based on the complex mixture of substances present. Largely, these will be various hydrocarbons, among the most useful being polyaromatic hydrocarbons. In addition, both oxygen and nitrogen heterocyclic hydrocarbons, such as parent and alkyl homologues of carbazole, quinoline, and pyridine, are present in many crude oils. As a result, these compounds have great potential to supplement the existing suite of hydrocarbons targets to fine-tune source tracking of petroleum spills. Such analysis can also be used to follow weathering and degradation of crude spills.

Environmental Impact of the Oil Shale Industry

Kiviõli Oil Shale Processing & Chemicals Plant in Ida-Virumaa, Estonia

Environmental impact of the oil shale industry includes the consideration of issues such as land use, waste management, and water and air pollution caused by the extraction and processing of oil shale. Surface mining of oil shale deposits causes the usual environmental impacts of open-pit mining. In addition, the combustion and thermal processing generate waste material, which must be disposed of, and harmful atmospheric emissions, including carbon dioxide, a major greenhouse gas. Experimental in-situ conversion processes and carbon capture and storage technologies may reduce some of these concerns in future, but may raise others, such as the pollution of groundwater.

Surface Mining and Retorting

Land Use and Waste Management

Surface mining and *in-situ* processing requires extensive land use. Mining, processing and waste disposal require land to be withdrawn from traditional uses, and therefore should avoid high density population areas. Oil shale mining reduces the original ecosystem diversity with habitats supporting a variety of plants and animals. After mining the land has to be reclaimed. However, this process takes time and cannot necessarily re-establish the original biodiversity. The impact of sub-surface mining on the surroundings will be less than for open pit mines. However, sub-surface mining may also cause subsidence of the surface due to the collapse of mined-out area and abandoned stone drifts.

Disposal of mining wastes, spent oil shale (including semi-coke) and combustion ashes needs additional land use. According to the study of the European Academies Science Advisory Council, after processing, the waste material occupies a greater volume than the material extracted, and therefore cannot be wholly disposed underground. According to this, production of a barrel of shale oil can generate up to 1.5 tonnes of semi-coke, which may occupy up to 25% greater volume than the original shale. This is not confirmed by the results of Estonia's oil shale industry. The mining and processing of about one billion tonnes of oil shale in Estonia has created about 360-370 million tonnes of solid waste, of which 90 million tonnes is a mining waste, 70–80 million tonnes is a semi-coke, and 200 million tonnes are combustion ashes.

The waste material may consist of several pollutants including sulfates, heavy metals, and polycyclic aromatic hydrocarbons (PAHs), some of which are toxic and carcinogenic. To avoid contamination of the groundwater, the solid waste from the thermal treatment process is disposed in an open dump (landfill or "heaps"), not underground. As semi-coke consists of, in addition to minerals, up to 10% organics that may pose hazard to the environment owing to leaching of toxic compounds as well as to the possibility of self-ignition.

Water Management

Mining influences the water runoff pattern of the area affected. In some cases it requires the lowering of groundwater levels below the level of the oil shale strata, which may have harmful effects on the surrounding arable land and forest. In Estonia, for each cubic meter of oil shale mined, 25 cubic meters of water must be pumped from the mine area. At the same time, the thermal processing of oil shale needs water for quenching hot products and the control of dust. Water concerns are particularly sensitive issue in arid regions, such as the western part of the United States and Israel's Negev Desert, where there are plans to expand the oil shale industry. Depending on technology, above-ground retorting uses between one and five barrels of water per barrel of produced shale oil.*In situ* processing, according to one estimate, uses about one-tenth as much water.

Water represents the major vector of transfer of oil shale industry pollutants. One environmental issue is to prevent noxious materials leaching from spent shale into the water supply. The oil shale processing is accompanied by the formation of process waters and waste waters containing phenols, tar and several other products, heavily separable and toxic to the environment. A 2008 programmatic environmental impact statement issued by the United States Bureau of Land Management stated that surface mining and retort operations produce 2 to 10 U.S. gallons (7.6 to 37.9 l; 1.7 to 8.3 imp gal) of waste water per 1 short ton (0.91 t) of processed oil shale.

Air Pollution Management

Main air pollution is caused by the oil shale-fired power plants, which provide the atmospheric emissions of gaseous products like nitrogen oxides, sulfur dioxide and hydrogen chloride, and the airborne particulate matter (fly ash). It includes particles of different types (carbonaceous, inorganic ones) and different sizes. The concentration of air pollutants in flue gas depends primarily on the combustion technology and burning regime, while the emissions of solid particles are determined by the efficiency of fly ash-capturing devices.

Open deposition of semi-coke causes distribution of pollutants in addition to aqueous vectors also via air (dust).

There are possible links from being in an oil shale area to a higher risk of asthma and lung cancer than other areas.

Greenhouse Gas Emissions

Carbon dioxide emissions from the production of shale oil and shale gas are higher than conventional oil production and a report for the European Union warns that increasing public concern about the adverse consequences of global warming may lead to opposition to oil shale development.

Emissions arise from several sources. These include CO_2 released by the decomposition of the kerogen and carbonate minerals in the extraction process, the generation of the energy needed to heat the shale and in the other oil and gas processing operations, and fuel used in the mining of the rock and the disposal of waste. As the varying mineral composition and calorific value of oil shale deposits varies widely, the actual values vary considerably. At best, the direct combustion of oil shales produces carbon emissions similar to those from the lowest form of coal, lignite, at 2.15 moles CO_2/MJ, an energy source which is also politically contentious due to its high emission levels. For both power generation and oil extraction, the CO_2 emissions can be reduced by better utilization of waste heat from the product streams.

In-situ Processing

Currently, the in-situ process is the most attractive proposition due to the reduction in standard surface environmental problems. However, *in-situ* processes do involve possible significant environmental costs to aquifers, especially since *in-situ* methods may require ice-capping or some other form of barrier to restrict the flow of the newly gained oil into the groundwater aquifers. However, after the removal of the freeze wall these methods can still cause groundwater contamination as the hydraulic conductivity of the remaining shale increases allowing groundwater to flow through and leach salts from the newly toxic aquifer.

References

- Public Health England. 25 June 2014 PHE-CRCE-009: Review of the potential public health impacts of exposures to chemical and radioactive pollutants as a result of shale gas extraction ISBN 978-0-85951-752-2

- Bautista H. and Rahman K. M. M. (2016). Review On the Sundarbans Delta Oil Spill: Effects On Wildlife and Habitats. International Research Journal, 1(43), Part 2, pp: 93-96. doi:10.18454/IRJ.2016.43.143

- Lotman, Silvia. "Op-ed: Don't let Estonian shale firm do to Utah what it has done to Estonia". The Salt Lake Tribune. Retrieved 2016-06-14.

- "Dispersant makes oil 52 times more toxic - Technology & science - Science - LiveScience - NBC News". msnbc. com. Retrieved 20 April 2015.

- Brady, Jeff (December 18, 2014). "Citing Health, Environment Concerns, New York Moves To Ban Fracking". National Public Radio. Retrieved 6 January 2015.

- "Strategic Environmental Assessment for Further Onshore Oil and Gas Licensing" (PDF). Department of Energy and Climate Change. June 2014. p. ?. Retrieved 11 November 2014.

- Vidic, R.D.; et al. (May 17, 2013). "Impact of Shale Gas Development on Regional Water Quality" (PDF). Science. 340 (1235009): 826. doi:10.1126/science.1235009. PMID 23687049. Retrieved 29 September 2014.

- Nicot, Jean-Philippe (2 Mar 2012). "Water Use for Shale-Gas Production in Texas, U.S." (PDF). Environmental Science and Technology. Retrieved 1 Nov 2014.

- "Developing Onshore Shale Gas and Oil – Facts about 'Fracking'" (PDF). Department of Energy and Climate Change. Retrieved 14 October 2014.

Permissions

Index

www.ingramcontent.com/pod-product-compliance
Lightning Source LLC
Chambersburg PA
CBHW061317190326
41458CB00011B/3828